Agricultural Science as International Development

For more than fifty years, international aid for agricultural research has been shaped by an unusual partnership: an ad hoc consortium of national governments, foreign aid agencies, philanthropies, United Nations agencies, and international financial institutions, known as CGIAR. Formed in 1971 following the initial celebration of the so-called Green Revolution, CGIAR was tasked with extending that apparent transformation in production to new countries and crops. In this volume, leading historians and sociologists explore the influence of CGIAR and its affiliated international research centres. Traversing five continents and five decades of scientific research, agricultural aid, and political transformation, it examines whether and how science-led development has changed the practices of farmers, researchers, and policymakers. Although its language, funding mechanisms, and decision-making have changed over time, CGIAR and its network of research centres remain powerful in shaping international development and global agriculture. This title is also available as Open Access on Cambridge Core.

Helen Anne Curry is Melvin Kranzberg Professor in the History of Technology at the Georgia Institute of Technology and an Honorary Senior Research Fellow at the University of Cambridge.

Timothy W. Lorek is Assistant Professor of History and director of the program in Global Sustainability and Justice at the College of St. Scholastica.

Agricultural Science as International Development
Historical Perspectives on the CGIAR Era

Edited by

Helen Anne Curry
Georgia Institute of Technology

Timothy W. Lorek
College of St. Scholastica

CAMBRIDGE
UNIVERSITY PRESS

CAMBRIDGE
UNIVERSITY PRESS

Shaftesbury Road, Cambridge CB2 8EA, United Kingdom

One Liberty Plaza, 20th Floor, New York, NY 10006, USA

477 Williamstown Road, Port Melbourne, VIC 3207, Australia

314–321, 3rd Floor, Plot 3, Splendor Forum, Jasola District Centre, New Delhi – 110025, India

103 Penang Road, #05–06/07, Visioncrest Commercial, Singapore 238467

Cambridge University Press is part of Cambridge University Press & Assessment, a department of the University of Cambridge.

We share the University's mission to contribute to society through the pursuit of education, learning and research at the highest international levels of excellence.

www.cambridge.org
Information on this title: www.cambridge.org/9781009434669

DOI: 10.1017/9781009434713

First published 2024

A catalogue record for this publication is available from the British Library.

Library of Congress Cataloging-in-Publication Data
Names: Curry, Helen Anne, editor. | Lorek, Timothy, editor.
Title: Agricultural science as international development : historical perspectives on the CGIAR era / edited by Helen Anne Curry, Timothy W. Lorek
Description: New York, NY : Cambridge University Press, 2025 | Includes bibliographical references and index
Identifiers: LCCN 2024006077 | ISBN 9781009434669 (hardback) | ISBN 9781009434706 (paperback) | ISBN 9781009434713 (ebook)
Subjects: LCSH: Consultative Group on International Agricultural Research. | Agriculture – Research – International cooperation – History. | Agricultural development projects – History.
Classification: LCC S540.8.C66 A47 2025 | DDC 630.72–dc23/eng/20240412
LC record available at https://lccn.loc.gov/2024006077

ISBN 978-1-009-43466-9 Hardback
ISBN 978-1-009-43470-6 Paperback

Contents

Figures

Tables

Contributors

Derek Byerlee was formerly a professor at Georgetown University, following a career in academia, the International Maize and Wheat Improvement Center, and the World Bank. His current interests are in the history of agricultural science in the twentieth century.

Helen Anne Curry is the Melvin Kranzberg Professor in the history of technology at the Georgia Institute of Technology. She currently leads the Wellcome Trust–funded project "From Collection to Cultivation: Historical Perspectives on Crop Diversity and Food Security" at the University of Cambridge.

Greg Edmeades is an independent scholar and native of New Zealand. He was formerly a physiologist/agronomist at the International Maize and Wheat Improvement Center, based in Mexico and Ghana, and a physiologist/breeder at Pioneer Hi-Bred International, based in Hawaii.

Marianna Fenzi is a research fellow at the Institute of Geography and Sustainability at the University of Lausanne in Switzerland. Her research focuses on environmental history, especially in relation to the Green Revolution and crop diversity conservation.

Courtney Fullilove is Associate Professor in the Department of History, the College of the Environment, and the College of Science and Technology Studies at Wesleyan University. She is author of *The Profit of the Earth: The Global Seeds of American Agriculture* (2017).

David J. Jefferson is a senior lecturer (above the bar) at the University of Canterbury Faculty of Law. His research and teaching cover a range of issues related to biodiversity conservation, biotechnology regulation, intellectual property in the agricultural and food sectors, ecosystem rights laws, and the protection of Indigenous knowledge systems.

Prakash Kumar is Associate Professor of South Asian History and History of Science at Pennsylvania State University. He is the author of *Indigo Plantations and Science in Colonial India* (Cambridge University Press, 2012) and is currently writing a book on the history of India's Green Revolution.

Gabriela Soto Laveaga is Professor of the History of Science and Antonio Madero Professor for the Study of Mexico at Harvard University. She is the author of *Jungle Laboratories: Mexican Peasants, National Projects, and the Making of the Pill* (2009). She is currently writing a book on histories of the Green Revolution in Mexico and India.

Sabina Leonelli is Professor of Philosophy and History of Science and director of the Centre for the Study of the Life Sciences (Egenis) at the University of Exeter, where she leads the European Research Council-funded project "A Philosophy of Open Science for Diverse Research Environments" and the "Data Ethics, Governance and Openness" strand of the Institute for Data Science and Artificial Intelligence.

Timothy W. Lorek is Assistant Professor of History at the College of St. Scholastica in Duluth, Minnesota, where he is also director of the Alworth Center for the Study of Peace and Justice, and director of the program in Global Sustainability and Justice. He is the author of *Making the Green Revolution: Agriculture and Conflict in Colombia* (2023).

Harro Maat is Associate Professor at the Knowledge, Technology, and Innovation group of Wageningen University. In a recent project he looks at historical linkages between rice-farming practices of Maroon groups in Suriname and rice economies in Africa and Asia.

Lucas M. Mueller is a senior research associate at the University of Geneva. He is currently writing a book on the global history of food contaminants since 1960.

Wilson Picado-Umaña is Professor at the Escuela de Historia, Universidad Nacional, Costa Rica. His research focuses on the history of the Green Revolution in Latin America.

James Smith is Professor of African and Development Studies and vice principal at the University of Edinburgh. He has researched the inter-relationships among science, innovation, and Africa for the last twenty years, and has worked with the International Livestock Research Institute on aspects of their governance during that period.

Rebekah Thompson is a senior policy manager at the Department of Health and Social Care, United Kingdom. She completed her Ph.D. in African Studies at the University of Edinburgh and was affiliated as a graduate fellow to the International Livestock Research Institute during her doctoral studies.

Acknowledgments

We began the work that would become this volume in late 2019, only to see its development slowed in the tumult of the pandemic years that followed. We are deeply grateful to our contributing authors, who stuck with the project despite disrupted research plans, zoomified workshop meetings, and all the other delays and upheavals the pandemic entailed. The initial workshop was planned as part of Helen's Wellcome Trust Investigator Award (grant no. 217968/Z/19/Z). We thank the Wellcome Trust for its support of her research program, including research and editorial work on this volume, and Dr. Jessica J. Lee, who helped organize the virtual workshop held in June 2021. We also extend thanks to the B&B Stern Foundation for its support of the Kranzberg professorship at Georgia Tech, funds from which enabled this volume to be published Open Access. We are especially grateful to Robert Hampson of Georgia Tech, Lucy Rhymer of Cambridge University Press, and their colleagues who facilitated the Open Access publication process. We also thank the summer stipend program through the National Endowment for the Humanities, as well as the Faculty Development Grant Committee and the Alworth Center for the Study of Peace and Justice at the College of St. Scholastica, for contributing to the funding of Tim's research efforts.

This book had its earliest origins in our meeting at the Agrarian Studies seminar at Yale University in 2019 and discovering then that we had mutual interests in the history of CGIAR. We have both been influenced beyond measure by the Agrarian Studies community. We acknowledge our indebtedness to this community, and to Kalyanakrishnan Sivaramakrishnan and James Scott in particular. The painting by Scott of migrant farm laborers on this edition's cover pointedly humanizes a global food system that is all-too-often dehumanizing. It serves as a reminder of the need for scholarship that does the same, in the service of creating more just and equitable ways of feeding the world.

Abbreviations

AAT	African animal trypanosomiasis
ALAD	Arid Land Agricultural Development program (Ford Foundation)
AR4D	agricultural research for development
ASOCAÑA	Colombian Sugarcane Growers Association
ASSINSEL	International Association of Plant Breeders for the Protection of Plant Varieties
CAP	Colombian Agricultural Program (Rockefeller Foundation)
CBD	Convention on Biological Diversity
CGIAR	Consultative Group on International Agricultural Research
CIAT	International Center for Tropical Agriculture (Colombia)
CIMMYT	International Maize and Wheat Improvement Center (Mexico)
CIP	International Potato Center (Peru)
CRPs	CGIAR Research Programs
CSIRO	Commonwealth Scientific and Industrial Research Organization (Australia)
CVC	Cauca Valley Corporation (Colombia)
ECF	East Coast fever
EUCARPIA	European Association for Research on Plant Breeding
EURISCO	European Plant Genetic Resources Search Catalogue
FAO	United Nations Food and Agriculture Organization
FARC	Revolutionary Armed Forces of Colombia
GFAR	Global Forum for Agricultural Research
GRAIN	Genetic Resources Action International
IARI	Indian Agricultural Research Institute
IBPGR	International Board for Plant Genetic Resources (Italy)
IBRD	International Bank for Reconstruction and Development
ICA	Colombian Agricultural Institute

ICARDA	International Center for Agricultural Research in the Dry Areas, Syria (1977–2012) (Lebanon)
ICRISAT	International Crops Research Institute for the Semi-Arid Tropics (India)
IDRC	International Development Research Centre (Canada)
IFPRI	International Food Policy Research Institute
IICA	Inter-American Institute for Cooperation on Agriculture (Costa Rica)
IITA	International Institute of Tropical Agriculture (Nigeria)
ILCA	International Livestock Centre for Africa (Ethiopia)
ILRAD	International Laboratory for Research on Animal Diseases (Kenya)
ILRI	International Livestock Research Institute (Kenya)
INCAP	Nutrition Institute of Central America and Panama
INCORA	Colombian Institute for Agrarian Reform
INIA	National Institute for Agricultural Research (Mexico)
IPGRI	International Plant Genetic Resources Institute (Italy)
IRAT	Institut de Recherches Agronomiques Tropicales et des Cultures Vivrières (France)
IRRI	International Rice Research Institute (Philippines)
IWMI	International Water Management Institute (Sri Lanka)
MAP	Mexican Agricultural Program (Rockefeller Foundation)
MENA	Middle East and North Africa
NERICA	New Rices for Africa
NGO	nongovernmental organization
OAS	Organization of American States
OECD	Organisation for Economic Co-operation and Development
OPEC	Organization of the Petroleum Exporting Countries
ORSTOM	Office for Overseas Scientific and Technological Research (France)
PCCMCA	Central American Cooperative Program for the Cultivation and Improvement of Food Cultivars
PLO	Palestine Liberation Organization
PROFRIJOL	Cooperative Regional Project on Beans for Central America, Mexico, and the Caribbean
PVS	participatory varietal selection
RAFI	Rural Advancement Foundation International
SDC	Swiss Agency for Development and Cooperation
TAC	CGIAR Technical Advisory Committee
TRIPS	Trade-Related Aspects of Intellectual Property Rights
UNCTAD	United Nations Conference on Trade and Development

UNDP	United Nations Development Programme
UNICEF	United Nations Children's Fund
UPOV	International Union for the Protection of New Varieties of Plants
USAID	United States Agency for International Development
WARDA	West Africa Rice Development Association (Ivory Coast)
WHO	World Health Organization
WIEWS	World Information and Early Warning System on Plant Genetic Resources for Food and Agriculture
WTO	World Trade Organization

Introduction
Past, Present, and Future
Histories of CGIAR

Timothy W. Lorek and Helen Anne Curry

Since 1971, international aid for agricultural research has been shaped by an unusual and ambitious partnership: an organization founded as an ad hoc consortium of national governments, foreign aid offices, philanthropies, United Nations agencies, and international financial institutions that is known today as CGIAR. At its founding, the Consultative Group on International Agricultural Research was tasked with fostering scientific research that would help "developing nations ... increase and improve the quality of their agricultural output."[1] Representative of an era of broad multilateral cooperation, and reliant on complex international funding networks, CGIAR assumed the profoundly localized mission of reshaping farmers and fields across diverse cultural, economic, and environmental contexts. The tensions arising as researchers and institutions navigated the demands and expectations of these distinct scales form the crux of CGIAR history. They have affected the changing disciplinary orientations of research centers, the ecologies prioritized in breeding, the expectations for intellectual property management, and even the words used to describe crops.

CGIAR was and remains a dynamic entity. Its organization, policies, and mission have morphed multiple times in response to changing international circumstances across its fifty-year history.[2] It took shape as the United Nations' "development decade" of the 1960s transitioned to a period characterized by Cold War détente, aid multilateralism, and increasingly decentralized neoliberal restructuring of government agencies in host countries – and its original contours reflect the assumptions and priorities of that time. Its instigators included influential administrators at the Rockefeller and Ford Foundations and the World Bank, and it brought together representatives of donor nations and other organizations under the sponsorship of the World Bank, United Nations Food and Agriculture

---

[1] International Bank for Reconstruction and Development, "New International Research Group Formed," May 20, 1971, https://hdl.handle.net/10947/259.

[2] John Lynam, Derek Byerlee, and Joyce Moock, "The Organizational Challenge of International Agricultural Research: The Fifty-Year Odyssey of the CGIAR," *Food Policy* 124 (2024): 102617.

Figure 0.1 Representatives of leading agencies and CGIAR bodies preside over a July 1975 CGIAR meeting in Washington, DC. The individuals seated at the table from left to right are a UNDP representative, the CGIAR executive secretary, the TAC secretary, the chairman (perhaps of the panel, affiliation unclear), an FAO representative, and a World Bank representative. © World Bank Group. License: CC BY-NC-SA. 4.0.

Organization (FAO), and the United Nations Development Programme (UNDP) (Figure 0.1). This group imagined replicating the bumper harvests of wheat and rice recently experienced in Mexico, India, and the Philippines – the products of the so-called Green Revolution – with new crops and new countries. And they placed international support for and coordination of research at the center of this vision, necessary to "reinforce national efforts" in agricultural science that they assessed as failing to address mounting needs in food production.[3]

The initial model adopted by CGIAR leadership in the pursuit of this goal prized two elements: expert oversight and institution building. A Technical Advisory Committee (TAC) – a select group of "distinguished

[3] International Bank for Reconstruction and Development, "New International Research Group Formed." On the "development decade," see *The United Nations Development Decade: Proposals for Action*, Report of the Secretary General (New York: United Nations, 1962), United Nations Dag Hammarskjöld Library, https://research.un.org/en/docs/dev/1960-1970.

international experts" – prioritized areas to be addressed with resources from CGIAR donors, as well as the best ways to carry out research on these "priority problems."[4] Influenced especially by the recent history of wheat and rice research, which had been undertaken at international institutions that targeted these crops, the committee identified the creation of further international research centers with clear mandates as the go-to route for enhancing agricultural science in the name of development. Establishing these centers and organizing them as an interlinked system became the second key element of CGIAR strategy. Its network of research centers mushroomed from a founding four in 1971 to thirteen in 1983 and eighteen a decade later.[5]

Much of CGIAR's institutional growth took place in postcolonial spaces, locations enmeshed in the legacies of formal or informal empire. As Courtney Fullilove observes in her analysis of CGIAR's move into the Middle East, its administrators and scientists operated in rural landscapes that were sometimes "the fields of empire, recast in the aftermath of World War II as buffers against communism." In the transition to a post–Cold War world, CGIAR's geography largely remained intact, but its globalizing ambitions turned away from geopolitical jostling and the so-called battle for hearts and minds. As development strategies pivoted towards market-based interventions, postcolonial states and other nations targeted for aid "became the grist for a globalized vision of market-led development, terrain imagined rather than realized in the winds of change" (see Fullilove, Chapter 1, this volume). The shift kept CGIAR in step with the United Nations' unrolling of the Millennium (2000–15) and then Sustainable Development Goals (2015–present) as the twentieth century transitioned to the twenty-first.

By 2023, there were fourteen CGIAR research centers, and their activities substantiate CGIAR's claim to being "the world's largest global agricultural innovation network"[6] (Figure 0.2). More than 9,000 scientists and staff sustain a program of research that has changed over the intervening decades, expanding from an early emphasis on growing ever-bigger piles of grain to incorporating such issues as agroforestry, water

---

[4] Consultative Group on International Agricultural Research (CGIAR), "CGIAR First Meeting, Washington, DC, May 19, 1971: Summary of Proceedings," June 9, 1971, https://hdl.handle.net/10947/260.

[5] Derek Byerlee and John Lynam, "The Development of the International Center Model for Agricultural Research: A Prehistory of the CGIAR," *World Development* 135 (2020): 105080; Selçuk Özgediz, *The CGIAR at 40: Institutional Evolution of the World's Premier Agricultural Research Network* (Washington, DC: CGIAR Fund, 2012), https://openknowledge.worldbank.org/handle/10986/23845.

[6] CGIAR, "Research Centers," www.cgiar.org/research/research-centers/. Numerical data were updated in June 2023.

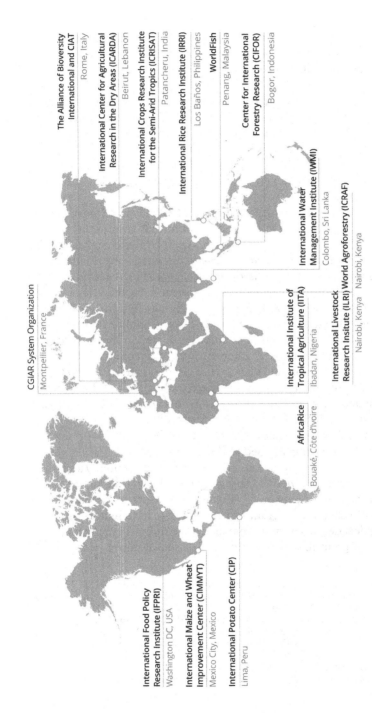

CGIAR System Organization
Montpellier, France

The Alliance of Bioversity International and CIAT
Rome, Italy

International Center for Agricultural Research in the Dry Areas (ICARDA)
Beirut, Lebanon

International Crops Research Institute for the Semi-Arid Tropics (ICRISAT)
Patancheru, India

International Rice Research Institute (IRRI)
Los Baños, Philippines

WorldFish
Penang, Malaysia

Center for International Forestry Research (CIFOR)
Bogor, Indonesia

International Water Management Institute (IWMI)
Colombo, Sri Lanka

International Institute of Tropical Agriculture (IITA)
Ibadan, Nigeria

International Livestock Research Insitute (ILRI) World Agroforestry (ICRAF)
Nairobi, Kenya   Nairobi, Kenya

AfricaRice
Bouaké, Côte d'Ivoire

International Food Policy Research Institute (IFPRI)
Washington DC, USA

International Maize and Wheat Improvement Center (CIMMYT)
Mexico City, Mexico

International Potato Center (CIP)
Lima, Peru

Figure 0.2 In 2021, the CGIAR system comprised fourteen international research centers. At the time of writing, the organization of CGIAR – including its constituent institutes – was taking new shape under One CGIAR. CGIAR, *Harvesting Research and Innovation for Impact* (Montpellier: CGIAR System Organization, 2022), p. 34. By permission of CGIAR.

management, social inclusion, and climate change adaptation.[7] According to CGIAR accounts, the labor and knowledge of scientists employed at these centers have changed the lives of the world's farmers for the better. Its list of "best scientific breakthroughs," prepared for its fiftieth anniversary, features many discrete research products whose uptake beyond the research facility can be traced and quantified: a vaccine for East Coast cattle disease, the cassava variety KU50, and a digital tool for banana disease identification, among others. But it also points to outcomes far more diffuse, like "enhancing food safety at all levels of the value chain," fostering a whole-landscape approach to natural resource management, and addressing inequality through "gender transformative research."[8] Not surprisingly, these claims to impact – whether discrete or diffuse – have been, and continue to be, vigorously contested.[9] Studies of research for agricultural development routinely question the scalability of interventions, neglected equity implications, sidelining of local and national perspectives in international agenda setting, and more.[10]

More than fifty years on from its founding – and despite its influential and enduring role in shaping the agendas, infrastructure, and labor force of agricultural research, as well as the crops tended by farmers around the world – CGIAR remains an enigmatic historical presence. Histories of twentieth-century agriculture and international development make frequent reference to CGIAR research centers, especially the most prominent of these, and occasionally to CGIAR itself. Yet if one sets institutional

---

[7] Uma Lele and Sambuddha Goswani, "CGIAR," in Uma Lele, Manmohan Agarwal, Brian C. Baldwin, and Sambuddha Goswani, eds., *Food for All: International Organizations and the Transformation of Agriculture* (New York: Oxford University Press, 2021), pp. 707–806.

[8] CGIAR, "Innovation Explorer: CGIAR's 50 Years of Innovations That Changed the World," www.cgiar.org/cgiar-at-50/innovation-explorer/.

[9] James Sumberg, John Thompson, and Philip Woodhouse, "Why Agronomy in the Developing World Has Become Contentious," *Agriculture and Human Values* 30 (2013): 71–83; James Sumberg and John Thompson, eds., *Contested Agronomy: Agricultural Research in a Changing World* (London: Routledge, 2012). See also James E. Sumberg, ed., *Agronomy for Development: The Politics of Knowledge in Agricultural Research* (London: Routledge, 2017).

[10] E.g., Nina de Roo, Jens A. Andersson, and Timothy J. Krupnik, "On-Farm Trials for Development Impact? The Organisation of Research and the Scaling of Agricultural Technologies," *Experimental Agriculture* 55, no. 2 (2019): 163–184; Marcus Taylor and Suhas Bhasme, "Model Farmers, Extension Networks and the Politics of Agricultural Knowledge Transfer," *Journal of Rural Studies* 64 (2018): 1–10; Linus Karlsson, Lars Otto Naess, Andrea Nightingale, and John Thompson, "'Triple Wins' or 'Triple Faults'? Analysing the Equity Implications of Policy Discourses on Climate-Smart Agriculture (CSA)," *Journal of Peasant Studies* 45, no. 1 (2018): 150–174. See also Mitch Renkow and Derek Byerlee, "The Impacts of CGIAR Research: A Review of Recent Evidence," *Food Policy* 35, no. 5 (2010): 391–402.

accounts aside, there are surprisingly few historical treatments that take these institutions as their primary concern.[11] It is difficult for a student or scholar who encounters a mention of CGIAR – let alone its constituent centers past and present, such as the International Water Management Institute (IWMI) or the International Livestock Research Institute (ILRI) – to locate accounts attentive to colonial and postcolonial experiences, national and international political histories, or cultural histories of science and technology.

The exceptions to this pattern are two centers that predated the formation of CGIAR and were made famous by their role in disseminating the headline-generating wheat and rice varieties of the 1960s: the International Maize and Wheat Improvement Center (CIMMYT) in Mexico and the International Rice Research Institute (IRRI) in the Philippines. In examining these institutions, historians have developed persuasive accounts of the role of Cold War geopolitics in shaping an approach to development focused on containing unrest among rural people in Latin America and Asia through interventions consonant with the interests of transnational agribusiness.[12] Following critiques of the Green Revolution that have circulated since the 1970s, they have typically highlighted the shortcomings of a vision of agriculture limited to technical interventions, especially novel plant varieties, capital-intensive mechanization, and petrochemical inputs, and ill-suited to the circumstances of the most socially and economically marginalized farmers.[13]

It is striking to note that even in the case of undeniably influential institutions like CIMMYT and IRRI, the years around 1970 tend to

---

[11] Examples of institutional accounts include Warren C. Baum and Michael L. Lejeune, *Partners against Hunger: The Consultative Group on International Agricultural Research* (Washington, DC: World Bank, 1986); Özgediz, *The CGIAR at 40*; Derek Byerlee, *The Birth of CIMMYT: Pioneering the Idea and Ideals of International Agricultural Research* (Mexico City: CIMMYT, 2016); John Lynam and Derek Byerlee, *Forever Pioneers – CIAT: 50 Years Contributing to a Sustainable Food Future … and Counting*, CIAT Publication No. 444 (Cali, Colombia: CIAT, 2017), http://hdl.handle.net/10568/89043.

[12] Key accounts along these lines include John H. Perkins, *Geopolitics and the Green Revolution: Wheat, Genes, and the Cold War* (Oxford: Oxford University Press, 1997); Nick Cullather, *The Hungry World: America's Cold War Battle against Poverty in Asia* (Cambridge, MA: Harvard University Press, 2010); Jonathan Harwood, *Europe's Green Revolution and Others Since: The Rise and Fall of Peasant-Friendly Plant Breeding* (London: Routledge, 2016); Marci Baranski, *The Globalization of Wheat: A Critical History of the Green Revolution* (Pittsburgh: University of Pittsburgh Press, 2022).

[13] E.g., Nick Cullather, "Miracles of Modernization: The Green Revolution and the Apotheosis of Technology," *Diplomatic History* 28, no. 2 (2004): 227–254; Elta Smith, "Imaginaries of Development: The Rockefeller Foundation and Rice Research," *Science as Culture* 18, no. 4 (2009): 461–482; Raj Patel, "The Long Green Revolution," *Journal of Peasant Studies* 40, no. 1 (2013): 1–63; Glenn Davis Stone and Dominic Glover, "Disembedding Grain: Golden Rice, the Green Revolution, and Heirloom Seeds in the Philippines," *Agriculture and Human Values* 34, no. 1 (2017): 87–102.

mark the endpoint in many existing accounts. From the perspective of the founding of CGIAR in 1971, the same period could be considered a starting point in which these went from singular institutions to fulfilling their promise as model operations for a globalized agricultural research infrastructure.

A number of recent accounts offer routes into facets of CGIAR influence beyond its association with "miracle" rice and wheat, and begin to explore its later history. Historians of Mexican science and politics have led the way in reinscribing Mexican ambitions in a history of global maize and wheat research at CIMMYT that has more typically ignored Mexican agronomists' contributions.[14] New studies have revealed the routes by which CGIAR came to be centrally involved in the management of global crop diversity and managed to maintain this position through decades of controversy regarding the ownership of plant genetic materials.[15] Histories situated outside the traditional geographic frame of Green Revolution histories, such as in Colombia and South Korea, temper stories of global influence with far more complex narratives of local experiences.[16] Histories of agricultural research adjacent to the work of CGIAR – whether livestock breeding in revolutionary Cuba, crop science in Mao's China, or Taiwanese development programs in Vietnam – decenter the dominant historiographic framework of a Western and capitalist Green Revolution and establish important boundaries to claims of CGIAR's novelty and influence.[17] CGIAR's reinvention of its goals for

[14] Netzahualcóyotl Luis Gutiérrez Núñez, "Entre lo inesperado y lo imprevisto: La sequía y los proyectos de mejoramiento de maíz y sorgo en el Bajío, 1943–1970," *Historia Mexicana* 70, no. 1 (2020): 207–258; Gabriela Soto Laveaga, "Beyond Borlaug's Shadow: Octavio Paz, Indian Farmers, and the Challenge of Narrating the Green Revolution," *Agricultural History* 95, no. 4 (2021): 576–608.

[15] Marianna Fenzi, "'Provincialiser' la Révolution Verte: Savoirs, politiques et pratiques de la conservation de la biodiversité cultivée (1943–2015)," Ph.D. dissertation, L'Ecole des Hautes Etudes en Sciences Sociales (2017); Helen Anne Curry, "From Working Collections to the World Germplasm Project: Agricultural Modernization and Genetic Conservation at the Rockefeller Foundation," *History and Philosophy of the Life Sciences* 39, no. 2 (2017); Helen Anne Curry, *Endangered Maize: Industrial Agriculture and the Crisis of Extinction* (Oakland: University of California Press, 2022).

[16] Timothy W. Lorek, *Making the Green Revolution: Agriculture and Conflict in Colombia* (Chapel Hill: University of North Carolina Press, 2023); Tae-Ho Kim, "Making Miracle Rice: Tongil and Mobilizing a Domestic 'Green Revolution' in South Korea," in Hiromi Mizuno, Aaron S. Moore, and John DiMoia, eds., *Engineering Asia: Technology, Colonial Development, and the Cold War Order* (London: Bloomsbury, 2018), pp. 189– 208.

[17] Reinaldo Funes-Monzote and Steven Palmer, "Challenging Climate and Geopolitics: Cuba, Canada, and Intensive Livestock Exchange in a Cold War Context, from the 1960s to the 1980s," in Andra B. Chastain and Timothy W. Lorek, eds., *Itineraries of Expertise: Science, Technology, and the Environment in Latin America's Long Cold War* (Pittsburgh: University of Pittsburgh Press, 2020): 137–158; Sigrid Schmalzer, *Red Revolution, Green Revolution: Scientific Farming in Socialist China* (Chicago: University

agricultural transformation and the objects of agricultural research have also come in for scrutiny, including its introduction of agendas for improving the nutritional profile of crops and addressing gender inequalities.[18] The emergent view of CGIAR's history nonetheless remains fragmented and partial. Opportunities for further analyses and richer historical understanding abound.

The contributions to this volume seize on that opportunity. Here leading historians and sociologists of agricultural research and international development explore the influence of CGIAR and its network of research centers on agriculture, science, and policy since the 1970s. Seeking to extend beyond the early years of CGIAR, and beyond the two most prominent centers, these chapters ask whether and how science- and center-led development changed the practices of farmers, researchers, and policymakers in the years that followed. They traverse five continents and five decades of scientific research, agricultural aid, and political transformation. They pose – and begin to answer – questions about CGIAR informed by the critical historiographies of science, agriculture, and development.

By gathering new critical historical scholarship on CGIAR in a single work for the first time, we hope to make crucial cross-cutting themes visible and bring new research questions to the fore. CGIAR is a sprawling enterprise whose history encompasses five decades of significant transformations in food, agriculture, and industry over diverse geographies and cultures. We do not – and could not – critically address all facets of its story in a single volume. This volume includes case studies of several CGIAR centers but leaves out multiple past and present centers, such as the International Food Policy Research Institute (IFPRI), IWMI, and WorldFish, as well as the research domains in which they specialized. It gives only limited attention to some of the highest-profile centers, such as IRRI and the International Potato Center (CIP). Our focus on CGIAR institutes omits the record of comparable efforts at research for development emanating from the Soviet Union or China. This institutional frame also diverts attention from the effects of CGIAR activity on individual farmers and communities. Rather than see these lacunae as a barrier to presenting a history of CGIAR, we offer the necessarily incomplete view

of Chicago Press, 2015); James Lin, "Martyrs of Development: Taiwanese Agrarian Development and the Republic of Vietnam, 1959–1975," *Cross-Currents: East Asian History and Culture Review* 33 (2022): 53–83.

[18] Sally Brooks, *Rice Biofortification: Lessons for Global Science and Development* (London: Routledge, 2010); Margreet van der Burg, "'Change in the Making': 1970s and 1980s Building Stones to Gender Integration in CGIAR Agricultural Research," in Carolyn E. Sachs, ed., *Gender, Agriculture and Agrarian Transformations* (London: Routledge, 2019).

as an opportunity. Above all, we hope this volume takes stock of the existing scholarship and sets promising agendas for the future. The contributions collected here cluster around three big themes – the role of geopolitics, the pursuit of research as development strategy, and the coordination and centralization of research within a system – but highlight many further analytical possibilities.

Geopolitics

Existing critical assessments of CGIAR and the international research centers typically emphasize their geopolitical functions, highlighting their association with Cold War security imperatives and the desire to shape a capitalist world agro-economy. This emphasis is not misplaced. CGIAR emerged after a World Bank–financed Commission on International Development (the Pearson Commission) called in 1969 for greater coordination of food and agricultural research, in line with the perspectives of World Bank president Robert McNamara and commission members representing the United States, Canada, Western Europe, and Japan.[19] But it is possible to give far more nuance to this dominant account.

As the chapters by Prakash Kumar on India's pursuit of the International Crops Research Institute for the Semi-Arid Tropics (ICRISAT) and Gabriela Soto Laveaga's study of CIMMYT as a Mexican institution highlight especially well, national political aspirations and demands produced ostensibly international institutions that were also constitutive of nation-building projects. Consider Kumar's description of ICRISAT in the age of Indira Gandhi: "It was partly because Gandhi could bring herself to see the international as opposed to the American face of ICRISAT, and because she could bring her constituents to believe in this international image as well, that ICRISAT was accepted even as popular anti-American sentiment in India was peaking." Kumar emphasizes that Gandhi's political balancing act between international and national objectives helped define ICRISAT in its early years. Other chapters in this section demonstrate similar processes for other institutes and regions.

Were, and are, CGIAR centers best understood, to paraphrase Kumar, as Indian (or Mexican, or Colombian, or Filipino, or Nigerian)? Or were they international? How did policymakers, scientists, and national

---

[19] The Pearson Commission, "Partners in Development," 1969. See the Records of the Commission on International Development (Pearson Commission), World Bank Archives, https://archivesholdings.worldbank.org/records-of-the-commission-on-international-development-pearson-commission.

politicians understand and communicate the purpose of institutions grounded in local soil yet always drawing funding and personnel from across the globe? As Timothy W. Lorek's Colombian contextualization of the International Center for Tropical Agriculture (CIAT) shows, sometimes both national and international agendas could – and did – run roughshod over local realities. In other cases, politicians' lofty aspirations for science-driven development to improve lives and livelihoods were beset by conflict, including armed conflict, as Courtney Fullilove explains in her account of the International Center for Agricultural Research in the Dry Areas (ICARDA) in Syria and Lebanon.

These chapters illustrate the tensions between international objectives or funding and local experiences at their most intense. Fullilove aptly characterizes "a globalized vision for agricultural development that made poverty alleviation into a single project and poverty itself into a uniform condition. While international research organizations have made increasing claims to operate at a global scale, on behalf of universal interests, the landscapes they traverse are more complex in agro-ecological and historical terms." The imagination of ostensibly ecology- and geography-specific centers, including CIAT and the International Institute of Tropical Agriculture (IITA) in Nigeria, each mandated to improve agriculture in the global tropics, as well as ICRISAT and ICARDA, which respectively aimed to produce knowledge and solutions for the semi-arid tropics and arid regions, flattened the heterogeneity of small-scale farming. It overlooked microclimates, cultural expectations, and local political contexts. The long-term negotiation between internationally oriented institutions and highly subjective – and always changing – localized contexts is at the heart of our geopolitically oriented chapters and indicates an important topic for future research.

Research as Development Strategy

A key theme in the history of CGIAR is its prioritization of scientific research as an instrument of development. This has inevitably demanded that CGIAR leaders and staff formulate problems that can be considered "solvable" through research. Such problems have ranged from the very general "second-generation development problems" – namely, inequality and unemployment generated by agricultural intensification – highlighted by Lucas M. Mueller in his examination of the CGIAR founders' formulation of early institutional missions. They have also encompassed more specific concerns, like the perceived "protein gap" in Latin America that drove bean research and extension programs at CIAT even as war and poverty generated more immediate threats to nutritional security, as Wilson Picado-Umaña documents. Regardless of the degree of specificity involved, the imagination of problems set boundaries around the expertise, tools, and labor considered relevant to CGIAR programs and development more broadly. Mueller's analysis shows that solving "second-generation development problems" demanded the know-how of agricultural economists, individuals practiced in thinking about market-based interventions, much as it did the labor of breeders who, for example, could produce groundnut varieties for new export-oriented agricultural programs at ICRISAT.

In some cases the articulation of research objectives mapped poorly onto the real needs of farmers or rural communities. This might be a product of the mismatch between a realizable scientific aspiration and the nature of the "problem" to be addressed. Picado-Umaña describes how breeders' efforts to produce a higher-yielding bean for Central American growers faltered in light of eaters' regular consumption of a wide variety of different beans and the complicated agro-ecologies of the region's bean fields. The quest to identify an "ideal type" on which to focus, in the interest of maximizing the impact of research, ran counter to local expectations for dietary diversity, not to mention strongly embedded socioeconomic inequalities and political conflict. The idealized rice "target environments" devised at the West Africa Rice Development Association (WARDA) and discussed by Harro Maat offer a different scenario, one in which research agendas remained mired in the political and economic objectives of old authorities, despite stated intentions of serving the needs of newly independent states and citizens. Locating the founding of WARDA within longer colonial histories, Maat shows that until the 1990s "WARDA continued to focus on the rice farming areas defined in the colonial period, addressing European commercial interests rather than the concerns of West African rice farmers." Maat's analysis offers a compelling reminder that understanding CGIAR-era research goals demands detailed knowledge of the projects on which they were built – and not just the national projects discussed by Soto Laveaga and Kumar, but in many cases colonial forerunners as well.

Perhaps the challenge for research as development has been deeper than simply the challenge of pinpointing a relevant problem, given the constraints imposed by disciplines, inherited assumptions, available resources, and other obstacles. In their study of ILRI and its predecessors, Rebekah Thompson and James Smith suggest that the challenge may lie in an unresolvable contradiction between doing science and pursuing development. They describe how a drive for "scientific excellence" in livestock research within CGIAR has created challenges for researchers caught between the often incommensurable imperatives of helping poor farmers *and* publishing peer-reviewed research. They urge scholars to remember that it "is important to recognize how institutions and funding bodies conceptualize excellence, as this shapes the way in which knowledge is produced and how research impact is ultimately perceived." This, too, points towards a topic ripe for further study, both within CGIAR and its constituent institutions and across the larger landscape of development initiatives.

Coordination and Centralization

One of the most difficult elements of CGIAR history to tackle is the extent to which it has – or has not – functioned beyond individual research programs and centers. Is there, or was there ever, meaningful system-wide activity that warrants closer historical examination of CGIAR as the network it aspired to be? Our contributors offer several possibilities, from center-based projects that operated via institutional interdependencies, such as the international coordination of maize research by CIMMYT, detailed here by Derek Byerlee and Greg Edmeades, to explicitly multi-center structures, including the oversight of the conservation of plant genetic materials across CGIAR by Bioversity and its predecessors, as analyzed by Marianna Fenzi, to events constitutive of CGIAR as a singular entity, such as consolidation of legal services to facilitate intellectual property management, as noted by David Jefferson.

In these domains and others, the vision of a global network of institutions drove efforts for system-wide coordination of research tools, objectives, and administration – to mixed effect. Byerlee and Edmeades claim successes in CIMMYT's international maize-breeding program, especially in its generating varieties suitable for drought-prone regions with low soil fertility in Africa. Yet they also chart a gradual decentralization of research, which allowed breeders in diverse contexts to produce lines suitable for their locales, and ever-growing interdependencies with private industry. These transformations belie the notion, in many ways foundational to CGIAR, of strong centralized research programs producing clear public goods. Fenzi's account similarly charts the "success" of CGIAR in establishing oversight over the world's major crop gene banks and expanding the extent of plant genetic resources conserved in these institutions. Her emphasis, however, is on the narrow worldview behind this approach to conservation, in which farmers' varieties were only considered useful as the raw materials for professional breeders, and therefore highlights the limitations on whom CGIAR gene banks ultimately served.

Harmonizing activities across a system was predictably difficult. Not everyone within an organization diverse in culture, geography, and discipline agreed on how best to deliver "global public goods," for example, as shown by Jefferson in his study of CGIAR's evolving approach to intellectual property. Jefferson charts three broad positions taken by various CGIAR stakeholders in response to the expansion of intellectual property protections in agriculture since the 1980s and the ever-increasing demands on CGIAR research to align with norms in the private sector. He also charts the often acrimonious debates in which these positions were sketched out. That many stakeholders still do not agree on what the appropriate position of CGIAR on intellectual property should be, or the usefulness of such protections to local research objectives, is evidenced in the disparate pursuit of intellectual property claims across the centers.

Coordinating across CGIAR institutions, whether in the interest of reducing costs, aligning stated objectives, or creating explicitly "system-wide" capacities, also produced new knowledge and expertise. In some cases, this expertise could simply be recruited. As Jefferson indicates, a demand for legal expertise to help manage intellectual property concerns drove the development of new centralized advisory capacities in CGIAR – presumably staffed by legal experts possessing comparable experience gained elsewhere. In other cases, expert knowledge essential to system-wide coordination had to be created. The history of crop descriptors examined by Helen Anne Curry and Sabina Leonelli presents one such example. Tasked with overseeing CGIAR centers' management of plant genetic materials, Bioversity and its predecessor institutes found themselves not only preparing lists of the agreed-upon attributes and terms to use in describing specific crops (ostensibly to facilitate the circulation of breeding materials), but also devising the rules that would govern the creation of such lists and overseeing their circulation and upkeep.

Crop descriptors are a research product in and of themselves and, like the many other CGIAR research products touched upon in this volume, must be viewed in light of the political and economic motivations that produced and perpetuated them. For Curry and Leonelli, this exercise leads back to the aspirations of CGIAR as axis. As they write, exerting oversight over crop descriptors "provided an opportunity for CGIAR to instantiate and consolidate its central position in a larger web of international agricultural research initiatives" and "served to advance CGIAR's identity as an essential resource for globalized development." Their observation points to yet another key research avenue opened up by this volume: identifying and understanding specific means by which

CGIAR, as distinct from its constituent institutes, made itself a dominant actor in the crowded domain of international development.

Even as we categorized contributions as speaking to one of these themes – geopolitics, research as development, or coordination and centralization – we recognized other points of intersection and multiple avenues opening up for further inquiry. Clearly, attention is needed to address the role of transnational agribusiness in the history of CGIAR, the turn to farming systems and resource management research, and the incorporation (or lack) of attention to gender and other equity concerns. We hope that readers will identify many more issues ripe for attention in the chapters that follow.

In 2021, its fiftieth anniversary year, CGIAR transitioned to a new mode of organization: One CGIAR.[20] In a bid for the "integration of CGIAR's capabilities, knowledge, assets, people, and global presence for a new era of interconnected partnership-driven research," CGIAR shifted from a model of networked but independent institutions into an imagined program of far more centrally planned and coordinated activities. Many of the drivers of this shift are predictable: to cut costs, exert greater control over research, and maintain relevance amidst changing global priorities. The outcomes are, of course, not knowable at all. The shift to One CGIAR reflects another inflection point in the longer history captured in this volume, wherein CGIAR and its constituent parts fluctuated between centralized and decentralized coordination models, international and national (or regional and local) objectives and orientation, and public versus private power.

The completion of this volume therefore coincides with another turning point in the organization and strategy of the very institutions it aims to parse. It is a fitting moment for the analyses presented here. We hope the contributions to this volume will offer many scholars and students an opportunity to consider through the lens of CGIAR what it has meant, historically, to conduct research in the name of development – and to consider critically what this pursuit has meant for scientists, farmers, and citizens over the past fifty-plus years.

[20] CGIAR, *CGIAR 2030 Research and Innovation Strategy: Transforming Food, Land, and Water Systems in a Climate Crisis* (Montpellier, France: CGIAR, 2021).

Part I

Geopolitics

1 Locating ICARDA
The Geopolitics of International Agricultural Research in the Middle East and North Africa

Courtney Fullilove

In the summer of 2012, armed gangs began raiding the headquarters of the International Center for Agricultural Research in the Dry Areas (ICARDA) in Tal Hadya, Syria, stealing vehicles, computers, and other equipment at night. Within a few months, ICARDA's field trials were abandoned and the experiment station dismantled. In November, a video uploaded to YouTube showed a group of armed men in front of ICARDA's vacant headquarters, declaring the institution a fallen bastion of Bashar al-Assad.[1] An international research organization applauded for its advances in the interest of struggling farmers was recast as an instrument of oppression and corruption. In the following years, the facility remained occupied by the anti-Assad groups al-Nusra and Ahrar al-Sham. By November, ICARDA had relocated operations to Amman, Tunis, and Beirut. The seizure, evacuation, and uncertain symbolism of ICARDA, however marginal to the story of the Syrian civil war, was a stark reminder of the embeddedness of international public organizations in nation-states, and of the sometimes fraught relationship of international research to global geopolitics.

This chapter explains how ICARDA came to be located in Syria by examining the broader geopolitical logic of international agricultural research. Ironically, Beirut had been the intended site of ICARDA's headquarters in the early 1970s, but planners deemed Lebanon's political situation too volatile as the country erupted into war in 1975. Ultimately ICARDA, then the newest of the Consultative Group on International Agricultural Research (CGIAR) centers, took root thirty-two kilometers from Aleppo in the village of Tal Hadya. Its mandate was to improve the livelihoods of resource-poor farmers in dry areas through research, working within national agricultural research systems and directly with farmers. Over the next thirty years, ICARDA became a research hub and home to a major international gene bank[2] (Figure 1.1).

---

[1] الكاردا ياجح رير تحري ادلب-الشام اجرار بتائك ("Idlib Liberation Roadblock ICARDA – Freedom Brigades of al Sham"), November 24, 2012, http://youtu.be/TzE49m1Tbzk (no longer available).

[2] This chapter originated in research conducted: at ICARDA in Aleppo, Syria in 2010; during a series of seed-collecting missions in the Southern Caucasus and Central Asia

Figure 1.1 A view of ICARDA's facilities in Tal Hadya, Syria, 2007.
Photo by Global Crop Diversity Trust/Cary Fowler. By permission of
Global Crop Diversity Trust.

Born of the Cold War, ICARDA emerged from exercises of European
imperialism, Great Power rivalries, and the concomitant restructuring of
a patchwork of modern nation-states in Western Asia and North Africa.
In the aftermath of World War II and European withdrawal from formal
governance, newly independent nation-states became battlegrounds of
the Cold War. The southern rim of Asia, which provided a buffer to the
Soviet Union, became a focus of US strategies of containment from the
1950s. Scholars have attended to the global movement of soldiers, arms,
and aid that fueled Cold War conflict in the "killing fields" of the Asian
rim.[3] They have paid less attention to the ways in which the institutional
development of international research organizations served Cold War
objectives.[4] The founding of ICARDA was part and parcel of the

between 2010 and 2015, which were facilitated by ICARDA staff; and at a site visit to the
new headquarters of the Genetic Resources Unit in Rabat, Morocco.

[3] Paul Thomas Chamberlin, *The Cold War's Killing Fields: Rethinking the Long Peace*
(New York: HarperCollins, 2019).

[4] For exceptions see, e.g., Nick Cullather, *The Hungry World: America's Cold War Battle
against Poverty in Asia* (Cambridge, MA: Harvard University Press, 2010);
Daniel Immerwahr, *Thinking Small: The United States and the Lure of Community
Development* (Cambridge, MA: Harvard University Press, 2018). The phrase "killing

American effort to domesticate Western Asia and North Africa according to the geopolitical terms of the Cold War, bringing Syria from the sphere of Soviet influence and into the American one. Designated the "Middle East and North Africa" (MENA), the region ultimately became synonymous with the extraction of oil resources.[5]

The framing of ICARDA in relation to the postwar MENA region grafted a political geography onto a broad range of ecological areas. Under the charge of CGIAR, agronomists characterized these regions in the vocabulary of ecology, establishing them as a terrain for "dryland" agricultural science.[6] Planners, drawing on climatic models, classified the region in agro-ecological terms devised in reference to the tropics. Functionally, their logic shored up a focus on rainfed, or unirrigated, agriculture in semi-arid and arid lands. But this rendering of dry areas masked the geopolitical framing of international agricultural research in the postwar period. The remainder of this chapter charts the imperial origins of international agricultural research in Syria, the Cold War on hunger, and CGIAR's classification of arid regions, towards an account of how dryland agricultural science became the ground for technological and political intervention in decolonized lands.

Imperial Prehistories of International Agricultural Research

Orientalism suffused the disciplines of environmental science as they developed in the late nineteenth and early twentieth centuries, as the historian Diana Davis has noted, "hiding power relations in specific

---

fields" was also invoked in Latin America, and Chapters 3 and 8 in this volume by (respectively) Timothy W. Lorek and Wilson Picado-Umaña track the institutional development of international research in such Cold War spaces.

[5] Like other entities administering, or hoping to administer, post-independence projects in the region, ICARDA planners referred to its intended domain as the "Near East and North Africa" or the "Middle East and North Africa." It is the work of this chapter to understand how international agricultural research contributed to the construction of the MENA region as a geopolitical category within West Asia and North Africa. I use each term advisedly.

[6] Diana K. Davis, *Resurrecting the Granary of Rome: Environmental History and French Colonial Expansion in North Africa* (Athens: Ohio University Press, 2009); Diana K. Davis, *Arid Lands: History, Power, Knowledge* (Cambridge, MA: MIT Press, 2016). The geographer Omar Tesdell has written that "from the perspective of Palestine, what might be called the 'global drylands assemblage' emerges as an uneven field of political and technical activity constituted in the Middle East, and also through its relations within North America and beyond"; see Omar Loren Tesdell, "Shadow Spaces: Territory, Sovereignty, and the Question of Palestinian Cultivation," Ph.D. dissertation, University of Minnesota (2013).

forms of knowledge production."[7] Europeans invoked biblical rhetoric to portray the dry lands of the Near East and North Africa as barren, desolate places of trial and suffering, in need of imperial intervention to reverse centuries of deforestation and desertification. European observers attributed environmental conditions to human degradation of the natural environment. In fact, the extent of both deforestation and desertification were exaggerated and often misrepresented a regional history of coping with the high temperatures and low rainfalls. As a region, Western Asia and North Africa can be characterized by thousands of years of sophisticated water control systems and agricultural practices adjusted to the natural environment.[8]

Europeans reiterated myths of environmental degradation to justify imperial projects. Across the region, narratives of overgrazing and excessive irrigation facilitated imperial goals of improvement and resource management. In Algeria, the French rendered themselves the heirs of Rome, there to restore a deteriorated environment to its rightful state.[9] In Egypt, British colonizers saw land that needed to be made productive and irrigated for cotton production in the late nineteenth century.[10] Meanwhile the French invested heavily in the Eastern Mediterranean, including the port of Beirut, railroads, and industry in the coastal region.[11]

The collapse of the Ottoman Empire further emboldened orientalist and biblically inflected interpretations of the landscape. In Palestine, the British justified control of Bedouin populations with a mandate to counter deforestation and blamed environmental deterioration on Arab land use and Ottoman mismanagement. Reforestation projects, and the broader commitment to "make the desert bloom," motivated early Zionists in the same region.[12] International wheat-breeding initiatives and a focus on Palestine as a site of domestication helped remake drylands as targets of colonization.[13] Iraq, in turn, figured as a battered and

---

[7] Diana K. Davis, ed., *Environmental Imaginaries of the Middle East and North Africa* (Athens: Ohio University Press, 2013), 22.

[8] Davis, *Arid Lands*; Davis, *Environmental Imaginaries*; Alan Mikhail, *Under Osman's Tree: The Ottoman Empire, Egypt, and Environmental History* (Chicago: University of Chicago Press, 2019).

[9] Davis, *Resurrecting the Granary of Rome*.

[10] Jennifer L. Derr, *The Lived Nile: Environment, Disease, and Material Colonial Economy in Egypt* (Redwood City, CA: Stanford University Press, 2019).

[11] Elizabeth R. Williams, *States of Cultivation: Imperial Transition and Scientific Agriculture in the Eastern Mediterranean* (Redwood City, CA: Stanford Ottoman World Series, 2023).

[12] Shaul Ephraim Cohen, *The Politics of Planting: Israeli–Palestinian Competition for Control of Land in the Jerusalem Periphery* (Chicago: University of Chicago Press, 1993).

[13] Omar Tesdell, "Wild Wheat to Productive Drylands: Global Scientific Practice and the Agroecological Remaking of Palestine," *Geoforum* 78 (2017): 43–51.

degraded Babylonia, waiting to be restored to its former glory as a cradle of civilization.[14]

Syria's construction as a modern nation-state was the collateral damage of World War I, as European powers jockeyed for control of the former lands of the Ottoman Empire. British and French designs led to an array of shoddy plans to divide the region into spheres of influence, ultimately resulting in the interwar ordering that placed Palestine, Trans-Jordan, Mesopotamia (Iraq), Egypt, and the Gulf within a British zone of influence, and Syria (including present-day Lebanon) within a French zone of influence. The mandate system established by the League of Nations in 1922 placed Palestine under British control and Syria under French control. These agreements, which fragmented traditional trade networks and cultural continuities, were accompanied by often disingenuous gestures towards Arab independence. The British and French "mandate" was a fig leaf for renewed imperial designs in a moment of political and economic disarray. Ostensibly installed to usher their charges into a new era of self-government, neither the British nor the French had any interest in stewarding national independence.[15]

Ultimately, the modern nation-states of Syria and Lebanon were carved from the broader region stretching from the north of the Arabian desert through contemporary Israel–Palestine and Lebanon to the Tigris–Euphrates river system. The area, alternately christened Suri (old Babylonian), the Levant (Italian traders), and *Bilad al-Sham* (the "country of Damascus"), comprises the lands of the so-called "Fertile Crescent": a term coined on the eve of World War I to describe the birthplace of agriculture in Western Asia. Amid a concerted policy of fragmentation, the French made sluggish and inconsistent gestures towards Syrian independence. This flip-flop exacerbated tensions between various groups who aspired to government and who expressed markedly different visions for Syria's future as a nation-state. In 1946, the French, hobbled by war, formally withdrew from Syria, leaving a nation-state mauled by European invasion and mismanagement. Lebanon, too, emerged as an independent state with borders that had been drawn by the French mandatory government.

---

[14] Priya Satia, "A Rebellion of Technology Development: Policing and the British Arabian Imaginary," in Davis, ed., *Environmental Imaginaries*, pp. 23–59.

[15] My account of Syrian history throughout this chapter benefits from concise histories by David W. Lesch, e.g. *Syria* (Cambridge: Polity Press, 2019); Patrick Seale, *The Struggle for Syria: A Study of Post-War Arab Politics, 1945–1958* (London: Tauris, 1987); and Philip Shukry Khoury, *Syria and the French Mandate: The Politics of Arab Nationalism, 1920–1945* (Princeton, NJ: Princeton University Press, 2016).

In the wake of World War II, as Britain and France ceded their spheres of influence in the former Ottoman Empire, the MENA region emerged as a theater of conflict between the USSR and the United States. The United States regarded the MENA region primarily as an oil-producing zone, with a handful of independent nation-states supplying newly insatiable Western European and American appetites. In 1956, fearing a loss of access to oil reserves in the likelihood of a Soviet invasion of the region, the United States repudiated a secretive British-Franco-Israeli invasion to reverse Nasser's nationalization of the Suez Canal. In the wake of the Suez crisis, the United States entered a perceived vacuum of power in the Middle East, courting new governments as building blocks in its nascent strategy to contain global communist influence. As Nikita Khrushchev made overtures to support Arab states against lingering British colonialism and contain Israeli influence, the United States, under the direction of President Dwight D. Eisenhower and Secretary of State John Foster Dulles, promised aid to any country requiring assistance to thwart communist infiltration. Anti-communism provided the conceptual language through which Americans framed their interest in the region. In practice, the Eisenhower Doctrine led the United States into successive machinations and interventions to impede Soviet influence and stave off pan-Arab realignment.[16] Syria and Lebanon found themselves tangled in these superpower rivalries, which in turn complicated regional relationships.

In the background, diplomats and their technical advisors reframed the region as the terrain of international development. Depictions of Western Asia and North Africa as degraded and in need of restoration continued during the postwar period, with little reference to the role of European invasion in their de-development. These characterizations, sketched by the United States Agency for International Development (USAID) and other foreign assistance agencies, were shored up by novel social and economic theories. Modernization theorists such as Walt Whitman Rostow posited that all civilizations proceeded through one path of

---

[16] In the broader context of an "Arab Cold War," the United States, for example, recognized the short-lived union of Syria and Egypt as the United Arab Republic (UAR) in 1958, then intervened militarily on behalf of the Maronite Christian leadership in Lebanon to prevent a potential expansion of the UAR. See, e.g., Peter L. Hahn, "Securing the Middle East: The Eisenhower Doctrine of 1957," *Presidential Studies Quarterly* 36, no. 1 (2006): 38–47; Maurice M. Labelle, "A New Age of Empire? Arab 'Anti-Americanism', US Intervention, and the Lebanese Civil War of 1958," *The International History Review* 35, no. 1 (2013): 42–69; Richard J. McAlexander, "Couscous Mussolini: US Perceptions of Gamal Abdel Nasser, the 1958 Intervention in Lebanon and the Origins of the US–Israeli Special Relationship," *Cold War History* 11, no. 3 (2011): 363–385; Douglas Little, "His Finest Hour? Eisenhower, Lebanon, and the 1958 Middle East Crisis," *Diplomatic History* 20, no. 1 (1996): 27–54.

development and looked, at the end, eerily like the United States. In this reading, agriculture was a pit stop between nomadism and industrialization in the progress of civilizations.[17] The need to restore land could justify a wide range of interventions, from agricultural and economic reforms to sedentarization and military force. Nor were these arguments the sole province of colonizers. As Egyptian President Gamal Abdel Nasser's leadership would demonstrate, these same reforming projects could be reclaimed for nationalistic ends.[18] Nasser's land reclamation projects were of a piece with his nationalization of the Suez Canal, and he played American and Soviet interests against one another.

Superpower rivalry and regional competition over the future of Arab nationalism exerted further pressure on Syria's weak and dysfunctional government. In this climate, and amid successive coups, the secular, socialist Ba'ath Party took power in March 1963, with a slogan of "freedom, unity and socialism." Within the party, traditionally marginalized minorities such as Alawite and Druze had entered positions of power, to the resentment of the Sunni majority. Two such figures were Salah Jadid and Hafez al-Assad, who would jockey for power within the Ba'ath Party. In 1970, following the disastrous 1967 Arab–Israeli war, Assad, then minister of defense, wrested power from Jadid. Assad's seizure of power was an outcome of long rivalry between an urban mercantile class dominated by Sunni Muslims, French, Islamists, and fascists, and a younger generation of Marxists (soon to form the Ba'ath Party) who rejected accommodations to imperial rule. As Assad faced growing isolation within the Middle East, global recession, and persistent sectarian and economic division within Syrian society, pragmatism over idealism was to be his governing strategy.[19]

The rivalries that brought Assad to power superimposed a deep divide between urban and rural Syria. Prior to World War I, the southerly city of Damascus had been linked to Beirut, Haifa, Jerusalem, and Baghdad, all

[17] W. W. Rostow, *The Stages of Economic Growth: A Non-Communist Manifesto* (Cambridge: Cambridge University Press, 1960), esp. chapter 2. See also Mark Mazower, *Governing the World: The History of an Idea* (New York: Penguin Books, 2012); David C. Engerman, *Staging Growth: Modernization, Development, and the Global Cold War* (Amherst: University of Massachusetts Press, 2003); Michael E. Latham, *Modernization as Ideology: American Social Science and "Nation Building" in the Kennedy Era* (Chapel Hill: University of North Carolina Press, 2000); Michele Alacevich, *The Political Economy of the World Bank: The Early Years* (Stanford, CA: Stanford University Press; World Bank, 2009). On USAID in Egypt, see Timothy Mitchell, "Afterword," in Davis, ed., *Environmental Imaginaries*, pp. 265–274.

[18] Jeannie Sowers, "Remapping the Nation, Critiquing the State: Environmental Narratives and Desert Land Reclamation in Egypt," in Davis, ed., *Environmental Imaginaries*, pp. 158–191.

[19] Lesch, *Syria*, pp. 87–111.

of which fell within the British zone of influence. Meanwhile, Aleppo, in the north, shared with its Turkish, Armenian, and Kurdish neighbors an orientation towards Central Asia along the path of the Silk Road, as well as to the Iraqi city of Mosul. Modern-day Syria is made up of semi-arid and arid land (the Syrian desert), along with a narrow coastal plain on the Mediterranean Sea. Populous urban centers constitute a vertical line from north to south, linking Aleppo, Hama, Homs, and Damascus. Syria's agricultural sector consisted of cotton, wheat, barley, sugar beet, and olive production. Rainfed agriculture predominated, as it does to this day. The gulf in wealth between the cities of the west and the rural land to the east contributed to longstanding tensions in Syrian society, compounded by the balkanization of historical trade routes to constitute French and British spheres of influence.

The CGIAR network came into being as Assad seized power; legacies of empire, Cold War development, and Arab nationalisms shaped its agenda. Withdrawing from formal empire, Europe and the United States competed to be the dominant exporters of food, then of agricultural inputs, based on a model of input-intensive industrial agriculture. As several contributors to this volume chart, the 1950s and 1960s saw the export of high-yielding seeds and agricultural methods, attributed to American agronomists and celebrated as the Green Revolution (see Prakash Kumar, Chapter 2, and Gabriela Soto Laveaga, Chapter 4, this volume). Aiming to build on the alleged successes of the Green Revolution, the United Nations Food and Agriculture Organization (FAO) supported programs of agricultural modernization and the free exchange of germplasm between countries for the use of breeders. By the 1970s, decolonized lands were the sites of modernization projects premised on genetically uniform, high-yielding monocultures and the prospective hosts of CGIAR centers for research on food security, rural poverty, and sustainable development. It fell to CGIAR's technical advisors to justify their designs.

At the inaugural meeting of its scientific advisory body, the Technical Advisory Committee (TAC), in November 1971, the FAO director general and Dutch agronomist Addeke Hendrik Boerma applauded the "new international approach to agriculture" for its promise to build on Green Revolution successes in "other regions of the world maintained on a global basis."[20] On behalf of the International Bank for Reconstruction and Development (IBRD), Director of Development Services Richard

---

[20] CGIAR Technical Advisory Committee, "Report of the First Meeting of the Technical Advisory Committee. 29 June–2 July 1971," November 5, 1971, 2, https://cgspace.cgiar .org/handle/10947/1422.

Demuth envisioned an application of the model "developed by the international cereals institutions" to other crops and livestock.[21] Focused initially on the increased production of cereal crops in the "Third World," the TAC attended to regions not yet served by CGIAR's four established research centers. After a series of TAC, working group, subcommittee, and donor meetings, ICARDA was established, with the Canada-based International Development Research Centre (IDRC) as executing agency. In January 1977, ICARDA assumed operations to pursue research into the agricultural systems of the MENA region. The technocratic process through which the center was founded and administered obscured the extent to which the war on hunger in which CGIAR centers participated was an aspect of an anti-communist project. Its architects and technical advisors linked the perceived successes of the Green Revolution to a vision of international development that would make Asia and North Africa after a Western European and American image.

Classification of Dry Areas

In institutional terms, the CGIAR TAC's recommendations determined the site and remit of ICARDA. As it mapped priority areas onto the world, the committee flagged the semi-arid and arid regions of the Near East and North Africa as "a major research problem which had not yet received adequate study."[22] It anticipated that a single center could not address the diversity of conditions of the region but nevertheless speculated that centralized research could accelerate agricultural development in low-rainfall areas.[23] By identifying low rainfall as the primary source of low agricultural productivity in the region, the committee incidentally disregarded institutional and political conditions, including colonial and post-colonial fragmentations of landholding and technocratic projects to exert greater control over agricultural resources.[24]

Although the TAC flagged the MENA region as understudied, the region was already populated by international organizations. The International Maize and Wheat Improvement Center (CIMMYT) had outreach programs there, as well as links with the FAO Near East Wheat and Barley Program. The Ford Foundation–funded Arid Land Agricultural Development program (ALAD) operated in the Beka'a Valley in Lebanon. Meanwhile, FAO was in the midst of a survey of existing

[21] Ibid.   [22] Ibid., 12.   [23] Ibid.
[24] Rafaelle Bertini and Abdallah Zouache, "Agricultural Land Issues in the Middle East and North Africa," *The American Journal of Economics and Sociology* 80, no. 2 (2021): 549–583; Williams, *States of Cultivation*.

research organizations in the Near East. As the TAC planned an exploratory mission, observers noted an upcoming meeting in New York attended by FAO, United Nations Development Programme (UNDP), USAID, and Ford and Rockefeller Foundation representatives to discuss "responsibilities and means of improving collaboration" among organizations pursuing agricultural projects in the region.[25] All of these activities marked the longer history of Euro-American involvement in the MENA region, and the persistent interest in its development.

To evaluate the research needs and priorities, the TAC commissioned a team led by Professor Dunstan Skilbeck of Wye School of Agriculture, University of London to visit countries in the MENA region in spring 1973.[26] The Skilbeck Committee, reporting in June 1973, recommended the establishment of a new center, internationally supported and multidisciplinary in approach, to serve the needs of the region. It recommended that the center assume global responsibility for select staple crops, including barley and durum wheat, and that it take a holistic approach to the needs of farmers on arid lands.[27]

The Skilbeck mission's report designated the region as a coherent one for reasons that were equal parts environmental and political. While the Near East and North Africa shared some problems of development with other regions, it also had a unique "agricultural environment and consequent research needs" arising "partly from its geographical location and partly from its long and sometimes turbulent history."[28] Moreover, these conditions were ones of degradation and marginality with social and cultural roots:

As a result of historical processes rather than any strong evidence of climatic change, much of the region, which was once the granary of ancient civilization, is now barely able to support a low population density at the subsistence level and there is extensive deforestation and degradation of natural grazing reflected in serious erosion and desert encroachment. Once fertile land has been abandoned, ancient irrigation systems have silted up or fallen into disuse and there is widespread salinity. The proportion of arable land to total area (only 6.3 percent for the Region as a whole) is lower than that in other developing regions, but the

---

[25] CGIAR Technical Advisory Committee, "Report of the Fourth Meeting of the Technical Advisory Committee, 2–4 August 1972: Draft," September 1972, 32, https://cgspace.cgiar.org/handle/10947/1434.

[26] CGIAR Technical Advisory Committee, "Report of the Fifth Meeting of the Technical Advisory Committee, 30 January–2 February 1973: Draft," March 1973, 41, https://cgspace.cgiar.org/handle/10947/1411.

[27] CGIAR Technical Advisory Committee, "Report of the Sixth Meeting of the Technical Advisory Committee, 25 July–2 August 1973: Draft," September 1973, https://cgspace.cgiar.org/handle/10947/1451.

[28] Ibid., 4.

balance is not largely composed of grazings or forests as in Latin America or Africa, but of unusable desert and wasteland.[29]

The committee attributed low yields, even in irrigated areas, to "social and structural rigidities and the persistence of traditional cultural practices."[30] The vague reference to Ottoman institutions and folkways omitted a more granular discussion of French and British interventions to restructure local landholdings and productivity.[31] Instead, the committee leapfrogged over recent colonial and postcolonial history to assert that population growth and urbanization exerted further pressure on resources. In the face of growing deficits, attempts to expand cultivation intensified conflicts between farmers and pastoralists shepherding sheep and goats.[32]

The region's agro-ecology defied reduction to climate classifications, combining a Mediterranean climate zone with harsh, arid conditions and searingly hot summers. While the majority of the target region, apart from Sudan and the southern Arabian peninsula, was within a Mediterranean climate zone characterized by rainfall in the winter and early spring, its environment was harsh and arid rather than temperate, marked by severe winters and unreliable precipitation.[33] The designation of a Mediterranean climate zone misrepresented agro-climatic conditions. In spite of this climatic diversity, there were many common features of the countries surveyed: chiefly, "searing summer temperatures" with low precipitation, making irrigation "pre-requisite for the production of most summer crops."[34] In the view of the committee, these common conditions should be made into a single culture of production: that is, "the ecological complementarities between zones of different production potential must be translated into production complementarities if the overall productive capacity of the region's agriculture is to be mobilized to meet its socio-economic goals."[35] This oblique analogy of production and ecology omitted a discussion of politics in structuring production, including the historical roots of inefficiencies and the capacity of states, land legislation, or specific configurations of land tenure to manage access to resources.

The struggle to name the center signaled a mismatch between the political and ecological orderings of the landscape. In its November 1973 meeting, the TAC adopted the working title International Center

---

[29] Ibid., i.   [30] Ibid., 5.
[31] Williams, *States of Cultivation*; Bertini and Zouache, "Agricultural Land Issues in the Middle East and North Africa."
[32] CGIAR Technical Advisory Committee, "Report of the Sixth Meeting of the Technical Advisory Committee," i–iv.
[33] Ibid., 4.   [34] Ibid.   [35] Ibid., 11.

for Research in Arid Lands.[36] It was a short-lived designation. In successive meetings, members of both the Skilbeck mission and the TAC Working Group on the Research Needs of the Near East and North Africa, which was appointed to review the mission's findings, objected to the inclusion of the terms "Arid Lands" or "Arid Zone" in the title. On the one hand, it portended conflict or redundancy with CGIAR's International Crops Research Institute for the Semi-Arid Tropics (ICRISAT), which had been founded the year prior to address the requirements of arid lands in the tropics (see discussion in the chapters by Prakash Kumar, Chapter 2, and Lucas M. Mueller, Chapter 5, this volume), as well as the Institute for Arid Zones in New Delhi managed by the Indian government. Moreover, many contended that the designation "arid" was "inaccurate as applied to the agro-ecological areas under consideration," inasmuch as these areas encompassed a diverse range of climate conditions in the Mediterranean. While some suggested the inclusion of "Mediterranean" in the title, others objected that parts of the region were outside the Mediterranean climate zone and that such a designation applied a "narrowly regional connotation to a centre whose work might have much wider application." The committee eventually accepted the term "dry areas" as the "most descriptive of the probable focus of the centre's work," and the title "International Research Centre for Agriculture in Dry Areas" was suggested.[37]

Although the TAC had noted that a single center could not address the diversity of ecological conditions in the region, the Skilbeck mission nevertheless reiterated the TAC's preference for a centralized organization to address major problems in low-rainfall areas. Following the Skilbeck mission, the TAC appointed a subcommittee to make recommendations for the location and staffing of the prospective center. The subcommittee sought "proximity to a broadly representative range of ecological conditions."[38]

Since it defined water as the limiting factor of agricultural production in the region, it elected to prioritize climate in choosing a site. Its agro-ecological mapping borrowed the Troll climate classification, which had also been used in the mission that established the parameters for

---

[36] CGIAR Technical Advisory Committee, "Report of the TAC Working Group on the Research Needs of the Near East and North Africa," November 1973, 14, https://cgspace.cgiar.org/handle/10947/1171.

[37] CGIAR Technical Advisory Committee, "Report of the Seventh Meeting of the Technical Advisory Committee, 4–8 February 1974: Draft," February 1974, 10, https://cgspace.cgiar.org/handle/10947/1439.

[38] CGIAR Technical Advisory Committee, "Location of the Proposed International Centre for Research in the Near East and North Africa," June 1974, 2, https://cgspace.cgiar.org/handle/10947/730.

ICRISAT.[39] Carl Troll and Karlheinz Paffen's 1965 classification of the "Seasonal Climates of the Earth" (Figure 1.2) divided tropical climates by the number of humid months, where humid months were defined as those in which mean rainfall exceeded potential evapotranspiration.[40] The tropics of South America and Africa provided the reference point for the model.

The Troll classification imperfectly represented the broad range of climate conditions in the MENA region. These included tropical dry climates with 2 to 4.5 humid months (in summer) and tropical dry climates with 2 to 4.5 humid months (in winter). The zones included a narrow strip of true Mediterranean climate and the semi-arid zone, and constituted "a certain degree of uniformity over the bulk of the area defined under this classification, which stretched from Afghanistan to Morocco."[41] In a quinquennial review of ICARDA, conducted by the TAC in 1984, the review panel found the division by altitude oversimplified, preferring distinctions which had relevance for crops and livestock: for example, between areas suitable for autumn sowing of wheat versus those with winter and spring plantings, or between those areas where livestock could graze in winter and those where they must be protected and fed.[42] ICRISAT agro-climatologists, too, later disputed the applicability of the Troll classification to the climate of India, framed as it was in relation to the tropics of Africa and the Americas.[43]

However imprecise, the very capaciousness of the Troll classification recommended it as an umbrella for the broad range of climate conditions in Western Asia and North Africa. ICARDA would ultimately address agricultural practices in littoral areas at altitudes of up to 1,000 meters, which had a Mediterranean climate of cool, moist winters and hot, dry summers, as well as areas with altitudes of 1,000 to 2,000 meters, which had extreme winter cold and summer heat and snow cover for up to five months of each year. The precipitation of the latter ranged from 200 to 600 mm rainfall equivalent per year. As the quinquennial review committee later described, ICARDA's work involved a "spectrum of

---

[39] CGIAR Technical Advisory Committee, "Report of the Fourth Meeting of the Technical Advisory Committee," 34, 42.

[40] Carl Troll, "Karte der Jahreszeiten-Klimate der Erde [The Map of the Seasonal Climates of the Earth]," *Erdkunde* 18, no. 1 (1964): 5–28.

[41] CGIAR Technical Advisory Committee, "Report of the Fourth Meeting of the Technical Advisory Committee," 34.

[42] CGIAR Technical Advisory Committee, "Report of the Quinquennial Review of the International Centre for Agricultural Research in the Dry Areas (ICARDA)," Report, January 1984, 6, https://cgspace.cgiar.org/handle/10947/1390.

[43] International Crops Research Institute for the Semi-Arid Tropics, ed., *Climatic Classification: A Consultants' Meeting, 1980* (Andhra Pradesh, India: ICRISAT, 1980).

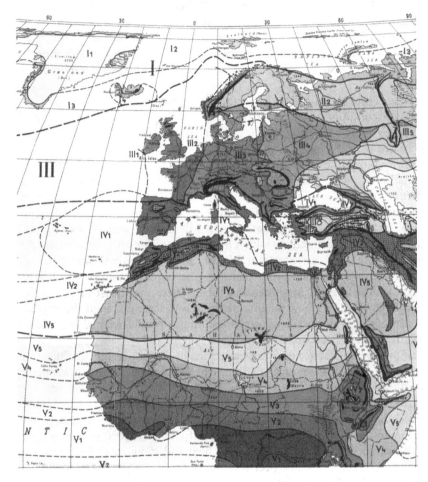

Figure 1.2 Detail of Troll and Paffen's "Seasonal Climates of the
Earth." C. Troll, "Karte der Jahreszeitenklimate der Erde, mit einer
farbigen Karte von C. Troll und K. H. Paffen," *Erdkunde* 18 (1964): 5–
28. By permission of *Erdkunde*.

environments," including the "warm winter littorals of North Africa, the
medium altitude environments, such as the Algerian steppe, Syria and
Iraq, and the true high altitude highlands of Afghanistan, Iran and
Turkey."[44]

[44] CGIAR Technical Advisory Committee, "Report of the Quinquennial Review of
ICARDA," 6–7.

In the absence of agro-ecological uniformity, the subcommittee charged with recommending a location for the new center asserted that political criteria were likely to outweigh ecological ones, although the exact nature of the former remained unspecified. The subcommittee identified "no single country in which an International Research Centre would be able to conduct a programme representative of the entire range of climate, soil, and resultant agricultural usage in the Near East and North Africa."[45] Since multiple substations were bound to be required for coverage, the subcommittee concluded that no "technical evaluation [would] produce a more definitive answer."[46] Rather, the decision was likely to be made according to "other criteria involving political, ethnical and other factors; external accessibility; working and living conditions; local availability of research infrastructures, and facilities such as universities; and the adequacy of land and water resources to support a major station."[47]

The Skilbeck report had also readily acknowledged that site selection for the center was likely to be made for reasons other than environmental ones, but these reasons went largely unnamed. There were exceptions, as when Z. H. K. Bigirwenkya, secretary general, East African community, suggested to the acting regional vice president of Europe, Middle East, and North Africa for IBRD that "assuming that there are no scientific reasons to the contrary, establishing the center in Iran could facilitate tapping of the oil financial resources to benefit all the three states [Algeria, Lebanon, and Iran]."[48] That financial interests, and the region's identification with oil resources, were not routinely named does not indicate their absence. On the contrary, the effort to fund ICARDA accompanied multiple overtures to Iran to join CGIAR as a member state. For some time, Abdul Majid Majidi, minister of plan and budget in Iran, expressed Iran's interest conditional to its designation as headquarters for ICARDA.[49] A number of donors also expressed the conditionality of

[45] CGIAR Technical Advisory Committee, "Location of the Proposed International Centre for Research in the Near East and North Africa," 7.
[46] Ibid.   [47] Ibid., 8.
[48] Z. H. K. Bigirwenkya to Martijn J. W. M. Paijmans, September 12, 1974, "Re: Cable from Price on ICARDA," Folder 1761726, CGIAR – G-10 – International Center for Agricultural Research in the Dry Areas (ICARDA) – Correspondence 72/74–01, Records of the Consultative Group on International Agricultural Research (CGIAR), World Bank Group Archives.
[49] Folder 1761726, CGIAR – G-10 – International Center for Agricultural Research in the Dry Areas (ICARDA) – Correspondence 72/74–01, Records of the Consultative Group on International Agricultural Research (CGIAR), World Bank Group Archives.

their funds based on contributions from oil-rich states.[50] CGIAR also made overtures to Saudi Arabia and Kuwait to join as donor members during site visits in search of a host country for ICARDA.[51] The financial underpinnings of international agricultural research and its geopolitical constitution were a matter of ongoing discussion, and this was especially the case in the context of the Organization of the Petroleum Exporting Countries (OPEC) crisis of 1973.

The subcommittee tasked with siting the new center nevertheless proceeded through prospective locations by process of elimination, presenting its recommendations "excluding political considerations, except for an assumption that the headquarters should be in an Arab country."[52] Iraq was eliminated because Kurdistan, which had the best ecological profile, was comparatively inaccessible. Iran and Turkey were eliminated because they were not Arab countries. The subcommittee saw merit in an Algerian or Tunisian location, but in the subsequent discussion regarded the Maghreb as a secondary research area to the Near East. Syria, like Tunisia, did not possess the full range of climate conditions, but it did have representative soils and both irrigated and rainfed agriculture. Aleppo provided a good site with a strong university and was home to Ford Foundation–funded development. Damascus boasted a new Arab League–funded center that planned to research agriculture in arid lands.[53] Although air access was poor, Beirut was a five-hour drive from Damascus. Lebanon had far and away the easiest access, living conditions, facilities, schooling, and university system, as well as a cooperative government interested in hosting a center. Many governments were amenable to hosting an international center, but established relations of European and American organizations with the Lebanese government made its availability apparent from the outset. However, Lebanon was considerably smaller, lacked irrigated land, and only represented a certain range of growing conditions. None of the options considered, with the possible exceptions of Iran and Algeria, had conditions ecologically representative of the entire region, largely because of the absence of cold plateau areas. Ultimately the subcommittee offered as recommendations

---

[50] CGIAR ICARDA Subcommittee and Daniel Ritchie, "ICARDA Subcommittee: Draft Minute of October 29, 1975 Meeting," November 19, 1975, 4, https://cgspace.cgiar.org/handle/10947/607.

[51] Warren C. Baum, "Memorandum on Progress in the Establishment of ICARDA," May 7, 1975, 2, https://cgspace.cgiar.org/handle/10947/859.

[52] CGIAR Technical Advisory Committee, "Location of the Proposed International Centre for Research in the Near East and North Africa," 2.

[53] The Arab League–funded Arab Center for the Studies of Arid Zones and Dry Lands (ACSAD), founded in 1971, served all member states.

several groupings of headquarters and substations in Algeria, Syria, Lebanon, Tunisia, Turkey, and Iran. Although the subcommittee ranked Algeria first among options for headquarters, the working group chose Lebanon.[54] Located in the most prosperous region of early twentieth-century greater Syria on the Eastern Mediterranean, Lebanon's independence had been recognized with the removal of the last French troops in 1946. Dominated by a Christian Maronite government, the country was an ally of the United States, and among the most diverse and prosperous of the new nation-states of the Middle East. By the early 1970s, Beirut had become a destination for tourists, banks, and diplomats.[55] It prevailed as a choice of headquarters largely because of the maturity of existing research networks and ease of living, including issues of staffing, communication, and international transportation. ALAD had a station in Lebanon's Beka'a valley; ICARDA would take over the station's operations. The committee recognized "that conditions in the valley, although offering a fairly wide range of elevation and rainfall (from 200 to 600 mm), were not typical of the area, particularly with regard to rainfed farming systems and cropping patterns." Therefore, it determined to establish a subsidiary site, probably near Aleppo, Syria, where land was abundant and the agro-ecological conditions were more typical of the region as a whole.[56] The committee also emphasized the need for a station to address the conditions of winter rainfall and snowy areas typical of mountainous Iran, Turkey, and Afghanistan, with Iran as a prime candidate due to its robust research network, abundant capital, and a government amenable to cooperation.

[54] The merits and demerits of various host countries were debated exhaustively following multiple site visits: CGIAR Technical Advisory Committee, "Report of the Eighth Meeting of the Technical Advisory Committee, 24 July–2 August 1974: Draft," September 1974, https://cgspace.cgiar.org/handle/10947/1408, 36–42; CGIAR Technical Advisory Committee, "Report of the Ninth Meeting of the Technical Advisory Committee, 3–7 February 1975: Draft," May 1975, 57–68, https://cgspace.cgiar.org/handle/10947/1436. A decision in favor of Lebanon, with substations planned for Syria and Iran, was announced by Warren Baum in May 1975: Baum, "Memorandum on Progress in the Establishment of ICARDA," 1.

[55] My account of the Lebanese civil war draws on the following: Chamberlin, *The Cold War's Killing Fields*, pp. 366–392; Paul Thomas Chamberlin, *The Global Offensive: The United States, The Palestine Liberation Organization, and the Making of the Post-Cold War Order* (Oxford: Oxford University Press, 2015); Itamar Rabinovich, *The War for Lebanon, 1970–1983* (New York: Cornell University Press, 2019); and Jonathan C. Randal, *The Tragedy of Lebanon: Christian Warlords, Israeli Adventurers, and American Bunglers* (Charlottesville, VA: Just World Books, 2016).

[56] CGIAR Technical Advisory Committee, "Report of the Seventh Meeting of the Technical Advisory Committee," 5; CGIAR Technical Advisory Committee, "TAC Quinquennial Review of ICARDA, 1984," 28.

Between Beirut and Aleppo

So it was that ICARDA was planned in 1974 as a tripartite center, with its headquarters in Beirut, a main station in Lebanon's Beka'a valley, a substation for low-altitude research in Aleppo, Syria, and a substation for high-altitude research in Tekmeh Dash (Tabriz) in northwest Iran – and a possible third substation in the Maghreb. However, the Iranian Revolution of 1979 summarily terminated CGIAR plans to establish a substation of ICARDA in Iran and scuttled plans to bring the country to CGIAR as a member state.[57] Plans to locate in Lebanon, too, would run aground, leaving only two small stations at Terbol and Kfardane in the Beka'a Valley.

Beirut's prosperity and diversity concealed the extent of rural poverty and the fragility of its representative government. On the basis of decades-old census data, the form of government granted Maronite Christians a permanent majority over Sunni Muslims. Territorial redistribution during the French mandate period had added large Muslim and Druze populations to the Maronite Christian communities on the coast, but the latter continued to enjoy the greatest political and economic power. The persistent privilege of Maronite Christians sowed discontent that ultimately undermined the fledgling nation's stability. Tensions had already once boiled over into civil war in 1958, leading the United States to intervene on behalf of the standing government. The move had deepened the alliance between the two countries, contributing to its large community of European and American foreign service personnel and its advancement as a likely base for an international research organization in 1974.

In 1975, the country plunged again into civil war. Jordan's expulsion of the Palestine Liberation Organization (PLO) from its borders in 1970 hastened its relocation to Lebanon. Lebanon's 100,000 Palestinian refugees tipped the country's demographic balance, lending credence to claims that new census data would require a new representative government acknowledging a Muslim majority. Commitments to Palestinian liberation and pan-Arab nationalism more broadly threatened to destabilize the Maronite Christian government and remake the political landscape of the Middle East. Closer to home, tensions with the PLO demonstrated the weakness of governments assembled according to European imperial designs. In 1975, Maronite and Palestinian forces

---

[57] CGIAR Technical Advisory Committee, "Report of the Seventh Meeting of the Technical Advisory Committee," 5. ICARDA struggled to find a suitable location for a high-altitude station, ultimately pursuing Pakistan (Baluchistan), the headquarters of its high-altitude program: CGIAR Technical Advisory Committee, "TAC Quinquennial Review of ICARDA, 1984," 92.

clashed. Leftist and pan-Arabist groups joined the Palestinians. The commercial heart of Beirut near the port was destroyed within months as the country descended into sectarian violence. The PLO came to patrol the "Green Line," a buffer zone between the Muslim West and the Christian East. Syria, which entered the conflict to check the growing power of the PLO, emerged as a guarantor of security in Lebanon.

Lebanon had become a failed state and a symbol of sectarian violence in the postcolonial world. The Iranian Revolution in 1979 only reinforced Western fears of sectarianism and religious fundamentalism. By comparison, Ba'athist Syria, a secular regime with progressive aspirations, appeared to move away from its Marxist origins in the direction of accommodation to international capital. With Arab unity as its guiding principle, Ba'ath leadership had initially rejected the existence of Israel as a Western puppet and paid lip service to the Palestinian Arab cause as married to its own. This doctrinaire foreign policy concealed rivalries within the Ba'ath party between the older urban elite and the younger rural population, who had been empowered by their military training to rise through the ranks of the party. Hafez al-Assad's seizure of power in an intra-Ba'athist coup ushered in a retrenchment away from radical foreign policy predicated on anti-imperial, anti-Western, pro-Palestinian, and pan-Arab nationalist ideology. His pragmatism ultimately translated into a more moderate foreign policy. Rather than throw Syria's lot in with the cause of Palestinian Arabs and pan-Arab nationalism, he broached a more moderate stance towards Israel and the West.

It was in this context that Assad invited ICARDA to set up operations in Syria. An agreement with the Syrian Arab Republic was signed in 1975 for the establishment of a principal station in Syria, with a separate agreement for a long-term loan of land in Aleppo province, spanning "a rainfall transect from an average of 200 mm per year in the south-east to 600 mm per year in the northwest." In November 1981, ICARDA's headquarters were formally moved from Beirut to Aleppo, leaving only the Terbol station operational (Figure 1.3).

But what was Syria to ICARDA, and ICARDA to Syria? A frank answer to this question requires more attention to the "turbulent history" to which the Skilbeck report alluded. For if it was too little to attribute the destruction of the "granary of ancient civilization" to "historical processes" or "social and structural rigidities," neither was it sufficient to assert that agricultural development projects could translate "ecological complementarities [into] production complementarities" without reference to political power.

The political economy of Syria as a modern nation-state was a collage of state-centered and imperial colonial Ottoman, French, and British

Figure 1.3 A row of greenhouses at ICARDA's research station and "temporary headquarters" (since 2012) in Terbol, Lebanon, 2018. Photo by Michael Major/Crop Trust. By permission of Global Crop Diversity Trust.

projects to make agrarian networks in the Eastern Mediterranean amenable to extraction. As the competition of Ottoman and French visions for agricultural productivity was made inert by the post–World War I fracture of the land into French and British mandate regions, the succeeding order facilitated intensified extraction according to technocratic and capitalistic forms of land and resource management.[58] The cumulative destabilization of land tenure systems, fragmentation of trade networks, and reintegration of territory into a nation-state divided between Mediterranean coastal plain and Syrian desert had created the preconditions for political, economic, and sectarian crisis. These same conditions provided the justification for renewed attention to agricultural development projects, framing the problem as one of low productivity with climatic and cultural causes.

Hafez al-Assad came to power as an outcome of sectarian and parliamentary rivalries; and ICARDA implanted itself into a Greater Syria developed by imperial reconstructions. Each provided the overall structure in which ICARDA would operate in Syria proper, as a polity marked by ethnic rivalry and economic inequality entered a period of rapprochement to Western capital. Assad was no liberal progressive. He ruled through military force, a secret police apparatus, and patronage arrangements rife with corruption that placed the merchant class, and anyone else who would

[58] Williams, *States of Cultivation*, esp. p. 14; Bertini and Zouache, "Agricultural Land Issues in the Middle East and North Africa," esp. p. 22 on "Colonial Legacies and the Legislative *Millefeuille.*"

deign to do business, in the pocket of the state. When Assad faced a challenge from the Islamist forces of the Muslim Brotherhood, he responded by shelling the city of Hama for days, sending in infantry and tanks to finish the job and pick through the rubble for surviving militants. Initial diplomatic reports understated the death toll at around 1,000. The actual loss was in the tens of thousands.[59]

In many respects, the embrace of Assad's Syria as a home for an international research organization mirrored US support of authoritarian governments that furthered national interests. After the outbreak of civil war in Lebanon, US President Gerald Ford, with Henry Kissinger at the helm of foreign policy, had concluded that Assad's defeat would threaten geopolitical security in the Middle East. Kissinger further held that Syria's intervention in Lebanon had the added benefits of challenging PLO leader Yasser Arafat's supremacy and widening a rift between Syria and the USSR, which disapproved of Assad's attack on the PLO. Syria's antipathy to Israel and ongoing conflict over the Golan Heights gave Kissinger little pause. US support of the Assad regime was consistent with Kissinger's realpolitik approach to managing Cold War rivalries and served a broader project of containment of Soviet influence in the Middle East and southern Asia.

For Assad, international organizations had other potential benefits. In a narrow sense, the presence of an international agricultural research institute offered practical solutions for an agricultural sector besieged by successive droughts, offering to bridge the gap between the urban elite and the impoverished rural areas to the east. More broadly, it offered the prospect of foreign investment and alliances to counter Syria's increasing isolation in the Arab world. In either sense, it was a potential source of capital for a deeply divided country. Even so, the terms of the agreement did not always favor local people. Apart from the potential for foreign investment and benefit to the agricultural sector, the exchange rate for ICARDA was set at 3.9 Syrian lira per USD, instead of roughly 5.4 SL per USD as of the time of the quinquennial review in 1984. This agreement had a detrimental effect on local staff and purchasing arrangements

---

[59] Estimates of fatalities provided by witnessing journalist Robert Fisk, Amnesty International, the Syrian Human Rights Committee, and the Syrian Network for Human Rights range from 10,000 to 40,000. Robert Fisk, *Pity the Nation: Lebanon at War*, updated edn. (Oxford: Oxford University Press, 2001); "SHRC.Org | Massacre of Hama (February 1982) Genocide and a Crime against Humanity | 2005 Reports," archived May 22, 2013, https://web.archive.org/web/20130522172157/http://www.shr c.org/data/aspx/d5/2535.aspx; "Syria: 30 Years on, Hama Survivors Recount the Horror," Amnesty International, February 28, 2012, www.amnesty.org/en/latest/news/ 2012/02/syria-years-hama-survivors-recount-horror.

within Syria.[60] Could such an institution heal the rift between the urban west and the rural east inscribed by Ottoman, European imperial, and postcolonial designs on the landscape?

Farming Systems

While the FAO director general had applauded CGIAR for its promise to build on Green Revolution successes, members of the Skilbeck mission and other observers saw no such revolution forthcoming. Those who had studied the agricultural conditions of the MENA region warned that social and political factors, rather than technical ones, constrained production, even as they rarely specified the nature of those factors. Asserting stagnation due to traditional cultural practices, they counseled a primary emphasis on farming systems as a whole, rather than genetic improvement of selected crops. In his commentary on the Skilbeck mission, Professor M. Nour of the FAO office in Cairo warned that "socio-economic realities must be taken into account if technical transplants were not to be rejected." He lamented a general neglect of rainfed areas, which constituted a "long history and tradition which would not be easily changed." Nour's commentary included only the most glancing reference to Ottoman, French, or British institutions or agronomic styles. He observed in passing that "in the past, there had been too much emphasis on technical solutions, and too little thought concerning the social and economic factors constraining and conditioning the successful use of technology."[61] The details of colonial and postcolonial territorial realignments, expropriation of land, fragmentations of landholdings and trade networks, nationalization of resources, and general evolution of land tenure went unremarked in the mission's report and the TAC's review.

By the time an official proposal was framed for the center, the priority areas for research were crop improvement, soil and water management, and animal production systems. ICARDA, in its mandate, sought "to develop appropriate technologies which, when integrated into improved farming systems, will increase the production of staple food commodities, especially cereals, food legumes, and sheep." Its research program consisted of four programs: Farming Systems, Cereal Improvement, Food Legumes Improvement, and Pasture and Forage Improvement. The broadest objective of research and training was to "increase and stabilize food production in the region." Specifically, ICARDA would "serve as an

---

[60] CGIAR Technical Advisory Committee, "TAC Quinquennial Review of ICARDA, 1984," 9.

[61] CGIAR Technical Advisory Committee, "Report of the TAC Working Group on the Research Needs of the Near East and North Africa," 10.

international center for research into and the improvement of barley, lentils and broad beans (*Vicia faba*)," as well as any other crops designated by CGIAR, and would serve a relay role for other international centers for research in other important crops in the region, such as bread wheat and chickpeas. In addition to crop improvement, ICARDA would "conduct research into and develop, promote and demonstrate improved systems of cropping, farming and livestock husbandry," facilitating connections between national, regional, and international researchers.[62]

The mandate was notable for its emphasis on farming systems over crop improvement. The Farming Systems Program (FSP) was a multidisciplinary, systems-oriented approach to agriculture, consisting of crop agronomy, agricultural economics, and livestock and soil science. Such an approach required attention to "the physical, biological and socio-economic problems which impose constraints on the widespread adoption of improved systems of cropping, farming and livestock husbandry." Rather than isolate plant or animal material, a farming systems approach considered the entire process of production, including attention to pre- and post-harvest factors and a robust program of research and training with a broad range of stakeholders. Broadly, the center's programs concerned both farming systems and genetic improvement, but planners viewed all "as components of improved farming systems, which should be the ultimate aim of the new Centre."[63]

Even as its mandate obscured colonial and postcolonial disruptions in rural lands, the insistence of ICARDA's architects on attending to socio-economic realities and farming systems, as opposed to single crops, distinguished the center from its predecessors. In priority and methodology, ICARDA took a new approach to the CGIAR mandate to reduce rural poverty, insisting on persistent and dynamic interaction between researchers and farmers. TAC members noted the extent to which a center in the MENA region would depart from "the narrowly defined commodity approaches of the earlier centers."[64] Five years into operation, the FSP was the largest and most complex of the research programs, often also taking the largest share of the budget.

---

[62] "Proposal for the Establishment of an International Centre for Agricultural Research in the Near East and North Africa," 139–146, Folder 1761726, CGIAR – G-10 – International Center for Agricultural Research in the Dry Areas (ICARDA) – Correspondence 72/74–01, Records of the Consultative Group on International Agricultural Research (CGIAR), World Bank Group Archives.

[63] CGIAR Technical Advisory Committee, "Report of the Seventh Meeting of the Technical Advisory Committee," 10; "Proposal for the Establishment of an International Centre for Agricultural Research in the Near East and North Africa."

[64] CGIAR Technical Advisory Committee, "Report of the Seventh Meeting of the Technical Advisory Committee," 7.

The commodity focus, and genetic improvement within it, nevertheless remained a principal aspect of the new center's mission. The interaction between the commodity programs of different CGIAR centers required careful management. Both barley and durum wheat were the province of CIMMYT, headquartered in El Batán, Mexico. The details of ICARDA's potential relationship to CIMMYT troubled early plans for the center. CIMMYT opposed removing barley as one of its mandate crops, even as the working group charged with establishing the dryland agricultural center concluded that CIMMYT could not meet the needs of the MENA region. For similar reasons, it recommended the possible transfer of the durum wheat program once the center was at full operating capacity.[65] A decade later, the quinquennial review reiterated the arguments for both transfers: 11 million hectares of barley were grown within the ICARDA region, as opposed to 700,000 in Latin America. Forty-four percent of the world's total durum area was within the ICARDA region, as opposed to 1.6 percent in Latin America, mostly in Argentina, where the crop was declining. Moreover, 97 percent of the durum wheat sown within the ICARDA region was rainfed, as opposed to irrigated.[66] This protracted negotiation indicates the extent to which a commodity focus remained central to CGIAR's overall program of research, troubling attempts to remake international agricultural development to serve the needs of local economies.

The Skilbeck Committee had recommended agro-climatological studies, a focus on rainfed agriculture, irrigated agricultural systems, and special problems of gypsiferous and saline soils. When pressed to reduce the mandate, members countered that each was a fundamental aspect of farming systems in the region. Rainfed and irrigated agricultural systems, for example, were closely intertwined. The final proposal nevertheless restricted the mandate to rainfed agriculture, noting that the distinguishing agro-ecological characteristic of the region was winter rainfall distribution. Rainfed agriculture was the practice that held the otherwise unwieldy mandate region together. Additionally, as of the quinquennial review in 1984, ICARDA had no lines of research in water management, which had been a component of the Skilbeck recommendation. The review panel found it "odd" that the mandate of ICARDA contained "so little reference to the study of soil-water relationship," citing the case made so strongly in the Skilbeck report for its centrality to

---

[65] Ibid., 4.

[66] CGIAR Technical Advisory Committee, "TAC Quinquennial Review of ICARDA, 1984," 24–25. After 1980, ICARDA began limited programs in supplementary irrigation, focused on winter-planted crops.

dryland agricultural systems.[67] Even so, the panel acknowledged that "the heterogeneity of the mandated area," between rainfed and irrigated agriculture and high and low elevation, for example, "posed a number of problems in interpretation of the mandate and program development."[68]

The limited scope of ICARDA's approach suggested the unwieldiness of a mandate region that was framed first in geopolitical terms and secondarily in agro-ecological ones. While ICARDA's commitment to farming systems did partial justice to the aspirations of its planners to address the roots of rural poverty, the winnowing of its mandate to exclude irrigated agriculture and soil and water management signaled its limited capacity to address some of the principal problems of agriculture in the region as a whole, and in Syria itself.

Economic liberalization failed to save Syria, even as it led to geopolitical realignment in Western Asia and the world. Hafez al-Assad had pursued liberalization to invest oil money flowing from the Gulf States in the form of aid and remittances from Syrian laborers in the Gulf. These windfalls proved short-lived. In the 1980s, oil prices plummeted. Aid and remittances declined accordingly. A global recession further hampered foreign investment, while Soviet withdrawal from the region deprived the country of substantial military and financial aid. Driven by economic necessity, Assad moderated the country's stance to the United States. This shift culminated in Syria's 1991 decision to join the coalition against Iraq's invasion of Kuwait and to participate in the Madrid Conference in pursuit of Palestinian–Israeli peace. To Assad's frustration, the process failed to return the Golan Heights to Syria. It nevertheless indicated his willingness to participate in international negotiations that could further territorial and economic interests.[69]

None of this geopolitical maneuvering was sufficient to address the underlying conditions of inequality and sectarianism within Syrian society. The Syrian civil war, begun in 2011, fulfilled the warnings of those who had alerted CGIAR planners to the political and socioeconomic roots of crisis in the region, even as they fell short of a full critique of imperial de-development of greater Syria. While decades of drought provoked significant public investment in irrigation in the early twenty-first century, it was not enough to reverse the impact on farmers in the country's rural center, who had long resented the political and economic supremacy of the western cities. As drought and increasing fuel prices plunged Syria's agriculture into crisis, Syrian farmers became some of the fiercest opponents of the regime of Bashar al-Assad, who had become

---

[67] Ibid., 6.   [68] Ibid., xxv.   [69] For a summary of these shifts, see Lesch, *Syria*, chapter 6.

president after his father's death in 2000. The son followed the example of his father, Hafez, in crushing dissent. The outcome was civil war.

Throughout these turbulent transitions, ICARDA was less important for its work in the field than as an institution that symbolized a Western vision for international agriculture. In the inaugural meeting of the TAC, the CGIAR chairman defined international research as "research which, while located in a specific country, was of wider concern regionally and globally, independent of national interest or control, and free from political dictates of anyone government whilst retaining appropriate links with national research systems to ensure necessary testing of results and feedback both of results and needs."[70] This vision of progressive, scientific agriculture, which provided the context for the proliferation of CGIAR centers across the globe, belied the fundamental commitments of donors to international commodity cultures and the formation of a coalition of states amenable to Western technical assistance. In establishing ICARDA, donor countries staked a claim to West Asia and North Africa as regions of influence, and a base and proxy for arid and semi-arid regions of strategic interest across the globe. ICARDA was part of a globalized vision for agricultural development that made poverty alleviation into a single project and poverty itself into a uniform condition. While international research organizations have made escalating claims to operate at a global scale, and on behalf of universal interests, the landscapes they traverse are more complex in agro-ecological and historical terms.

The post–Cold War history of ICARDA provides an instructive coda to this history. In the 1990s, ICARDA inaugurated plant genetic resources collecting expeditions and collaborations in the countries of the former Soviet Union, focusing on biodiverse regions of the Caucasus Mountains and Central Asia. Thus the end of the Cold War provided new geographies for the complex of international research and development within the geography of Central and West Asia and North Africa that had first characterized Eisenhower's southern rim strategy of containment. Ostensibly, these missions brought new regions into the fold of the CGIAR system, offering farmers membership in a network of international technical assistance. But it is also intentional that these missions were situated on the periphery of the Soviet Union, which had long structured the geography of the CGIAR network. Rural lands were not simply grounds of poverty: they were the fields of empire, recast in the aftermath of World War II as buffers against communism. In the collapse

---

[70] CGIAR Technical Advisory Committee, "Report of the First Meeting of the Technical Advisory Committee."

of the Soviet Union, they became the grist for a globalized vision of market-led development, a dream imagined rather than realized in the winds of change.

Since the outbreak of civil war in Syria in 2011, ICARDA's operations have decentralized across the Middle East and North Africa, with head-quarters back in Beirut. Decentralization realized the initial orientation of planners towards a broad range of territories and enabled renewed claims that international agricultural research can address the needs of small farmers across the globe. Orientation towards the world's farmers nevertheless requires us to be lucid about the ways in which the imperial prehistory of international scientific research continues to structure neocolonial power relationships, often concealing or abetting conflict.

2 US–India Entanglements and the Founding of ICRISAT in India

Prakash Kumar

In July 1971, agronomist Ralph Cummings wrote to the director of the Rockefeller Foundation's agricultural program in India and mentioned his upcoming preliminary visit to India "regarding the feasibility of the suggested Upland Crops Research Institute." On his docket were meetings with such key people as the agriculture minister, C. Subramaniam, the agriculture secretary, B. Sivaraman, and Planning Commission member Tarlok Singh. Cummings was also scheduled to meet agricultural scientists M. S. Swaminathan and B. P. Pal at such premier Indian institutions as the Indian Agricultural Research Institute (IARI) and the Indian Council of Agricultural Research, respectively. In all these encounters, Cummings hoped to "discuss the extent to which they wish[ed] serious consideration be given to India as [the] location for the Upland Crops Research Institute." This institute was to be a world resource center for research on crops in areas of low rainfall.[1] By the following July an institute with this mission was in place on the outskirts of Hyderabad. It was christened the International Crops Research Institute for the Semi-Arid Tropics, or ICRISAT, and funded through the recently established Consultative Group on International Agricultural Research (CGIAR).

Although historical accounts commonly put US foundations and agronomists like Ralph Cummings at the forefront in explanations of the history of global institutions such as ICRISAT in the 1970s, American political and scientific ambitions offer at best a partial understanding. As I show, the outreach, interest, and facilitation involved in the establishment of ICRISAT in India were enmeshed in the momentum of past agricultural programs and the contingencies related to choices made by key Indian political leaders in a changing political climate. The establishment of ICRISAT seemingly showcases departures from the high-yield agricultural

Acknowledgments: Gopi Swamy provided research assistance with Telugu archives in Hyderabad for this chapter.
[1] Ralph Cummings to Guy B. Baird, July 15, 1971, Ralph Cummings Collection, MC 312, Special Collections, North Carolina State University, Raleigh, NC (hereafter Ralph Cummings papers), Folder 3, Box 34, Series 33.

agenda of the 1960s that India embraced under American inspiration. The latter, commonly captured by the descriptor Green Revolution, was focused on wheat, foisted on fertile, irrigated lands in three key states in India's northwest, and promulgated by the government through economic incentives.[2] The inauguration of ICRISAT happened after a rupture in Indo-US diplomatic relations and a consequent end to the ongoing collaborations on agrarian programs between the two nations. The upcoming institution would focus on a different set of crops that thrived in a different agro-climatic context and also seemed hitched to a different agrarian program. That said, this chapter argues that ICRISAT – despite the pronouncements of important political actors in India to the contrary – represented an institutional continuation of a Green Revolution vision that had become consolidated in the preceding decade.

The lightning speed with which an institution as large as ICRISAT went from a plan on paper to bricks and mortar demands a history that considers the context of prior agrarian projects in India, including their American lineage, as well as scientific and political developments in India in the 1970s that made an "international" institute palatable to many of the country's leaders. I begin my analysis by sketching the agendas that brought American agricultural scientists to India in the 1950s and 1960s, before turning to events more immediately surrounding the creation of ICRISAT in association with CGIAR. Lastly, I discuss the Indian context in which proposals for ICRISAT were received, showing how the successful launch of the institute in 1972 depended on reconciling the claims of two key political leaders in India with the rhetoric of international research on dryland crops.

Deepening US–India Entanglements in the 1950s and 1960s

The 1950s and 1960s saw a growing entanglement of American and Indian interests in projects of rural uplift and agricultural yield enhancement that were undertaken in India. President Harry Truman's project of exporting technical aid to developing nations through his Point Four program opened the gate for the arrival of American technical experts to

---

[2] I use the phrase "Green Revolution" to refer to the 1960s agricultural transformation in India using high-yielding variety seeds that was made possible through the state provision of subsidies on seeds, fertilizers, and pesticides, the creation of infrastructure, and the state's guarantee of a remunerative purchase price. See Akhil Gupta, *Postcolonial Developments: Agriculture in the Making of Modern India* (Durham: Duke University Press, 1998); Glenn Davis Stone, "Commentary: New Histories of the Indian Green Revolution," *Geographical Journal* 185, no. 2 (2019): 243–250; for debates on the global Green Revolution, see Prakash Kumar et al., "Roundtable: New Narratives of the Green Revolution," *Agricultural History* 91, no. 3 (2017): 397–422.

India.[3] These experts made an entry in multiple sectors, with the majority working on agricultural and rural development projects.

The footprints of American experts were first noticeable with the launch of India's ambitious community development projects on October 1, 1952. Earlier that year, India and the United States had signed an agreement to allow for the arrival of American experts. Among these were sociologists, agricultural experts, and "communitarians" who provided key advice to Indian officials and helped set up training centers for the staff of community projects. They made a major contribution specifically in the training of village-level workers, the cadre of the community development project, who bore the primary responsibility for educating farmers in myriad tasks, ranging from seed selection to well-digging to rural education and welfare. Aside from the US State Department, the Ford Foundation was also involved in the training of village-level workers. While the American expertise was crucial, the Indian government showed extraordinary commitment to the idea of bottom-up development that was implicit in the communitarian idea. Within a few years, the government had created a vast, nationwide infrastructure of community centers and blocks. Financial and bureaucratic support for this work was made readily available under Prime Minister Jawaharlal Nehru's leadership, not least due to his extraordinary interest in the project.[4]

A parallel American interest in augmenting India's agricultural production also progressed in India in the 1950s and 1960s. Historians of American foreign relations have argued that US-led efforts to mitigate hunger in India and across Asia in the 1950s and 1960s were motivated by the Cold War objective of countering communism. They were sustained by the notion that agricultural atrophy and low productivity in overpopulated nations like India generated breeding grounds for radical ideologies and communist takeovers. The Rockefeller Foundation in particular was at the forefront of this global drive to fight hunger, most famously by contributing to the expanded cultivation of high-yielding varieties of wheat and rice.[5]

---

[3] For a synoptic view of the United States' Point Four program and its push for development in postcolonial nations, see Steven Macekura, "The Point Four Program and US International Development Policy," *Political Science Quarterly* 128, no. 1 (2013): 127–160.

[4] A history of the community development project in India and a description of its various activities appear in Prakash Kumar, "'A Big Machine Not Working Properly': Elite Narratives of India's Community Projects, 1952–58," *Technology and Culture* 60, no. 4 (2019): 1027–1058.

[5] For the Rockefeller Foundation's early forays into agricultural aid, see Tore C. Olsson, *Agrarian Crossings: Reformers and the Remaking of the US and Mexican Countryside* (Princeton, NJ: Princeton University Press, 2017); for the spread of its programs to

The Rockefeller Foundation's initial outreach in India started at IARI, with the agronomist Ralph Cummings as the foundation's director of field staff in the country. Based in India from 1956, Cummings oversaw a project to start a postgraduate teaching program in agricultural sciences at IARI and another to jump-start research on three principal crops of maize, sorghum, and millet.[6] The Rockefeller Foundation dispatched Kenneth O. Rachie, an expert from its Mexican program – the site of its earliest efforts to develop high-yielding varieties – to "assist primarily in the development and execution of the research on improvement of sorghums."

After the middle years of the 1960s, as the high-yielding varieties of wheat and rice spread in India, the Rockefeller Foundation pursued a parallel interest in the development of sorghum and millet in India. Thus, late in 1968, Sterling Wortman of the Rockefeller Foundation, the hero of its Mexican Agricultural Program, wrote to India's agriculture minister suggesting that he meet with Wortman's emissary, Ralph Cummings, to discuss intensifying research efforts on sorghum and millet. The foundation had been pursuing work on those crops for more than a decade, but Wortman felt that the existing programs still did not meet all the world's needs "at this point in history." Wortman thought that India would be "the logical location for an [additional] intensified effort." He knew that the Indian government was actively considering the launch of all-India coordinated research schemes on sorghum and millet and that its preeminent scientists, such as M. S. Swaminathan, B. P. Pal, and A. B. Joshi, were positively inclined towards pursuing development of these crops. He thus invited the minister to meet with Ralph Cummings to discuss future projects in India in which the Rockefeller Foundation could participate. Wortman referred to Subramaniam's prior visit to Mexico, where he had seen him, and it seemed the two knew each other well. Clearly Wortman was using his prior acquaintance with the minister in pushing for the expansion in India of the Rockefeller Foundation's work on sorghum and millet.[7] In short, then, the foundation's interest in India

Asia, see Nick Cullather, *The Hungry World: America's Cold War Battle against Poverty in Asia* (Cambridge, MA: Harvard University Press, 2010).
[6] "Statement Relative to Tentative Plan of Operation under Rockefeller Foundation – Government of India Agreement on Agricultural Research and Education," Ralph Cummings papers, Folder 3, Box 34, Series 33. The IARI postgraduate institution was set up on the recommendation of a 1955 Indo-US team. The Rockefeller Foundation stepped in to provide funds. M. S. Randhawa, *Agricultural Research in India: Institutes and Organizations* (New Delhi: ICAR, 1958), p. 45; Hadley Read, *Partners with India: Building Agricultural Universities* (Urbana: University of Illinois, 1974), p. 26.
[7] Sterling Wortman to M. S. Swaminathan, November 18, 1968, Ralph Cummings papers, Folder 3, Box 34, Series 33.

was hardly limited to wheat and paddy (rice) – the showpieces of the Green Revolution – but, rather, from the 1950s extended to crops like sorghum and millet. These nascent interests anticipated the later interest of CGIAR and Indian stakeholders that concretized around the effort to start ICRISAT.

The New Multilateralism and a Concrete Interest in "Upland Crops"

By 1970, US philanthropies led by the Rockefeller and Ford Foundations had been active in India for almost two decades, as had the United States Agency for International Development (USAID) and its prior incarnations. The staff of these organizations worked with and through the very large infrastructure of Indian agricultural institutions and engaged with many Indian scientists. The momentum of the work carried out by this binational community of agronomists, crop specialists, and other scientists foretold the collaborative spirit that later helped to launch ICRISAT.

The concrete idea of an institute that would embrace crops grown in areas of low rainfall took shape amidst deliberations at the Rockefeller Foundation in 1970. Its scientists had spent some time identifying and discussing distinct crop geographies of the world. An internal report compiled by Clarence Gray, a specialist in the foundation's agricultural sciences division, focused on rainfed, or unirrigated, areas in tropical Asia and Africa and justified the need for setting up a new organization to focus on the crops of these regions. Its beginning premise was that advances made in crop yields in Asia in the preceding decade had bypassed areas with low rainfall. "While there have been impressive gains in wheat and rice, a large production problem still exists in Asia," Gray argued. His report called the "contemporary yield-increasing technologies" of the 1960s vintage inadequate measures which "had little applicability and relevance in the unirrigated, rainfed uplands." With this logic, the report funneled attention towards a category denoted as "upland crops."[8] This categorization picked up a salience and reflected a broader pattern wherein the Rockefeller Foundation contributed to the founding of new, geography-specific institutes, such as the International Center for Tropical Agriculture in Colombia (CIAT) and the International Institute of Tropical Agriculture (IITA) in Nigeria, both in 1967.

Gray's consequential report borrowed ideas from geographers and was influenced by contemporary perspectives that established linkages among

---

[8] Clarence C. Gray, "Discussion Paper for an International Upland Crops Program," Rockefeller Foundation, June 1971, 59, https://cgspace.cgiar.org/handle/10947/474.

climate, rainfall distribution patterns, and agro-ecology. Gray relied on those connections to shape an imagined geography that made available a specific global space for action. The descriptor "upland" was commonly used by geographers to designate landscapes of higher altitudes that were located above the floodplains. These were mostly areas of low rainfall.[9] Gray took an imaginative leap from here to address "upland crops" globally, describing them, and then specifying them as lands of uncertain productivity. Gray's account took a particular lead from the climatologist Carl Troll's study of global seasonal rainfalls to identify and group rain-deficient regions in the tropics. While admitting that Troll's broad categorizations of global rainfall patterns could not hold true year after year, Gray nonetheless presented "agro-climatic situations" for which he believed specific food production and crops research policies could be devised. He identified lands with definitive deficiencies: ones that went without rains for more than seven months every year. The wetter or irrigated areas where high-yielding varieties of rice and wheat had spread thus far due to the Green Revolution measures of the 1960s had been the privileged ones, cornering the bounties of scientific innovation. The drier regions now needed attention.

Clarence Gray's new geography of low-rainfall areas implied that these areas that were populated by agrarian masses had not yet reached their potential to be the most productive. The world could leave them behind at its own peril. The Rockefeller Foundation's perception of the need for a new institution of agricultural research was pinned on turning attention to these subaltern lands, crops, and people. Gray considered a wide range of crops in these regions, but highlighted four to make his point: sorghum, millet, chickpeas, and pigeon peas. Pointing to "the inadequate state of tropical crop production technology" for such crops in Asia and Africa broadly, he made the case for establishing a new center that would serve as a world resource for research on these crops. This center would "develop and demonstrate improved cropping patterns and systems of farming which optimize the use of human and natural resources in low-rainfall, unirrigated, upland tropics."[10]

Gray's advocacy seemingly had an effect, as both the Ford and Rockefeller Foundations began to support the cause of upland crops. Their support crystallized at two preparatory meetings called by Ford and Rockefeller Foundation officials in the summer and fall of 1970 to which important scientists from many countries were invited. Among the agricultural specialist invitees from Asian nations to this meeting was B. P. Pal, the director general of the Indian Council of Agricultural

---

[9] Ibid.   [10] Ibid., v, 68.

Research. Others arrived from the Philippines, Pakistan, Malaysia, and Indonesia; these were the nations where the foundations had shown prior interest.

The foundations' effort to form an upland crops institute transpired almost in parallel with their effort to jump-start the global consortium CGIAR (as discussed by Lucas M. Mueller, Chapter 5, this volume), which emerged in May 1971 under the aegis of the World Bank and United Nations. The 1970s saw an opening for wider collaboration among existing multilateral institutions. The World Bank was looking to channel its funds towards an organization that would dispense with the need for US bilateral aid to individual countries. The World Bank's president, Robert McNamara, found particularly willing partners in this task in the Rockefeller and Ford Foundations.

A foundations-wide counterpart to multilateralism was building its own momentum with a different logic. In the 1960s, a preference for decentering aid for agriculture and rural welfare began to find favor among foundation leaders. In the summer of 1966, John D. Rockefeller III spoke with trepidation to the Far East American Council in New York, warning about an apparent "American overpresence" in Asia that might turn out to be counterproductive. The talk, copies of which were pre-circulated, was titled "Our Dilemma in Asia" and stressed that "our presence supports ... [Asia's] self-preservation but it bothers their self-respect." Rockefeller emphasized the need to restore balance, whereby greater Asian initiatives in security, finance, and development could be achieved.[11] As a corollary to building Asian solidarity and initiative, John D. Rockefeller III also suggested moving away from policies of American bilateral aid and towards multilateral aid that would reduce the American footprint in Asia while securing the same set of goals.

The coming together of the World Bank, United Nations, and the foundations led to the formation of CGIAR. This partnership had a specific outcome for the "upland crops" project that had germinated within the foundations. The foundations brought into CGIAR their existing research centers and programs, including their plans for an upland crops institute. This initiative got subsumed within CGIAR's emerging projects in India, where it was sold, thanks to the CGIAR umbrella, as a United Nations initiative. The Rockefeller Foundation's upland crops institute thus had its reincarnation in India as ICRISAT, an

---

[11] John D. Rockefeller III, "Our Dilemma in Asia," May 17, 1966, Ralph Cummings papers, Folder 3, Box 34, Series 33, p. 1. The idea of multilateralism in aid apparently had a broader constituency. Rockefeller cited two World Bank presidents, Eugene Black and George Woods, as well as the influential senator from Arkansas, James William Fulbright, who served on the powerful Foreign Relations Committee.

international research center sustained by multilateral funding that would target areas now designated as the "semi-arid tropics." The formal plan for ICRISAT concretized rapidly as CGIAR and its core decision-making body, the Technical Advisory Committee (TAC), were formed. At its very first meeting, the TAC formed a team to explore the feasibility of such an institution that comprised Louis Sauger (Senegal), Hugh Doggett (United Kingdom), John Comeau (Canada), and Ralph Cummings (United States). The team settled on Hyderabad in southern India as the future site for the new institution. The Ford Foundation was designated by CGIAR as the executing agency to negotiate details with the government of India. It was Ford Foundation officials who signed an agreement with India on March 28, 1972 on behalf of CGIAR. The institution was set up as an international entity under the United Nations Privileges and Immunities Act.[12] The governing board held its first meeting in Hyderabad on July 5, 1972.[13]

The Political and Diplomatic Context in India

Just as outreach on ICRISAT in India was an admixture of foundation and CGIAR efforts out of which different strands of influence can be teased, the local history of ICRISAT's establishment was threaded with continuities and disjunctures that require purposeful unravelling. In particular, the launch of ICRISAT coincided with a rupture in long-running USAID-assisted agricultural programs in India. To those loose ends was tied the umbilical cord of the new institution. The birth of ICRISAT as a wellspring of Indian agrarian visions in the 1970s was rooted in this moment of transition.

The decade of the 1970s was one of tremendous flux in US–India relations with respect to aid and in terms of overall diplomatic relations that suddenly turned sour. The historian Srinath Raghavan connects these changes to the new "Nixon Doctrine" that *prima facie* aimed to deal with the changing dynamics of the Cold War, the rise of a multipolar world, and the acceleration of globalization. On aid relationships, the Nixon Doctrine clearly preferred "a trimming of the American foreign aid program by turning away from bilateral, project-based aid and technical assistance and toward ... multilateral financial flows to developing countries." On the diplomatic front, the Nixon administration actively sought

---

[12] Statement by Deputy Minister of State for Agriculture Jagannath Pahadia, Rajya Sabha, May 24, 1972, 103–104, https://rajyasabha.nic.in/Documents/Official_Debate_Nhindi/Floor/80/F24.05.1972.pdf.

[13] "ICRISAT Presentation," July 27, 1976, Ralph Cummings papers, Folder 2, Box 69, Series 33, p. 1.

the intercession of Pakistan in establishing a better relationship with China. These two tendencies seemed to come to a point of explicit realization around the time that India entered the Bangladesh War, the 1971 armed conflict between Bengali nationalists and the Pakistani military that ultimately resulted in the birth of Bangladesh. Taking India's declaration of war against Pakistan on December 3, 1971 as an act that crossed the line, Nixon announced the halt of economic aid to India, including $87 million in USAID support already in the pipeline. The curtailing of USAID presence in India aligned with a prevailing mood among US and Indian officials about "reorienting economic ties" that aimed to reduce USAID's "footprint in India."[14]

In 1971, when Nixon's decision to cut off aid was announced, the hammer fell most notably on the initiative by five US land-grant universities that had been active in India on a charge from USAID on a myriad of agricultural programs.[15] Late in the summer of 1972, G. V. K. Rao, development commissioner in the state of Mysore, informed the University of Tennessee team working for the southern Indian states that all of its programs would be terminated on September 30, 1972. Rao gratefully acknowledged the role the Americans had played in institution- and program-building for Mysore's department of agriculture and in setting up the University of Agricultural Sciences in Bangalore.[16]

The top executives of the agricultural universities that were being set up with expertise from American land grants expressed regret at the

---

[14] Srinath Raghavan, *Fierce Enigmas: A History of the United States in South Asia* (New York: Basic Books, 2018), pp. 273–308, at 294–295. David Engerman's study of foreign aid in India discusses the longer arc of USAID efforts there. In the years of somewhat strained Indo-US relations under Indira Gandhi from 1966 to 1971, there was a certain trend towards "financialization of aid" that abetted the withdrawal of American personnel from field offices in India and instead leaned on providing support through dollars alone. India, for its part, also emphasized that, while it would be open to financial aid, it would slash the number of American experts actually present on the ground. David Engerman, *The Price of Aid: The Economic Cold War in India* (Cambridge: Harvard University Press, 2018), pp. 250–259, 328–337. After the 1971 interregnum, the United States much preferred to pump aid into India through the World Bank's Aid India Consortium.

[15] Between 1960 and 1971, twelve Indian "land-grant" state universities emerged, projecting implementation of a specifically Americanist "land-grant modernization" vision in India; Prakash Kumar, "Modalities of Modernization: American Technic in Colonial and Postcolonial India," in John Krige, ed., *How Knowledge Moves: Writing the Transnational History of Science and Technology* (Chicago: University of Chicago Press, 2019), pp. 120–148, esp. 134–140. See also Henry C. Hart, *Campus India: An Appraisal of American College Programs in India* (East Lansing: Michigan State University Press, 1961).

[16] G. V. K. Rao to William Ward, July 8, 1972, Agency for International Development and College of Agriculture Records, AR.0387, University of Tennessee, Knoxville, Special Collections Library (hereafter USAID/Tennessee papers), Folder 7, Box 9.

snapping of ties with their American counterparts. A letter from G. Rangaswami, the vice chancellor of the newly established Tamil Nadu Agricultural University, regretted "the decision taken at Delhi to terminate the USAID operations" and expressed remorse that his "hopes of receiving assistance to develop the University have been completely thwarted." He had been in contact with the USAID team since 1958 and was "very unhappy" over the recent turn of events.[17] The University of Agricultural Sciences vice chancellor, K. C. Naik, called this breakdown in collaborative engagements with USAID an "unfortunate development" and was "sad that a most profitable relationship developed over many years between the US universities and Indian agricultural universities has come to an abrupt end."[18] Naik spoke very approvingly of the results that he thought the USAID programs had wrought. Without them "the progress of Indian agriculture, including that of agriculture in Mysore state would have been trivial." Naik was referring broadly to the long-running American aid programs in rural India as he alluded to the "seeds sown by the TCM [the US Technical Cooperation Mission to India] and USAID and the help received from a few selected US universities" since the 1950s, and to American participants "who have worked with us intimately, as members of a family, for reorienting our educational system in agricultural sciences, to render effective service to our farmers."[19]

In the 1950s and 1960s, USAID-assisted agricultural projects in India had served as a magnet around which US and Indian collaborators coalesced. They formed an epistemic community in India within which common visions of agrarian progress developed and prospered. These forces ensured not only that the formal break in American aid would not spell the end of certain agricultural programs, but also that an initiative such as ICRISAT, which promised to continue prior visions of agricultural progress, would rise and be consolidated. Naik mused about the first joint Indo-American team of 1955 that had initially studied the prospect of the land-grant model for establishing agricultural universities in India. The team's recommendations had led to the first postgraduate teaching program in agricultural sciences at IARI in Delhi. This program was supported by the Rockefeller Foundation, and Ralph Cummings, who was appointed director of postgraduate

[17] G. Rangaswami to William B. Ward, June 26, 1972, USAID/Tennessee papers, Folder 7, Box 9.
[18] K. C. Naik to William B. Ward, June 22, 1972, USAID/Tennessee papers, Folder 7, Box 9.
[19] K. C. Naik, Statement, USAID/Tennessee papers, Folder 7, Box 9, p. 3.

teaching, was assigned the responsibility for defining and developing the program.[20] A second joint Indo-American team was constituted in 1959 – Naik was a member of this team – and its recommendations solidified the project to launch agricultural universities in India with the aid of US land-grant institutions.[21] Cummings was on this occasion appointed by the government of India to lead a committee in drafting the basic framework for agricultural universities in India. The Cummings Report of 1960, as it came to be later called, provided the blueprint for different states to draft legislation for their respective agricultural universities. Cummings was involved with these separate university projects in a supervisory role.

Speaking in 1972, Naik was being prophetic in hoping that the cessation of collaborative programs with the US universities was "temporary." He "look[ed] forward to the day when we may be able to pick up the threads and once again proceed on a path of cooperation and collaboration for the good of Indian agriculture."[22] The thick matrix of collaborative programs of the past and the relationships to which Naik alluded boded well for the future. The momentum of these programs was such that a cast of characters on the Indian and the American side was standing by and provided a propitious context for the birth of ICRISAT in Hyderabad. In a move that was telling of how old connections paved the way for the new institution, Cummings returned to India as the first director general of ICRISAT, steadying the institution in its initial years between 1972 and 1977. He was representative of those who straddled the worlds of the US State Department and private foundations, as well as CGIAR. Together, individuals like Cummings and Naik ensured that their agrarian visions survived in the face of blips or breakdowns in political and diplomatic relations.

The return of Ralph Cummings to the helm of ICRISAT testified to the resilience of these actors in mobilizing externally funded agrarian programs in India. Yet Cummings and others could not have achieved such outcomes without institutional and political mobilization within India as well, and it is to these mobilizations that I now turn.

---

[20] "Statement Relative to Tentative Plan of Operation under Rockefeller Foundation – Government of India Agreement on Agricultural Research and Education," Ralph Cummings papers, Folder 3, Box 34, Series 33.

[21] K. C. Naik and A. Sankaram, *A History of Agricultural Universities* (New Delhi: Oxford and IBH Publishing), pp. 20–23.

[22] K. C. Naik, Statement, USAID/Tennessee papers, Folder 7, Box 9, p. 4.

Priming the Pump

In February 1979, Ralph Cummings delivered the prestigious Lal Bahadur Shastri Memorial lecture – named after the country's second prime minister and icon of Indian farmers' prowess – in which he celebrated twenty-five years of scientific contributions to agricultural progress in India. The tone of the talk was slightly autobiographical, with Cummings drawing a straight line from his arrival in India in 1956 and his contribution to the launch of the country's first postgraduate teaching program at IARI to the 1970s, when the country had turned its focus to "semi-arid" crops.[23] Two years earlier, he had ended his tenure at ICRISAT and assumed the chairmanship of the CGIAR TAC. He was convinced of the need for constant attention to the application of science in bolstering agricultural yields. "You have to run as fast as you can to stay where you are. To get someplace else, you have to run even faster," he explained to colleagues in Indonesia, paraphrasing from Lewis Carroll's *Alice in Wonderland*.[24] In his new role of global ambassador of crop research, Cummings ensured that his message of acceleration and intensification – and his celebration of external aid as the means of bringing about change – could not be missed. However, what Cummings ignored in his account of the recent history of agricultural science in India was the synergy between ICRISAT's global objectives and India's national project in the 1970s. A study of Indian administrative records, political documents, and domestic context shows that Indian scientists and politicians, too, had come to embrace the idea of expanding the Green Revolution to the crops of "dryland" areas.

In India, the Fourth Five-Year Plan (1969–74), adopted under Prime Minister Indira Gandhi, had accelerated national efforts to promote research on "rainfed crops" – those grown without access to irrigation – a move that anticipated ICRISAT's focus on "upland" or "semi-arid" crops. In that sense, the Rockefeller Foundation outreach of 1971, which I described above, could not have been more opportune. Indeed, a case can be made that the watchful Rockefeller Foundation officials saw India's eagerness to move towards a focus on rainfed crops and decided the ground was propitious to bring their effort to India. The building of political will in India towards this agricultural agenda could be seen in parliamentary discussions and in concrete steps that solemnized new

---

[23] Ralph Cummings, "Science in Service to Agriculture – A Quarter Century of Progress," Lal Bahadur Shastri Memorial Lecture, New Delhi, February 2, 1979, Ralph Cummings papers, Folder 3, Box 72.

[24] Ralph Cummings, "Agricultural Research: Problems and Prospects," Address at Ujung Pandang, South Sulawesi, Indonesia, September 26, 1977, Ralph Cummings papers, Folder 1, Box 72, p. 1.

research programs. For example, the Indian Council of Agricultural Research initiated all-India coordinated projects for millet (1965), pulses (including chickpeas and pigeon peas) (1966), and sorghum (1969).[25] Another coordinated program on groundnut came later during the Eighth Plan. It was the Fourth Plan, which started in 1969, that apportioned definitive funds for these coordinated nationwide projects and enabled the setting-up of designated institutions. These were then supplanted by an integrated dryland agriculture development project that was also launched during the Fourth Plan.[26] The demand for research on these crops often came from constituents of rainfed regions and their representatives in parliament. In 1972, for instance, Rajya Sabha MP from Karnataka heckled the agriculture minister, asking repeatedly if the funds allocated to such crops were not "meager," considering that 70 to 80 percent of agricultural lands in the country were farmed under rainfed conditions, while also demanding to know why only a small portion of the allocated funds had been used.[27]

As these separate schemes developed, programmatic connections were established around specific crops. For instance, IARI's all-India coordinated project on sorghum had a subsidiary center in Hyderabad. The same city was also the seat of the Andhra Pradesh Agricultural University, which was being built with the help of Kansas State University (through the USAID land-grant program).[28] At the Andhra Pradesh campus Ralph Cummings and fellow Rockefeller Foundation agronomist Lee House coordinated the Rockefeller-sponsored sorghum program.[29] If anything, it was the connections and reciprocities between US aid programs, Indian political projects, and scientists that ultimately built a critical mass of

---

[25] Statement by Minister of State for Food, Agriculture, Community Development and Cooperation, Shri Annasaheb Shinde, Rajya Sabha Debate, March 21, 1969, 5165–5166, https://rajyasabha.nic.in/Documents/Official_Debate_Nhindi/67/F21.03.1 969.pdf; Minister Shri Shinde replied to a similar question the following year: Rajya Sabha Debate, February 27, 1970, 38–39, https://rajyasabha.nic.in/Documents/ Official_Debate_Nhindi/71/F27.02.1970.pdf. For a report on the centers and subcenters set up under the plan for sorghum and millets, see "Scientific Research on Coarse Grains," Rajya Sabha Debate, August 9, 1972, 53–54, https://rajyasabha.nic.in/ Documents/Official_Debate_Nhindi/81/F09.08.1972.pdf.

[26] Question by MP from Karnataka, Shri Veerendra Patil to Minister of State for Agriculture, Rajya Sabha Debate, December 6, 1972, 23, https://rajyasabha.nic.in/ Documents/Official_Debate_Nhindi/82/F06.12.1972.pdf.

[27] Sher Singh, "Development of Dry Farming," Statement, Rajya Sabha Debate, December 6, 1972, 20–24, https://rajyasabha.nic.in/Documents/Official_Debate_Nhin di/Floor/82/F06.12.1972.pdf.

[28] *Sixteen Years in India: A Terminal Report* (Manhattan: International Agricultural Program, Kansas State University, 1972).

[29] M. S. Swaminathan, "In the Beginning …," in Lydia Flynn, Ajay Varadachary, and Kate Griffiths, eds., *ICRISAT at 30: The Historic Journey to the Semi-Arid Tropics* (Patancheru: ICRISAT, 2002), pp. 1–6, at 1.

support for ICRISAT. Thirty years after its founding, M. S. Swaminathan (who was appointed director general of the Indian Council of Agricultural Research in 1972) reminisced that the germ of the idea of an international center in India for crops like sorghum came from Lee House, and he had enthusiastically welcomed it, suggesting that if such a center were to come up, it should additionally focus on millet.[30]

As ICRISAT came into being – with a focus on four mandate crops of sorghum, millet, chickpeas, and pigeon peas, to which groundnuts were added in 1976 (see Lucas M. Mueller, Chapter 5, this volume) – these synergies and associations between programs and institutions provided justifications to move forward. The minister of state for agriculture, Annasaheb Shinde, referred to those synergies when he announced the inauguration of the institute in the lower house of the Indian parliament. Shinde thought that ICRISAT would provide a "good opportunity to Indian agricultural scientists," as established experts would now be able to tap into ICRISAT's collections of globally accumulated genetic materials for its mandate crops.[31] As it moved forward with its programs, ICRISAT routinely drew on the resources of Indian agricultural institutions and the informal network of Indian scientists that the prior work of foundations in India had fostered. The government of India, for its part, appreciated the global resources of ICRISAT and made use of them to advance its own agendas. In the 1970s, Indian officials specifically asked for help to bridge the gap with developed nations on the quality and yield of its pulses – crops increasingly central to domestic political agendas.[32]

The Politics of the 1970s

On the morning of January 12, 1975, India's prime minister, Indira Gandhi, landed at Begumpet airport in Hyderabad, en route to lay the foundation stone of ICRISAT's campus (Figure 2.1). She was received at the airport by the state's governor and chief minister. Her entourage included the central minister for transport, Kamalapati Tripati, and All India Congress Committee Secretary P. V. Narasimha Rao, a future

---

[30] Ibid., p. 1.
[31] Written statement by minister of state for agriculture, in Lok Sabha Annasaheb P. Shinde, *Lok Sabha Debates*, Ninth Session (Fifth Lok Sabha), 5th Series, vol. XXXIII, no. 16, Lok Sabha Secretariat, December 3, 1973, 194–195, at 195, https://eparlib.nic.in/bitstream/123456789/1140/1/lsd_05_09_03-12-1973.pdf.
[32] "Biology of Yield on Pulse Crops," Statement of Minister of State for Agriculture and Irrigation Shri Shah Nawaz Khan, Rajya Sabha Debates, November 27, 1974, 27–30, https://rajyasabha.nic.in/Documents/Official_Debate_Nhindi/Floor/90/F27.11.1974.pdf.

Figure 2.1 Prime Minister Indira Gandhi inaugurating ICRISAT in 1975. Photo courtesy of ICRISAT and reprinted by permission.

prime minister. The arrival of Indira Gandhi and the company of high-level officials signaled the political salience of the new institution to the ruling party and to her brand of popular politics.

Reaching the ICRISAT site at Patancheru, twenty-five kilometers outside the city, Gandhi addressed the audience gathered there on the relevance of the new institute and its research program, emphasizing India's core problem of hunger. All other social and developmental issues could wait until the problem of hunger had been solved. She referred explicitly to the Green Revolution policies of the mid 1960s and the criticism her government had faced over how those interventions had aggravated economic disparity. She reminded the audience that, despite fomenting inequality, the Green Revolution had resolved recurrent famines and food shortages of the past. India was now at a new stage in its quest to solve the problem of hunger. The nation needed to extend the "modern methods" of the Green Revolution to areas practicing dryland agriculture. Some 70 percent of India's agricultural land was owned by smallholder farmers, many of whom were located in semi-arid zones. The new institution with its focus on small farmers and semi-arid crops could potentially address hunger and inequality simultaneously.[33] Gandhi's visit to ICRISAT relayed clearly her support for a strategy of expansion: "we have to follow modern methods in arid land as soon as possible."[34] In this, she drew a direct line from the prior Green Revolution to its anticipated dryland sequel. The emphasis on productivity that had been applied in wheat and rice now needed to be extended to crops of low-rainfall areas.

Gandhi's address to ICRISAT also amplified the "international" character of the new institute. Trying to counter the division of the world into "first-, second-, third-, or fourth-world countries," she idealized the institution as a global center in which scientists from multiple countries participated. Although she asserted pride in the contributions made by Indian scientists, she would not shy away from assistance offered by other countries. Praising ICRISAT's composite backing, Gandhi argued, "This international institute which is doing research on crops in semi-arid tropics is a model for conduct of international relations between nations."[35] Indian scientists were themselves copartners in the institution after all, she implied.

Out of political necessity, Gandhi overlooked the fact that CGIAR, although reflecting an emerging multipolarity, was still backed primarily by expertise and money from US-based sources. Perhaps the cover of CGIAR as being backed by the United Nations Food and Agriculture Organization (FAO), the United Nations Development Programme (UNDP), and the World Bank was sufficient for the Indian political class to accept the declaration of its international character. ICRISAT's American sheen had been diluted just enough to allow Indian politicians to sell the institution to their constituents as anything but American. Gandhi could instead highlight the "pooling of talents of scientists and technicians, regardless of nationality, race, or color" in ICRISAT. They were all unified, Gandhi stressed, "in [waging] this greatest of all wars, the war against hunger."[36] It was partly because Gandhi could bring herself to see the international as opposed to the American face of ICRISAT, and because she could bring her constituents to believe in this international image as well, that ICRISAT was accepted, even as popular anti-American sentiment in India was peaking in the wake of the Nixon

---

[33] "There Should Not Be Any Difference between Nations in Eradicating Hunger: PM," *Andhra Prabha* (January 12, 1975), 1, 2; "Scientists Should Help in Higher Food Production at Lower Cost: PM Request," *Andhra Patrika* (January 12, 1975), 1, 4. Translation from Telugu.
[34] "There Should Not Be Any Difference," 1, 2.
[35] "Mankind Should Put Efforts in Removing Hunger Problem: PM Indira Gandhi," *Andhra Bhumi* (January 12, 1975), 1, 2. Translation from Telugu.
[36] *ICRISAT at 30*, 29.

administration's policies in providing support to Pakistan during the Bangladesh War.[37]

ICRISAT's opening coincided with an era of momentous changes in India's national electoral politics as the reins of power passed from Indira Gandhi and the Congress Party to a rival political formation that brought to the helm two non-Congress Party prime ministers – Morarji Desai in 1977 and Chaudhury Charan Singh in 1979. Four years after Gandhi laid ICRISAT's foundation stone, Charan Singh arrived at ICRISAT as prime minister to "dedicate" the campus, signaling ICRISAT's full-fledged operationality. Gandhi and Singh were two polar opposites in the politically divisive 1970s. Paul Brass speaks of "the life-long struggle between ... [Charan Singh] and Indira Gandhi."[38] These rivalries came to the boil in the 1970s.[39] The presence of Gandhi and Singh on the ICRISAT campus four years apart represented a unique and rare convergence in their respective claim-making.

As Prime Minister Charan Singh arrived on the ICRISAT campus in August 1979, the diversity of political support enjoyed by the new institution shone through (Figure 2.2). This time the distinguished gathering included key ministers for agriculture, industry, and defense – individuals who were themselves of diverse political leanings and now part of a coalition government. Also in attendance was the Andhra Pradesh chief minister, Marri Chenna Reddy, who belonged to the Congress Party, a rival of the political conglomeration that had catapulted Singh to power in New Delhi. The joint appearance of these leaders at ICRISAT highlighted the issues on which a range of political parties could agree, despite standing in opposition along party lines at the center. Singh had established his political reputation as a peasant leader from the eastern state of Uttar Pradesh. At the national level, he had continued to speak for farmers. It is from this position that he spoke at ICRISAT. Singh exhorted the new center's scientists to "give utmost priority to removing inequalities in economic development of our nation." Clearly Singh was thinking in terms of India's predominantly agrarian economy,

---

[37] On the negative popular Indian sentiment towards the United States in the early 1970s, see Raghavan, *Fierce Enigmas*; Engerman, *The Price of Aid*.

[38] Paul Brass, *An Indian Political Life: Charan Singh and Congress Politics, 1967 to 1987* (New Delhi: Sage Publications, 2014), p. xiii.

[39] After Indira Gandhi lost the national election of 1977, Singh, as home minister of the Janata Party government, pursued legal cases against Gandhi and her son, Sanjay Gandhi. When Janata Party rule collapsed under the weight of its internal rivalries, Singh became prime minister as the new leader of the Janata Party (Secular). With a Machiavellian sleight of hand, at this moment Gandhi's party supported Singh, enabling him to conjure a majority and form a government on July 18, 1979. Gyan Prakash, *Emergency Chronicles: Indira Gandhi and Democracy's Turning Point* (Princeton, NJ: Princeton University Press, 2019), pp. 356–357.

Figure 2.2 Prime Minister Charan Singh at the dedication of ICRISAT in 1979. Photo courtesy of ICRISAT and reprinted by permission.

in which small farmers in rainfed areas had to be pulled up economically. He mentioned the dependence of 70 percent of the country's agriculture on uncertain rains and thus underlined the importance of ICRISAT in addressing this vast geography and "helping the farmers who are dependent solely on rainfall." Meanwhile the chief minister, Reddy, thanked ICRISAT for choosing his state as its location. An element of regional pride suffused Reddy's adulation as he emphasized the importance of Andhra Pradesh in Indian agriculture.[40] In short, a tenuous alliance seemed to exist over ICRISAT. This secured its place as an Indian, as well as an international, institution.

Conclusion

ICRISAT was welcomed in India in 1972 because its scientific goals looked appropriate to and in line with established state research programs

---

[40] *Andhra Jyoti*, August 31, 1979. Translation from Telugu.

and the nation's settled agenda for agrarian modernization. ICRISAT's focus on marginal lands and cultivators offered conjoint space to accommodate the politics of both Indira Gandhi and Charan Singh. Gandhi's rise from 1967 to 1972 was based on an effort to project herself as "a radical reformer." Her populist politics in the 1970s highlighted the goal of *garibi hatao*, the removal of poverty.[41] Charan Singh held different political positions, on account of his stature as a peasant leader, but ICRISAT still accommodated his advocacy on behalf of marginal farmers. The nature of programming at ICRISAT allowed both Indira Gandhi and Charan Singh to come to a consensus without compromising their different electoral politics.

Set up as an "international center," ICRISAT in India mirrored the dynamic world order of the 1970s and India's realization of its own priorities within new global patterns. The diminishing impact of the rigid bipolar divisions of the Cold War that scholars have identified as a trend in the 1970s is visible in the cast of multilateral organizations and national governments that stood behind the establishment of ICRISAT.[42] ICRISAT was the fifth international agricultural research center of the CGIAR system and – as Ralph Cummings emphasized in his role as ICRISAT's first director general – the first center to be established after the formal constitution of that system.[43] Cummings' emphasis on ICRISAT's origins within CGIAR was meant to highlight the new institution's composite backing. But in many ways the vision and work at ICRISAT reflected a continuation of prior trends in agricultural development, including India's earlier pursuit of the Green Revolution, that were earmarked now by a new stage in institutional evolution. The circumstances of postcolonial India allowed for the emergence of new forms of institutionalized expertise that developed outside the direct realm of the local state. ICRISAT arrived in a generative space where global visions sought the approval of Indian scientists and politicians, if not the Indian state per se.

---

[41] For Indira Gandhi's stamp on the turbulent Indian politics of the 1970s, see Ramachandra Guha, *India after Gandhi: The History of the World's Largest Democracy* (New Delhi: Macmillan, 2007), pp. 464–488; Prakash, *Emergency Chronicles*, pp. 136–139.

[42] Daniel Sargent, *A Superpower Transformed: The Remaking of American Foreign Relations in the 1970s* (New York: Oxford University Press, 2015).

[43] Ralph Cummings, "ICRISAT Presentation," July 27, 1976, Ralph Cummings papers, Folder 2, Box 69, Series 33, p. 1.

3   Conflicted Landscape
CIAT and Sugarcane in Colombia

Timothy W. Lorek

In a bunker far under the ice on the Norwegian island of Svalbard, 12 degrees latitude north of the Arctic Circle, seeds descended from maize landraces collected in Colombia are catalogued as part of the effort to preserve the world's agricultural genetic diversity and cultural patrimony. Here in the Svalbard Global Seed Vault, reserve seeds are filed in an abandoned coal mine for future breeding projects or in case of catastrophe. Nearly a million samples of seeds in the vault represent duplicates for over one-third of the genetic diversity contained in seed banks around the world.

The Svalbard Global Seed Vault opened in 2008 under a tripartite agreement between the Norwegian government, the Nordic Genetic Resource Center, and the Global Crop Diversity Trust, the latter a partnership between the United Nations Food and Agriculture Organization (FAO) and the Consultative Group on International Agricultural Research (CGIAR).[1] That year, the vault received its first batch of shipments of duplicate seeds from around the world, including Colombia, where CGIAR operates out of the International Center for Tropical Agriculture (CIAT), opened in 1967 outside the city of Palmira on the tropical floodplain of the Cauca River.[2]

Within months in late 2017 and early 2018, both the Svalbard Global Seed Vault and CIAT celebrated anniversaries. Scientists, philanthropists, government officials, and journalists arrived at Svalbard airport in Longyearbyen, Norway on the arctic ice and at Alfonso Bonilla Aragón International airport in Palmira, Colombia in the tropical heat. They

---

[1] On Svalbard and seed banks, see Helen Anne Curry, "The History of Seed Banking and the Hazards of Backup," *Social Studies of Science* 52, no. 5 (2022): 664–688; Helen Anne Curry, *Endangered Maize: Industrial Agriculture and the Crisis of Extinction* (Oakland: University of California Press, 2022); Courtney Fullilove, *The Profit of the Earth: The Global Seeds of American Agriculture* (Chicago: University of Chicago Press, 2017); and Xan Sarah Chacko, "Creative Practices of Care: The Subjectivity, Agency, and Affective Labor of Preparing Seeds for Long-Term Banking," *Culture, Agriculture, Food, and Environment* 41, no. 2 (2019): 97–106.
[2] "Svalbard Global Seed Vault Celebrates 10 Years," Press Release, February 27, 2018, International Institute of Tropical Agriculture (IITA), www.iita.org/news-item/svalbard-global-seed-vault-celebrates-10-years/.

celebrated these institutions and their histories and looked forward to their continued roles in tackling some of the twenty-first century's enduring issues, including food security, climate change, war, and natural disaster.[3] The stories these sites tell about themselves have changed since CIAT's founding in 1967; however, one constant remains – their global orientation.

This chapter introduces CIAT in Palmira, Colombia, contextualizing its founding and historical evolution and juxtaposing its research agenda and purported accomplishments against the lived reality of late twentieth-century rural Colombians, particularly those in Valle del Cauca, the political department in which Palmira and CIAT are located. Although CIAT is a key node in a global network that spans from Palmira to Svalbard, the global history perspective presented at anniversary celebrations and in official publications, whatever its other merits, misses or minimizes the Colombian side of the story.[4]

CIAT's founding role in the establishment of CGIAR in 1971, and its continuing significance today, situates the site within the circulation of scientists, seeds, and funding in a global network of agricultural science. But in Colombia, the social, political, and economic landscape in which CIAT operates may be discerned in the speaker list for the celebrations of CIAT's fiftieth anniversary in 2017: corporate executives, local and national politicians, a peace negotiator with the guerrilla, international aid workers, and land-grant academics. Their convergence in the agrarian history of the Cauca Valley binds the experiences of the Latin American Cold War and the multigenerational Colombian conflict to global processes of agricultural science, development, and capitalism.

In Palmira, CIAT's history has played out against the backdrop of a landscape indelibly marked by a parallel history of agricultural corporatization and monoculture, particularly via the ascendency of a sugarcane agro-industrial complex that has, since 1959, organized under the auspices of the politically powerful Colombian Sugarcane Growers Association (ASOCAÑA). The refined sugar industry has many by-products – candy, alcohol, and biofuels, but also land

---

[3] To date, the only withdrawal from the Svalbard Global Seed Vault came between 2015 and 2017 when the International Center for Agricultural Research in the Dry Areas (ICARDA) requested seeds to relocate its research and conservation work from war-torn Syria to new sites in Lebanon and Morocco. See "Svalbard Global Seed Vault Celebrates 10 Years." Also see Courtney Fullilove, Chapter 1, this volume.

[4] The history of CIAT has thus far remained the purview of official, institutional histories. See, for example, the history commissioned for CIAT's fiftieth anniversary: John Lynam and Derek Byerlee, *Forever Pioneers – CIAT: 50 Years Contributing to a Sustainable Food Future . . . and Counting*, CIAT Publication No. 444 (Cali, Colombia: CIAT, 2017), http://hdl.handle.net/10568/89043.

concentration and the Colombian government's commitment to private capital growth over land reform, environmental and public health issues related to nitrate leaching from chemical fertilizers and inputs including glyphosate herbicide, and societal consequences such as diabetes and obesity related to a steady rise in per capita refined sugar consumption. Collectively, these have played an important role in the country's long-running and multidimensional armed conflict.[5] In fact, one of the first victims of a guerrilla insurgency kidnapping in Colombia occurred in 1965, when the Revolutionary Armed Forces of Colombia (FARC) symbolically targeted, kidnapped, and murdered Harold Eder, heir to the largest and oldest sugarcane corporation in the valley.[6] Throughout much of the second half of the twentieth century, parts of the Cauca Valley remained a landscape of conflict. But it was also a landscape of conflicting agricultures, or at least agricultural visions. This valley was and is home both to a sprawling sugarcane monoculture and to a world-renowned international research entity dedicated to the improvement of staple food crops, most notably beans, rice, cassava, and pasture grasses. What does the jarring realization that CIAT exists in a sea of corporate sugarcane suggest about the conflicted landscapes of global food production, scientific research, and CGIAR?[7] (See Figure 3.1.)

[5] On the growth of private agribusiness, see Joan Marull, Olga Delgadillo, Claudio Cattaneo, María José La Rota, and Fridolin Krausmann, "Socioecological Transition in the Cauca River Valley, Colombia (1943–2010): Towards an Energy-Landscape Integrated Analysis," *Regional Environmental Change* 18, no. 4 (2018): 1073–1087. On environmental and public health issues, see Olga Deldadillo-Vargas, Roberto Garcia-Ruiz, and Jaime Forero-Álvarez, "Fertilising Techniques and Nutrient Balances in the Agriculture Industrialization Transition: The Case of Sugarcane in the Cauca River Valley (Colombia), 1943–2010," *Agriculture, Ecosystems and Environment* 218 (2016): 150–162. Details on glyphosate and public health from private communications with retired CIAT scientist Douglas Laing, August 2013. See also "'No puede ser que en Cali se sufra de recortes de agua': Douglas Laing," *El Tiempo* (September 9, 2014), www.eltiempo.com/archivo/documento/CMS-14509898. On sugar consumption rates, see US Foreign Agricultural Service, *Colombia – Sugar Annual – Colombian Sugar Industry Maintains High Production Levels*, Global Agricultural Information Network (GAIN) Report Number 1904 (April 12, 2019): https://apps.fas.usda.gov/newgainapi/api/report/downloadreportbyfilename?filename=Sugar%20Annual_Bogota_Colombia_4-12-2019.pdf, and Andrew Jacobs and Matt Richtel, "She Took on Colombia's Soda Industry. Then She Was Silenced," *New York Times* (November 13, 2017), www.nytimes.com/20 17/11/13/health/colombia-soda-tax-obesity.html. On violence in the sugarcane zone, see Michael Taussig, *Law in a Lawless Land: Diary of a Limpieza in Colombia* (Chicago: University of Chicago Press, 2003).

[6] Cali's most circulated daily ran a feature on Eder's murder as part of its coverage of the Colombian government's negotiation of a peace accord with the FARC from 2013 to 2016. Catalina Villa, "La historia de Harold Eder, uno de los primeros secuestrados de las FARC," *El País* [Cali, Colombia] (September 22, 2016), https://www.elpais.com.co/proceso-de-paz/la-historia-de-harold-eder-uno-de-los-primeros-secuestrados-de-las-farc.html.

[7] The imagery of a sea of sugarcane is elaborated in Carmen Cecilia Rivera, Luis Germán Naranjo, and Ana María Duque, *De María a un mar de caña: Imaginarios de*

Figure 3.1 CIAT land at headquarters in Palmira leased to corporate sugarcane producers surrounds the gates to the research center (seen in the side mirror), July 2022. Photo by Timothy W. Lorek.

In this chapter, I borrow the concept of deterritorialization from anthropology and political geography to suggest an important framework for writing the history of CGIAR centers from the ground up.[8] In these pages, I'm less interested in evaluating the global impact of CIAT, and even less in celebrating its successes or condemning its failures. Instead, my interests coalesce around the international center's relationship to the landscape, people, and politics just outside its gates. I advocate for seeing research centers in place and evaluating science and knowledge production as actions that take place. Despite their global orientation and circulation, CGIAR centers still rely on local, place-specific environments for soil and water, on local people to drive shuttles and buses, tend to fields, serve food, and clean the bathrooms, and on local businesses and regional and national politics for coordination, agreements, infrastructure, and even funding, whether applied directly to the center's operations or indirectly via the financing of roads, airports, and state-funded public university programs, to mention a few. Despite these ties, the beneficiaries of CIAT's research seem to consistently live beyond the Cauca Valley. Viewed from the Cauca Valley, I argue that CIAT has adopted a placeless research agenda, that is, one not specific to a particular location or

naturaleza en la transformación del paisaje vallecaucano entre 1950 y 1970, 2nd edn. (Cali, Colombia: Universidad Autónoma de Occidente, 2017).

[8] On the conceptualization of deterritorialization, see Gilles Deleuze and Félix Guattari, *A Thousand Plateaus: Capitalism and Schizophrenia* (London: Continuum, 1987). On theorizing territory and deterritorialization in Colombia, see Arturo Escobar, *Territories of Difference: Place, Movements, Life, Redes* (Durham, NC: Duke University Press, 2008). For a more recent example, see Marcela Velasco, "Territory and Territoriality in Colombian Politics," *Contextualizaciones Latinoamericanas* 8, Special Issue (May 2016): 1–19.

attuned to the nuances of local sociopolitical conditions. Instead, CIAT has offered scientific solutions for a placeless conceptualization of the global tropics, a phenomenon described in this volume for CGIAR more broadly, for example in Derek Byerlee and Greg Edmeades' description (Chapter 9, this volume) of CIMMYT's mega-environments in maize breeding, Prakash Kumar's account (Chapter 2, this volume) of the imaginative definition of upland crop zones, and others.[9] Immediately outside CIAT's gates, imagined solutions to the supposed challenges of the global tropics are minimally applicable amidst the reality of the sugarcane agro-industrial complex.

CIAT's advocates undoubtedly feel that it is not their institution's purpose to address local sociopolitical conditions. In this regard, CIAT scientists and their proponents adhere to a familiar myth, or, more generously, ideal, that views science as apolitical. Of course, CIAT, born of the hissing boilers and hot furnaces of the geopolitical Cold War, is nothing if not political.[10] As we shall see, CIAT leadership's historical and internally contested decision not to address local conditions was a political choice in the slippery language of anti-politics.[11]

At the same time, CIAT advocates are right to point out that the center did not create the inequalities of the Cauca Valley. The purpose here is not to blame CGIAR sites for creating or exacerbating rural inequalities.

[9] Also in this volume, Wilson Picado describes a disconnect between CIAT bean programs and on-the-ground realities of displacement during the Central American civil wars of the late 1970s and 1980s. Harro Maat (Chapter 6, this volume) and Rebekah Thompson and James Smith (Chapter 7, this volume) similarly describe a degree of geographic disconnection or placelessness in CGIAR research agendas in Africa.

[10] Recent assessments of the Cold War geopolitics of science and expertise in Latin America include Andra B. Chastain and Timothy W. Lorek, eds., *Itineraries of Expertise: Science, Technology, and the Environment in Latin America's Long Cold War* (Pittsburgh: University of Pittsburgh Press, 2020); Anne-Emanuelle Birn and Raúl Necochea López, eds., *Peripheral Nerve: Health and Medicine in Cold War Latin America* (Durham, NC: Duke University Press, 2020); and Eden Medina, Ivan da Costa Marques, and Christina Holmes, eds., *Beyond Imported Magic: Essays on Science, Technology, and Society in Latin America* (Cambridge, MA: MIT Press, 2014).

[11] The classic treatises on the anti-politics of the postwar development era include James Ferguson, *The Anti-Politics Machine: "Development," Depoliticization, and Bureaucratic Power in Lesotho* (Minneapolis: University of Minnesota Press, 1994); Timothy Mitchell, *Rule of Experts: Egypt, Techno-Politics, Modernity* (Berkeley: University of California Press, 2002); and, for Colombia, Arturo Escobar, *Encountering Development: The Making and Unmaking of the Third World* (Princeton, NJ: Princeton University Press, 1995). More recent examples for the Colombian case include Amy C. Offner, *Sorting out the Mixed Economy: The Rise and Fall of Welfare and Developmental States in the Americas* (Princeton, NJ: Princeton University Press, 2019) and a critique by Rebecca Tally, "How Not to Win Friends": Mundane Matters, Constant Critique, and the Rockefeller Foundation's Defense of Wheat Production in Colombia, 1950–1965," *Agricultural History* 97, no. 1 (2023): 84–120.

Rather, this chapter seeks to historicize CIAT in Cauca Valley soil and understand its creation and evolution as intertwined with, rather than separate from, the challenges of monoculture, land concentration, violence, and peace that preceded the center's establishment in 1967 and continue to play out beyond its gates after its fiftieth birthday.

CIAT, like the fourteen other centers of the present CGIAR consortium, exists in a place.[12] The places where international research agendas come into contact with specific agricultural economies reveal the contradictions and conflicted landscapes of global agricultural systems. Situating CIAT in place, then, is not an act in defiance or condemnation of the research center, much less its scientific community. Perhaps this chapter may even offer CIAT scientists an introduction to the complicated place where they do their important work. More boldly, it offers an attempt to contribute to the evolving mission of CIAT itself, to "help policymakers, scientists, and farmers respond to some of the most pressing challenges of our time, including food insecurity and malnutrition, climate change, and environmental degradation."[13] All of these profoundly affect the Cauca Valley in Colombia. If CIAT is to be a part of a sustainable peace in Colombia, then recognizing and reckoning with its history in place is an important start.

Local Precedents and Contexts for CIAT

CIAT is located along a highway that cuts across sugarcane fields for approximately thirty kilometers to connect the regional city of Palmira, Colombia (population approx. 300,000) with the large metropolis of Cali (approx. 2.2 million). Cali's international airport, which regularly delivers the world's scientific community to CIAT, is also located along this corridor just north of the research center. Coincidentally, this international airport (originally Palmaseca International airport) was inaugurated in 1971 in order for Cali to host the Pan-American Games, the same year that CIAT and the three other founding centers merged into

---

[12] This is historically indisputable, although it raises interesting questions for the twenty-first century, as CIAT and the One CGIAR system move towards greater coordination, digital operations, and, presumably, an enhanced future role of AI technologies. Notably, CIAT merged with Bioversity International as part of this reconfigured One CGIAR system, further detaching CIAT from its historic headquarters in Palmira, Colombia. See CGIAR Platform for Big Data in Agriculture, "The Platform for Big Data and the Digital Future of CGIAR" (December 29, 2021), https://bigdata.cgiar.org/blog-post/the-plat form-for-big-data-and-the-digital-future-of-cgiar. Also see Helen Anne Curry and Sabina Leonelli, Chapter 10, this volume.

[13] CIAT, *CIAT Today: An Overview* (Cali, Colombia: CIAT, 2018).

the global consortium known as CGIAR. Cali's growth and connection to the world offer important background for the establishment of CIAT.

The Cauca Valley is located in southwestern Colombia, between two divergent ranges of the Andes Mountains (Figure 3.2). In the political department of Valle del Cauca, the Cauca River flows north past the Cali metropolitan area. A fertile alluvial plain stretches east of the river at approximately 1,000 meters elevation. This valley has been the subject of grandiose proclamations of paradise and future agricultural bounty since at least the early nineteenth century. In fact, the rise of an industrial sugarcane zone here in the twentieth century was due substantially to the site's climate and geography – it is one of the few places on the planet that can support a year-round sugarcane harvest. Climatic advantages similarly contributed to the valley's long-standing tradition of hosting agricultural research centers. The valley's unique ability to produce two annual crops of maize attracted both Colombian and Caribbean agricultural scientists from the 1920s through the 1950s, eventually including Rockefeller Foundation scientists from the United States after 1948.

Figure 3.2 Monocultures of sugarcane viewed from the air dominate the fertile alluvial lands of the Cauca River valley around Palmira, Colombia, July 2022. Photo by Timothy W. Lorek.

Despite the region's long-advertised potential, industrial-scale agriculture took hold relatively recently. Before the 1930s, the Cauca Valley's most desirable soils hosted a patchwork of sprawling cattle ranches owned by regional elites, much to the chagrin of would-be agriculturalists. In the nineteenth century, these ranches fed and clothed the gold-mining zones of the Pacific Coast to the west and Antioquia to the north. The abolition of slavery in the mid nineteenth century ushered in a period of social unrest in the valley as newly freed Afro-Colombians from the mining districts migrated to the towns and cities of the valley and joined with other people of color in the region in opposition to the large ranching estates, or latifundia. Racially charged conflict erupted in the 1850s over fencing and landowners' enclosure of the commons, precipitating a series of national civil wars and an ongoing competition for the political allegiance of the popular classes.[14]

At around the same time, a new cohort of would-be industrialists with ties to international import-export circles purchased land and began to settle in the valley. One of these newcomers, James "Santiago" Eder, began assembling what would become the valley's first industrial-scale sugar operation with the opening of a steam roller mill imported from Scotland in 1900. That enterprise, Manuelita SA, remained a critical player in the expansion of the sugarcane industry throughout the twentieth century and is a major multinational player in the Cauca Valley today.

In the 1920s, the rural economy of the Cauca Valley pitted a traditional cattle-ranching elite against a growing industrialist class beginning to (slowly) coalesce its investments around sugar. Apart from both of these groups, a vast array of small- and medium-scale cultivators, some with land titles, others without, grew rice, cacao, plantains, sugarcane (milled in simple *trapiches* to produce unrefined sucrose, or *panela*), and coffee (in the foothills at the margins of the northern valley in particular). These rural valley residents included Afro-Colombians, as well as recently arrived colonists from Antioquia to the north, a group mythically associated with the expansion of a middle-class agricultural frontier in Colombia. With many groups and interests at odds in a relatively small but fertile valley, political energies focused around land tenure and space. The ranching elite and their political benefactors became the most common targets of charges of inefficiency and wasted space, particularly as Colombian cities such as Cali and Medellín grew and adopted heightened industrial ambitions and a need for robust food-producing hinterlands.

In this context, the departmental government funded a series of projects in the late 1920s aimed at quelling rural conflict and laying the

---

[14] James E. Sanders, *Contentious Republicans: Popular Politics, Race, and Class in Nineteenth-Century Colombia* (Durham, NC: Duke University Press, 2004).

infrastructure for future industrial growth and economies of scale.[15] In 1927, the department of Valle del Cauca and the national government joined forces to fund the Palmira Agricultural Experiment Station, one of three regional agricultural stations created to serve the country's different geographies and corresponding crop regimes. In the 1930s, following the national electoral triumphs of the Liberal Party, with its populist and reformist agenda, the young Palmira station enhanced its emphases on the scientific improvement of a diverse palette of crops, including hybrid rice, maize, sugar, citrus, and even experimental projects such as *Cannabis sativa*, pursued to explore hemp as a possible domestic fiber for coffee sacks. The focus on diverse and multiscalar agriculture at Palmira complemented both regional and national moods. A series of agricultural schools and colleges were founded around Valle del Cauca at this time to educate the sons of *campesinos*, as well as a future generation of expert agronomists. The Palmira station partnered in these education programs and aggressively touted its extension efforts. Similarly, at the national level, the minister of agriculture advocated for protectionist tariffs on imported foodstuffs and offered fireside chats on the radio to speak directly to the country's farmers. Taking cues from the agricultural bureaucracy of the US Department of Agriculture, as well as more climatically and culturally similar efforts in Mexico, Brazil, and Puerto Rico, actors at different levels within the Colombian state sought to foster a dynamic and self-sufficient agricultural sector.

Sugarcane was just one subject of many for researchers at Palmira in the 1930s. Initially, Colombian agronomists' efforts to promote disease-resistant hybrid canes received a lukewarm reception from valley cultivators. Only the few industrial-scale firms had the capital to invest in a risk like adopting new varieties, such as the so-called POJ lines circulating from the Dutch-operated Proefstation Oost Java (POJ). However, a severe breakout of the sugarcane mosaic virus in the mid 1930s devastated Colombian growers. Many of the small- and medium-scale *panela* producers lost land and market share as the large entities, like Manuelita, invested further in disease-resistant hybrids and hired foreign breeders. The Colombian state likewise moved to intervene in the mosaic crisis, wresting full control of the Palmira Agricultural Experiment Station from the Department of Valle del Cauca and amping

[15] On the history of agricultural science in the Cauca Valley and the life of the Palmira Agricultural Experiment Station, see Lorek, *Making the Green Revolution*. On the local Cauca Valley roots of Cold War international development programs, see Timothy W. Lorek, "Strange Priests and Walking Experts: Nature, Spirituality, and Science in Sprouting the Cold War's Green Revolution," in Chastain and Lorek, eds., *Itineraries of Expertise*, pp. 93–113.

up experimentation with sugarcane cultivars in a collaborative effort with industrial producers and the US Department of Agriculture Sugarcane Research Center in Canal Point, Florida. A newly strong sugar sector thus grew out of the mosaic crisis with firm collaborative investments from Bogotá and Washington, DC. Manuelita, for its part, built a massive new factory with a refinery that inaugurated production in 1952. By mid century, Valle del Cauca's agrarian populists and political boosters would have to look elsewhere to realize their dream of a bountiful and dynamic valley, at least one not exclusively reserved for sugarcane.

Violence and Development

As the industrial sugarcane sector grew stronger in the aftermath of the Colombian state's interventions against the mosaic virus, some of the proponents of the old Palmira Agricultural Experiment Station and its extension mission reached out to foreign experts and funders to keep their programs going.[16] As World War II ended and the Cold War dawned, the US government responded, establishing Point Four projects and collaborations with Cauca Valley partners, including a series of exchanges run through Michigan State University. The Rockefeller Foundation also responded, launching the Colombian Agricultural Program (CAP) in 1950, the first international expansion of the pilot Mexican Agricultural Program, considered by many to be the institutional birth of the Green Revolution. CAP partnered with the Palmira station and used that site as a base for some of its critical projects, such as maize breeding. Maize landraces collected by Colombian agronomists based in Palmira and Medellín on behalf of the Rockefeller Foundation would contribute genes to new high-yielding varieties in Asia and eventually make their way to the Svalbard Global Seed Vault.[17] During the 1940s and 1950s,

---

[16] On the rise of an industrial sugarcane sector in the Cauca Valley, see Adriana Santos Delgado and Hugues Sánchez Mejía, *La irrupción del capitalismo agrario en el Valle del Cauca: Políticas estatales, trabajo y tecnología, 1900–1950* (Cali, Colombia: Programa Editorial Universidad del Valle, 2010) and Hugues Sánchez Mejía and Adriana Santos Delgado, "Estado, innovación y expansión de la agroindustria azucarera en el valle del río Cauca (Colombia), 1910–1945," *América Latina en la Historia Económica* 21, no. 3 (September–December 2014): 201–230.

[17] On the maize collection program in Colombia, see L. M. Roberts, U. J. Grant, Ricardo Ramírez E., W. H. Hatheway, and D. L. Smith, in collaboration with Paul C. Mangelsdorf, *Races of Maize in Colombia* (Washington, DC: National Academy of Sciences – National Research Council, 1957). For the latest assessment of these maize collection programs in Latin America, see Curry, *Endangered Maize* and Diana Alejandra Méndez Rojas, "Los libros del maíz: Revolución Verde y diversidad biológica en América Latina, 1951–1970," *Letras Históricas* 24 (spring–summer 2021): 149–182.

sugar production grew in the Cauca Valley, but so did foreign technical assistance as the Rockefeller Foundation and others took the reins of Colombian projects and centers originally established in the 1920s. These bifurcating processes – sugarcane intensification and Cold War developmentalism – shared a place and time amidst great social and political turmoil. The period between 1946 and approximately 1958 is remembered in Colombia as *La Violencia*, or The Violence. This period of horror, especially acute in the Colombian countryside, is often attributed to partisan conflict between regional agents of the Liberal and Conservative political parties. However, as Mary Roldán and others have convincingly argued, the conflict had much deeper, locally situated motivations, not least a long history of unresolved grievances related to the control of land and water.[18] In the Cauca Valley, the rise of the Cauca Valley Corporation (CVC), a David Lillienthal-approved irrigation and electricity agency launched in 1954, further regulated and concentrated access to water, benefitting those with private property deeds and political connections. Members of the regional business elite, including the owners of sugar refineries, sat on CVC's board of trustees.[19] CVC emerged in the exact years of *La Violencia* in the Cauca Valley.

Moments of heightened tension and danger loomed over the irrigated sugarcane fields of the Cauca Valley in the 1950s. One observer in the municipality of Florida, some thirty kilometers south of Palmira, wrote to the national minister of government in Bogotá and described a "chaotic state of ruin and death that is bathing the soul of the country in blood." In the town of Corinto, just south of Florida in Cauca Department, murder occurred in town and country ("se mata en poblado y en despoblado"), when and how one pleased ("cuando se quiere y como se quiere").[20] A subcommander in the Valle del Cauca unit of the Colombian military described the effects on land and property of this "undeclared civil war." "Landowners abandon their properties, leaving them in the hands of unscrupulous usufructuaries or decide to sell them at a derisory price," he reported after one Cauca Valley massacre.[21] In this way, violence and the fear of violence affected land value and land tenure by intensifying

---

[18] Mary Roldán, *Blood and Fire: La Violencia in Antioquia, Colombia, 1946–1953* (Durham, NC: Duke University Press, 2002).

[19] On the Cauca Valley Corporation, see Offner, *Sorting out the Mixed Economy*.

[20] Rogerio Pulgarín to Minister of Government, October 7, 1959, Folder 2083, Box 222, Department of Valle 1959, General Secretary, Ministry of Government, Documents Received, Archivo General de la Nación, Bogotá (AGN).

[21] Jaime Rubiano Santoyo to Señor Mayor-Comandante Unidad Policía Valle, September 5, 1959, Folder 2083, Box 222, Department of Valle 1959, General Secretary, Ministry of Government, Documents Received, AGN.

land concentration throughout the valley. Anthropologist Michael Taussig collected memories of *La Violencia* as a participant observer in the valley in the 1970s. Residents described to him large landowners who took advantage of the "frightful insecurity of those times" to drive down land prices, accelerating processes of smallholder disadvantage relative to an expanding number of large sugar corporations.[22] Some of the largest landowners employed *pájaros* (literally "birds," slang for mercenaries) to protect landholdings and usurp new territory in the Cauca Valley, and CVC too resorted to armed protection as it expanded its irrigation projects.[23] Foreshadowing the rise of paramilitaries in Colombia at the end of the twentieth century, this mid-century militarization of private property and natural resources accelerated the process of industrializing and commercializing the landscape.

Some of the worst violence in the Cauca Valley, including the descriptions above, took place during what is sometimes referred to as the "Late Violence," the period following the formation of the National Front political alliance in 1958, which theoretically ended the partisan aspect of the conflict. As Robert Karl has described, the bipartisan agreement that produced the National Front in Colombian politics set the stage for martial law and the further suppression of land grievances in the name of national reconciliation. Peasant self-protection units formed during *La Violencia* evolved in this post–Cuban Revolution period into offensive-minded guerrilla insurgencies.[24] In the Cauca Valley, the most famous of these, the FARC, kidnapped and murdered Harold Eder of the Manuelita sugar corporation in 1965. Other businessmen in the sugar sector would be targeted by guerrilla groups in the ensuing decades.[25]

Access to land and water, pressure points in the Cauca Valley and largely unresolved since the abolition of slavery in the 1850s, became Cold War issues during the 1950s with the conjunction of CVC and its role in the expansion of the sugarcane industry, the developmentalism of the World Bank and Rockefeller Foundation projects, and *La Violencia*. CIAT would emerge out of this cauldron and, on the Colombian side,

---

[22] Michael Taussig, "Peasant Economics and the Development of Capitalist Agriculture in the Cauca Valley, Colombia," *Latin American Perspectives* 3 (1978): 62–91, at 68.

[23] Nazih Richani, *Systems of Violence: The Political Economy of War and Peace in Colombia* (Albany: State University of New York Press, 2002), G. Sánchez and D. Meertens, *Bandits, Peasants and Politics: The Case of La Violencia in Colombia* (Austin: University of Texas Press, 2001), and A. Reyes Posada, *Guerreros y campesinos, el despojo de la tierra en Colombia* (Bogotá: Editorial Norma, 2009). On CVC and armed protection, see Offner, *Sorting out the Mixed Economy*, pp. 71–72.

[24] Robert A. Karl, *Forgotten Peace: Reform, Violence, and the Making of Contemporary Colombia* (Berkeley: University of California Press, 2017).

[25] Sonia Milena Jaimes Peñaloza, *Familia, caña, y banano: Las actividades empresariales de Rodrigo Holguín* (Medellín: La Carreta Editores, 2012).

had roots in the land reforms of the high Cold War orchestrated under the National Front government of Liberal President Alberto Lleras Camargo (1958–62). Lleras Camargo had long championed interhemispheric diplomacy, having previously held the post, among others, of general secretary of the Organization of American States (OAS) (1950–54). The moderate agrarian reform of 1961 would keep Colombia close to the United States, positioning the country as a major partner in the anticommunist mission, ready to receive its first batch of Peace Corps Volunteers and Alliance for Progress aid, both Cold War projects of the Kennedy White House.

The 1961 effort was the second of two controversial land reforms passed by the Colombian government during the middle decades of the twentieth century, both under Liberal presidents and ostensibly designed to help small farmers. Among other things, an earlier 1936 land reform sought to ease tenure disputes and the grievances of squatters on uncultivated portions of latifundia (*colonos*) and farmers of vacant public lands (*baldíos*) by creating a series of land judges with jurisdiction to determine claims. This arrangement was quickly criticized by cultivators without titles and their political allies, decrying the bias and dealing between judges and large landowners.[26] President Lleras Camargo's 1961 land reform would similarly produce significant backlash, as it came to be seen as co-opted by large landed interests and a growing agribusiness sector.

The 1961 reform created two new national institutes. The first, the Colombian Agricultural Institute (ICA), was designed by National Front architects to integrate agricultural research with education and extension. Then and now, ICA operated locally out of the old Palmira Agricultural Experiment Station, today just east of the CIAT grounds. From its inception, ICA partnered with the Colombian Institute for Agrarian Reform (INCORA), the land reform agency created out of the 1961 law and designed to modestly distribute title to public lands and colonization zones without redistributing or nationalizing private property. In addition to INCORA, ICA worked with the National University and the Colombian national airline Avianca, which transported soil samples from farmers to research centers such as the one at Palmira. Animal science formed a key component of ICA's mission during the 1960s, especially dairy, along with crop improvement. Extension services also became a major emphasis of ICA, in particular providing information to "la familia campesina" regarding research and technology in order to raise

---

[26] See Catherine LeGrand, *Frontier Expansion and Peasant Protest in Colombia, 1830–1936* (Albuquerque: University of New Mexico Press, 1986).

living standards.[27] In this early phase, ICA operated six experiment sites (including Palmira) and five more subsidiary research stations across the country.

Recently, geographers and anthropologists have reflected upon the *deterritorializing* effects, if not outright strategies, of INCORA, the new land reform agency tied to ICA. As Juan Pablo Galvis described it, "land reform, as formulated in Colombia, was a deployment of state power that was instrumental in the historical production of marginal territories."[28] By the end of the 1960s, over 96 percent of new titles granted through INCORA were for public lands and areas of recent colonization – "new settlement regions." Many of these new settlement regions were in locations like Putumayo, an Amazonian department in southern Colombia, reflecting the Colombian state and international agencies' consensus that food production would increase and social conflict (e.g., communist revolution) decrease by relocating small farmers to peripheral territories, away from sites of friction with expanding large landowners and agribusiness in the fertile valleys.[29] This arrangement also involved pressure by landowning elites for INCORA to focus on distributing untilled land rather than break up private property.[30] After 1961, for example, the Cauca Valley's major landowners' organizations, including the sugar industry's ASOCAÑA, proposed the so-called Sugar Plan (Plan Azucarero) to INCORA, lobbying for the expansion of large-scale sugarcane cultivation in order to expand wage labor and the regional economy. INCORA could not agree to the plan outright but, in coordination with the semi-autonomous CVC, declared that land generated from that agency's reclamation projects would be slated for sugarcane production. CVC also cut deals with large landowners, exempting them from INCORA's caps on property size if they paid taxes or made investments to benefit CVC's land reclamation projects. These arrangements stimulated a mutually beneficial cycle of using taxes and investments as an official exemption strategy from state regulations, wherein those taxes and investments were specifically tied to the expansion of sugar.[31]

The internal dealings of large landowners, CVC, and INCORA dovetailed with the final phase of the Rockefeller Foundation's CAP. With the

---

[27] "Historia del Instituto Colombiano Agropecuario," *Republica* (June 2, 1968), Folder 76, Box 12, Series 311, RG 1.2, Rockefeller Foundation (RF), Rockefeller Archive Center (RAC).

[28] Juan Pablo Galvis, "Developing Exclusion: The Case of the 1961 Land Reform in Colombia," *Development and Change* 40, no. 3 (2009): 509–529, at 511.

[29] On resettlement in Putumayo, see Kristina M. Lyons, *Vital Decomposition: Soil Practitioners and Life Politics* (Durham, NC: Duke University Press, 2020).

[30] Frances Thomson, "The Agrarian Question and Violence in Colombia: Conflict and Development," *Journal of Agrarian Change* 11, no. 3 (July 2011): 321–356.

[31] Offner, *Sorting out the Mixed Economy*, pp. 74–78.

blessing of the Rockefeller Foundation and in partnership with the Alliance for Progress, ICA and INCORA represented an alternative model to Soviet or Chinese collectivization or Cuban land expropriation, providing modest land reform (or, more accurately, resettlement) and the coordinated organization of agricultural development operations towards international and capitalist Cold War objectives.[32] One representative of the US Department of State justified the United States' assistance to Colombia under the Alliance for Progress, declaring:

Present land tenure conditions appear to be the main deterrent to attainment of political stability in Colombia . . . without opportunities, accumulative discontent and frustration may well develop into a full-scale rebellion . . . Agrarian reform will help to promote social justice and preserve western culture . . . The social problems today in Colombia do not stop with the personal tragedy of the millions of "campesinos" involved. What happens to them now affects general hemispheric order and struggle to maintain free institutions everywhere. Cuba is not far away and the influence is being felt . . . The western world requires that what occurs in the Andes of South America must be different from what happened in the Sierra Maestra.[33]

Land reform ushered in the first occasion that the Rockefeller, Ford, and Kellogg Foundations cooperated on a single project in Colombia, which would be repeated and codified with the establishment of CIAT later in the decade. As a package, ICA and INCORA became an early poster child for the Alliance for Progress and the Peace Corps. They received funding from USAID, the United Nations Special Fund, FAO, the World Bank, the Inter-American Development Bank, and others, reflecting a new era in cooperative international development logistics and funding channels. The University of Nebraska served as prime contractor for the Mid-American State Universities Association and thereby put twelve staff members in Colombia. Proponents hoped to grow this figure to thirty staff within the ICA–National University system in Colombia to be joined by other individuals from the foundations, USAID, and other partner organizations. ICA offered a central repository in which international agencies and sponsors could deposit funds to Colombian agriculture.

---

[32] This process in Colombia paralleled a similar land resettlement project in Brazil. See Ryan Nehring, "The Brazilian Green Revolution," *Political Geography* 95, no. 1 (May 2022): 102574; Wendy Wolford, "The Casa and the Causa: Institutional Histories and Cultural Politics in Brazilian Land Reform," *Latin American Research Review* (2016): 24–42.

[33] Harold T. Jorgensen, "End-of-Tour Report Submitted by Mr. Harold T. Jorgensen, Agrarian Reform Advisor (Agricultural Advisor)," 1963. Quoted in Galvis, "Developing Exclusion," p. 519.

Much of this international funding centrally deposited to ICA and INCORA would be indirectly funneled to aid agribusiness. The Sugar Plan presented to INCORA, for example, identified the expansion of corporate sugar production as a strategy for luring foreign investment in the aftermath of the Cuban Revolution and the United States' resultant loss of one of its major sources of sugar.[34] On another plain, at the individual and family level, young Colombian agronomists had to choose between government-run research centers or the private sector. Colombian agribusiness consistently outbid the government, luring many newly trained agronomists to their corporate payrolls. A growing cadre of Colombian agronomists and geneticists emerged from the expanded university systems, including the National University's agricultural campus in Palmira and the Universidad del Valle in Cali, recipients of large sums of international funding. As Colombian agribusiness, including the sugarcane sector in the Cauca Valley, grew through the scientific achievements of this group, the government and its publicly oriented scientific stations struggled to keep up. More positions were filled by foreigners, and, not coincidentally, the research orientation of stations such as Palmira shifted towards an international Cold War agenda using science as a weapon against tropical poverty, population growth, and political volatility. The CAP Director's Report of 1959 foreshadowed this emerging situation:

> The lack of sufficient trained personnel remains the main bottleneck in the rapid advance of agricultural technology . . . the demand for agronomists by commercial companies and large farmers has increased in proportion to the rapid development of agriculture in Colombia. This demand by commercial concerns has caused salaries to be raised, and several of the well-trained agronomists have left Government employment to accept higher-paying positions elsewhere.

Exposing the program's underlying support for the growth of private industry and commercial agriculture, the report concluded: "However, this is in general a healthy situation and reflects a rapidly developing agricultural economy."[35]

The work of entrepreneurs like Humberto Tenorio, who opened the first privately owned hybrid seed company in the Cauca Valley, reflected this broader shift towards privatization. In this vein, the Rockefeller Foundation pivoted to sponsoring exchanges between Colombian scientists and private industry. For example, a promising breeder named Eduardo Chavarriaga received a Rockefeller Foundation travel grant to

---

[34] Offner, *Sorting out the Mixed Economy*, p. 74.
[35] Colombian Agricultural Program, Director's Annual Report, May 1958–April 1959, 2, Annual Reports, Agricultural Operating Programs, RF, RAC.

go to the United States and work with the Pioneer Hi-Bred Seed Company in Iowa. A generation after Henry A. Wallace, founder of Pioneer Hi-Bred, visited Palmira, Chavarriaga's studies in Iowa included developing models of seed distribution that might be applicable to the Cauca Valley.[36] "It is anticipated that the development and cooperation of private enterprise can substantially increase the effectiveness of corn improvement programs in general," the Rockefeller Foundation reported in 1962.[37] Other agronomists in training followed the growing connections forged through the Rockefeller Foundation relationship, including future leaders of the valley sugarcane industry such as Roberto Holguín who, similarly, obtained advanced agronomy degrees from Iowa State University in this era.[38]

During the 1960s, the Rockefeller Foundation partnered with ICA and its internationalist agenda for national agricultural research, education, and extension. While working to transfer project leadership to Colombians, the foundation turned its attention to more ambitious projects that would transform the sites of its host-country research programs into a global network of agricultural science. Such a network would detach program research from local contexts and contingencies and, as Norman Borlaug suggested, the "complications" of national politics, which, as we have seen, were considerable in rural Colombia. It would be coordinated and driven by a shared set of values to deliver science like interchangeable parts – specifically designed for broad geographic or climatic zones but otherwise transferable across national boundaries and cultural contexts.

CIAT and the Cold War

A 1966 report outlined the Rockefeller Foundation's collaborative spirit and global perspective. This report was the work of Lewis Roberts, a veteran of both the Mexican and Colombian Agricultural Programs, and Lowell Hardin, an agricultural economist at Purdue University recently hired as a senior agricultural specialist with the Ford Foundation. Roberts and Hardin described two ways to increase global food production: to obtain higher yields from land already in use, or to bring new land into cultivation. An international tropical agriculture

---

[36] Lewis M. Roberts Interviewed by William C. Cobb, NY, August 22–25, 1966, 46, Folder 3, Box 23, RG 13: Oral Histories, RF, RAC.
[37] Colombian Agricultural Program, Annual Report, 1961–1962, Annual Reports, Program in Agricultural Sciences, RF, RAC.
[38] On Holguín, see Jaimes Peñaloza, *Familia, caña, y banano.*

research institute would invest in the first method.[39] INCORA, through its policy of resettling untitled peasants in marginal territories, pursued the second.

Roberts and Hardin set the parameters for the organization of the international center which would complete the phasing out of the CAP.[40] The hot tropics, they thought, contributed little to global food production and struggled to keep pace with population growth. "Outside of Communist Asia and west Asia, most of the world's diet-deficit subregions are in the tropical belt between the Tropics of Cancer and Capricorn," they wrote.[41] Ignoring the long history of domestic agricultural science and inter-Latin American and Caribbean networks, the authors argued that such regions as the Cauca Valley had been largely bypassed by modern agricultural science, with only export crop technologies developed under the auspices of colonialism.[42] Export crops like rubber, sugar, bananas, cacao, tea, cotton, and spices, they thought, had received scientific attention, but not the staple food crops of the region, despite the Palmira Agricultural Experiment Station's work with rice and maize since 1927.[43]

Why did Roberts and Hardin present Palmira as their choice for this new international research center? They felt they needed to choose a location within the ecological zone of tropical agriculture, but one in which the climate would favor the maintenance of a germplasm collection. The Cauca Valley's comparatively mild tropical heat and modest rainfall met this condition, and it further offered distinct microclimates nearby to simulate different environments. Colombia was important geopolitically, and Cali and Palmira were geographically central within the country's transportation network, particularly as nearby Buenaventura continued its post–Panama Canal ascendency as the country's top port. By the mid 1960s, the Cauca Valley had extensive connections to locations throughout Colombia and beyond via a growing system of railroads and highways, as well as plans to expand the international airport in time to host the 1971 Pan-American Games.

[39] Lewis M. Roberts and Lowell S. Hardin, "A Proposal for Creating an International Institute for Agricultural Research and Training to Serve the Lowland Tropical Regions of the Americas," October 1966, 1, Folder 788, Box 112, Subseries 3, Series VI, Subgroup I, RG 6.7: New Delhi Field Office, RF, RAC.
[40] A detailed institutional history of the founding of CIAT and its subsequent achievements is Lynam and Byerlee, *Forever Pioneers*.
[41] Roberts and Hardin, "A Proposal for Creating an International Institute for Agricultural Research," 1.
[42] Ibid., i.   [43] Ibid., 12.

There was, of course, an institutional base already in place there as well, and the new site would be constructed adjacent to the National University's agronomy school and the now ICA-operated experiment station. The Universidad del Valle was nearby, with its important work in agricultural economics, public health, and nutrition, funded by the Ford and Rockefeller Foundations.[44] In addition, the Cauca Valley offered something important to well-to-do international researchers: "attractive living conditions are available in Cali," they noted. As in the past, Roberts and Hardin also praised the Colombian national and regional governments for their support in the form of promised land and a generally favorable attitude towards the proposed institute. The pair, biased by Roberts' experience with the CAP, did not identify any alternative sites that matched Palmira's advantages in these areas.[45] For Roberts and Hardin, the eyes of the Rockefeller and Ford Foundations, Palmira offered the right set of ingredients for their global development concoction.

The new institute at Palmira would be modeled on its forerunners, the International Rice Research Institute (IRRI, organized in 1960) in the Philippines and the International Maize and Wheat Improvement Center (CIMMYT, established in 1966) in Mexico. IRRI represented the Rockefeller and Ford Foundations' first collaborative attempt to enhance the world's tropical food supply through a coordinated scientific institution. As Gabriela Soto Laveaga describes in Chapter 4, this volume, CIMMYT then emerged from the Rockefeller Foundation's Mexican Agricultural Program (MAP). The new Palmira site, CIAT, was thus strategically designed to capitalize on the work already being done on staple grains at its partner institutions.[46] Like CIAT in Palmira, the International Institute of Tropical Agriculture (IITA) was formed in 1967 in Nigeria. Together these four sites comprised the original CGIAR network, formalized in 1971 with support from the Rockefeller Foundation, the World Bank, FAO, and other international entities. As such, CIAT and its sibling sites in the Philippines, Mexico, and Nigeria represented the original coordinating institutions of the collaborative Green Revolution. The group's clear Cold War mission could be discerned in the personnel responsible for its formation: Robert McNamara, serving as president of the World Bank, launched the Commission on International Development (the Pearson Commission) in 1968, which recommended the coordinated steps that led to the organization of CGIAR.

---

[44] Ibid., 56.   [45] Ibid., ix.   [46] Ibid., ii, 13, and 20.

The Rockefeller Foundation's CAP filtered its work, equipment, and personnel ("liquidating itself") into two distinct creations that represented new directions in agricultural science conducted in the country.[47] One of these became CIAT, oriented towards international tropical agriculture and the intensification of staple crop production. The other was the Colombian government's ICA and its land reform agency INCORA, pursuing domestic agricultural improvement and the expansion of cultivation into marginal territories.

The Rockefeller Foundation Board of Trustees appropriated $3 million to CIAT in April 1969 and began reassigning CAP staff. By then, eight foundation staff members from CAP had already pivoted to working in Cali and guiding CIAT even before its official opening in 1967. One of these, Ulysses Jerry Grant, director of the CAP maize-breeding effort before being reassigned to India, returned to the Cauca Valley to become the first director of the new international center. The Rockefeller Foundation also sold its staff residential retreat along the Magdalena River and transferred that sum to help finance CIAT.[48] After the 1970 closing of the foundation's agriculture-centric Bogotá Field Office, more personnel and capacities were transferred to CIAT or to the remaining Cali Field Office, which coordinated with the new international center, CVC, and the regional universities.[49]

CIAT was designed to be a "catalyst" for economic and agricultural development in the global tropics. To these ends, the institute and its scientists collaborated with local and national institutions in Colombia and partner organizations around the world, including IRRI and CIMMYT in particular. In the growing network that would soon become CGIAR, CIAT focused on the humid tropics below 1,000 meters.[50] From the outset, an interdisciplinary team of geneticists, agricultural economists, and engineers cooperated with research stations in key tropical regions of Latin America to enhance those few staple crops "vitally important from the standpoint of nutrition," including legumes, maize, rice, and animal products, as well as root crops, vegetables, and tropical fruits[51] (Figure 3.3).

---

[47] "Donation of Equipment – Colombian Agricultural Institute," June 21, 1968, Folder 88, Box 14, Series 311, RG 1.2, RF, RAC.
[48] "Donation of Staff Residence House to International Center of Tropical Agriculture," May 23, 1969, Folder 89, Box 14, Series 311, RG 1.2, RF, RAC.
[49] Folder 27, Box 3, Series I, RG 6.9: Cali Field Office, RF, RAC.
[50] "CIAT: Programas de Adiestramiento" (1970) pamphlet, 6, Folder 788, Box 112, Subseries 3, Series VI, Subgroup I, RG 6.7: New Delhi Field Office, RF, RAC.
[51] Roberts and Hardin, "A Proposal for Creating an International Institute for Agricultural Research," pp. v and vi. On CIAT's work with legumes in Central America, see Wilson Picado-Umaña, Chapter 8, this volume.

Figure 3.3 Tony Bellotti, entomologist in the CIAT cassava program (wearing the Twins baseball cap), works with Colombians in Palmira in this undated image. Special thanks to John Lynam for identifying Bellotti. Rockefeller Archive Center, Ford Foundation Photographs, Folder 510, Box 32, Series 1, CIAT 522. Photograph by James Foote, courtesy of Rockefeller Archive Center.

By 1973, CIAT had settled on tropical agricultural research in six main lines: beef, hogs, cassava, beans, maize, and rice. Each of these were truly international in procedure and scope. The beef line, for example, studied cattle and pasturage on Colombia's eastern plains and partnered with Texas A&M University in disease control. Its scientists served in consultations with Bolivia, Brazil, Colombia, Ecuador, Peru, Venezuela, and across the Caribbean. The rice line worked closely with IRRI in the Philippines and with producers in Central America and the Andean countries. The hog line, similarly, partnered with universities in Bolivia and Costa Rica. It also looked to integrate horizontally, partnering with the cassava and maize groups, for example, in the study of enhanced agricultural systems on small family farms.[52] The cassava line, for its part, worked closely with cassava researchers at IITA in Nigeria.

In 1975, CIAT employed approximately 150 scientists and related personnel from 13 countries. In addition to its research lines and regular international partnerships, the institute sent *científicos* for consultation work across Latin America.[53] By 1983, CIAT's roster had swelled to 1,200 employees, including 92 scientists hailing from 24 countries.[54] In addition to consultations, CIAT regularly distributed pamphlets and publications around the world and administered international seminars and outreach programs on select technical topics.[55] The institute focused on "comparative advantage" to apply its research and consultations to the specialization of its partners and other CGIAR institutions.[56]

Land tenure offered CIAT scientists and directors considerable conversation fodder from the outset. An ongoing debate within the institute centered on whether or not to embark on a rural development project in addition to crop science. Specifically, the institute considered "a major rural uplift program" in Valle del Cauca beyond the research center's property. Director Jerry Grant described this concept as working with small farmers to help design improved farming systems for them, as well as improving educational, health, and other services. Proponents hoped to emulate the Puebla Project (Plan Puebla) in Mexico, initiated in 1967 and targeting the intensification of rainfed smallholder (*minifundista*) agriculture and fertilizer distribution through the formation of cooperatives and other cost-sharing mechanisms. Grant and his allies at CIAT imagined such a project for the Cauca Valley, partly in response to the frequent question hurled at them from every direction: "How are your results going to help the small farmer?" However, other leadership, including David Bell, vice president of the Ford Foundation, viewed this as a "diversion of talent" at such an early stage of research and development. Bell and others suggested attention to land tenure and questions of equity would be more appropriately addressed by ICA,

[52] CIAT, *Informe Anual*, 1973, National Library (BNC), Bogotá.
[53] CIAT, *Informe Anual*, 1975, National Library (BNC), Bogotá, pp. xiv–xv .
[54] *CIAT Internacional* 2, no. 3 (November 1983): 1, National Library (BNC), Bogotá.
[55] CIAT, *Informe Anual*, 1973, p. 8.   [56] CIAT, *Informe Anual*, 1975, p. xv.

perhaps with technical support from CIAT.[57] ICA and INCORA, in accord with the lobbying interests of large landowners, CVC, and agribusiness, specifically focused on resettling untitled cultivators in marginal or peripheral lands. David Bell and CIAT's official position to defer to and provide technical support to ICA thus supported the deterritorializing move that helped declutter the Cauca Valley of peasant cultivators and redirect the burden of their long-standing political grievances away from the local expansion of agribusiness.

This debate continued several years into CIAT's operations. In 1973, leadership again examined the relevance of their work to small farmers. They organized a meeting that October with forty representatives from not only CIAT but the Universidad del Valle, ICA, and several international organizations. They discussed how to integrate scientific research and technology into small-farm agricultural systems, and they evaluated the impact of new technologies on the well-being of independent farming families. CIAT likewise pursued a program targeting "sistemas para pequeño agricultores" – the Small Farm Systems program launched with the Ford Foundation in 1973. This program funded observatory field work by CIAT personnel in the distant Colombian Llanos and Caribbean coastal regions, as well as collaboration with IITA in Nigeria.[58] This approach accepted on principle that committed small farmers would be left to the peripheries, while others would provide ample wage labor for the complex of valley soils and CVC waters allocated to produce sugarcane. As if confirming this unfolding reality, CIAT studies revealed the comparatively high market share of Valle del Cauca's large commercial farms in relation to those of neighboring departments such as Huila, Antioquia, and Nariño. CIAT's own studies underscored the inequalities embedded in Valle del Cauca agriculture and the department's intensifying concentration of credit and capital in large-scale operations. Nevertheless, and despite (or perhaps because of) these realities in its own backyard, its work in rural development and small-farmer systems concentrated on the global peasant and regions of Colombia far removed from Valle del Cauca. Seeing CIAT in the light of INCORA reveals the mirroring aspects of deterritorialization in Colombia. From this vantage, CIAT was never intended to help small farmers in the Cauca Valley. As INCORA facilitated the resettlement of small or untitled farmers in Amazonian Putumayo or the remote eastern plains of Casanare, CIAT offered the technical training and science for expanding cultivation

---

[57] David E. Bell to F. F. Hill (Inter-Office Memorandum), February 22, 1971 (Reporting on visit to CIAT, 2/10/71), Folder 698, Box 27, Subseries A, Office Files of David Bell, Office of the Vice President, Ford Foundation, RAC.

[58] CIAT, *Informe Anual*, 1973, pp. 8 and 243–245. Also Lynam and Byerlee, *Forever Pioneers*, pp. 30–33.

into these hot lowland territories. Meanwhile, in the Cauca Valley, ASOCAÑA members increased their share of land and water.

Conclusions

On November 8–9, 2017, CIAT celebrated fifty years of its footprint in global agriculture.[59] A contingent of distinguished guests commemorated the moment in Palmira. Colombian President Juan Manuel Santos spoke, as did Minister of Finance Mauricio Cárdenas Santamaría. Other speakers included Governor of Valle del Cauca Dilian Francisca Toro Torres, Mayor of Palmira Jairo Ortega Samboní, and Juan Camilo Restrepo Salazar, head of the Colombian government's peace negotiating team with the ELN (National Liberation Army) guerrilla group. The ambassador of France to Colombia joined the politicians assembled, as did leaders of FAO, the World Bank Group, and CGIAR. Officers and representatives of nongovernmental organizations (NGOs) such as the Wildlife Conservation Society and the Global Harvest Initiative assembled with corporate executives from the likes of DuPont Pioneer and professors from the usual assortment of land-grant universities and Ivy League institutions, including Columbia, Cornell, Michigan State, Minnesota, and Rutgers.

The attendees spoke triumphantly of "fifty years, fifty wins."[60] Speakers and the conference program largely organized their remarks around contemporary investment-generating phrases, including "building a sustainable food future," "the future of climate change research," and "aligning public and private interest to scale up and deliver impact."[61] The framing language may have changed with the times, but the institution's global ambitions have remained. For five decades, scientists, academics, politicians, and corporate executives have converged upon CIAT from afar, pulled by the institution's centripetal position in an orbit of tropical agricultural science. In turn, the research and technologies undertaken at CIAT have radiated out to the tropical world like a centrifugal force. Apply your buzzwords of choice, CIAT has made the Cauca Valley a critical node in a contemporary global food system.

In sharp contrast to the current rhetoric of small farmers and sustainable cultures presented by CIAT, but in overlapping timelines, the Cauca Valley sugar complex accumulated resources and technical advantage. Through its technical assistance to ICA and partnerships with INCORA

---

[59] "CIAT 50: 1967–2017: Celebrations at Headquarters, Cali, Colombia: 8–9 November 2017," https://alliancebioversityciat.org/ciat50.

[60] A triumphant narrative persists in the commissioned history of CIAT; see, for example, Lynam and Byerlee, *Forever Pioneers*.

[61] "CIAT 50: 1967–2017: Celebrations at Headquarters."

in Cold War Colombia, CIAT has accepted, if not embraced, that status quo. As a package, land governance and CIAT crop research in Colombia function to aid in the accumulation of the best land and resources for agribusiness and remove those with land grievances to start anew on the margins and bring new territories under cultivation.

Even amidst the fiftieth anniversary celebrations of 2017, the apolitical ideal of the 1960s remained. The commemorative institutional history published that year, for example, described it this way: "One of the core characteristics of the international centers was that they were apolitical, and the Roberts–Hardin proposal had reaffirmed this by stressing that CIAT would not be involved in the land reform question. Rather CIAT would give particular consideration to improving productivity of small farmers."[62] The language is key. CIAT was designed to focus on increasing the productivity of small farmers. But the institution's built-in collaboration with ICA and INCORA determined the geography of where those small farmers sowed. ICA and INCORA were the mechanisms for bringing new land under cultivation, often at the margins of Colombian territory – in the eastern tropical savannah of the Llanos, in the acid soils of the southern rain forests of Amazonia – and often by relocating farmers without title away from the most fertile valleys where land tenure issues had long simmered. These valleys increasingly became the objects of international development projects and aid, including CVC, for example, which regulated water and pursued land reclamation projects in consultation with the ASOCAÑA businessmen on its board. The agro-industrial sugar complex grew in the Cauca Valley; CIAT did not cause the exacerbation of inequality outside its gates, but it was part of the mechanics of shifting territory in Colombia.

The history of rural conflict in Colombia is as ironic as it is tragic: many of these recipients of INCORA grants or CIAT technical assistance have witnessed firsthand the paralyzing war between guerrillas, paramilitaries, drug traffickers, and the Colombian army. CIAT's stated objective to improve agrarian livelihoods around the tropical world has proven to be largely a mirage in its own Cauca Valley. More specifically, that objective was never intended for those in its own backyard. No wonder the "CIAT 50" pamphlet during the celebrations in Palmira featured a Southeast Asian family rather than a Cauca Valley one on its cover.[63]

---

[62] Lynam and Byerlee, *Forever Pioneers*, p. 30.
[63] The cover can be seen by accessing the full report: CIAT, *Building a Sustainable Food Future since 1967: Fifty Years and 50 Wins* (Cali, Colombia: CIAT, 2017), https://hdl.ha ndle.net/10568/89145.

4 CIMMYT's Early Years
Rooted in Mexican Experience, Designed to Be International

Gabriela Soto Laveaga

To understand the global impact of the International Maize and Wheat Improvement Center, or CIMMYT, we must first delve into the creation of the center. Though born a Mexican institution with an intended international focus, CIMMYT's roots are in binational cooperation and a longer practice of global germplasm exchange aimed at producing better crops. This chapter first examines the historical background of CIMMYT, and then considers the main shift in the center's mission over the span of its first forty years (1966–2006). To illustrate this shift, the chapter relies on brief overviews of how CIMMYT worked on the ground – training wheat breeders and working with farmers on specific projects – which serve to illustrate the broad aim and reach of the organization, before turning to its crafting a message for a world stage. The conclusion offers a reflection on the place of CIMMYT in global agriculture research today.

Historical Roots

Launched in the 1960s, CIMMYT is unique among other research centers of CGIAR (Consultative Group on International Agricultural Research) in that it traces its roots to an impactful 1940s agricultural development program known as the Mexican Agricultural Program (MAP).[1] As I discuss here, CIMMYT was designed to be international in scope but remained connected to Mexico and former MAP personnel

Acknowledgments: I am grateful to Helen Anne Curry and Timothy W. Lorek for their drive to get this finished during the pandemic. I am especially grateful for their patience and unwavering support.

[1] Among CGIAR research institutes, CIMMYT and the International Rice Research Institute (IRRI) in the Philippines have the most similar origin stories. Both were deeply influenced by the Rockefeller Foundation's mission and its personnel, were the first centers to emerge in the 1960s, paving the way for the others, and were initially focused on cereals research (wheat and rice, respectively). Yet there are significant differences as well, including in operating structure, funding sources, and research focus. See Lowell S. Hardin and Norman R. Collins, "International Agricultural Research: Organising Themes and Issues," *Agricultural Administration* 1, no. 1 (1974): 13–22.

to harness existing technology, a pool of trained scientists, and research experience.[2]

CIMMYT would eventually become part of CGIAR's first cohort of geographically diverse agricultural research centers and, arguably, its best known.[3] As Derek Byerlee and John K. Lynam maintain, centers such as CIMMYT were "the major institutional innovation of the 20th century for foreign assistance to support agricultural development and food security."[4] Byerlee and Lynam, speaking about CGIAR centers, echo historians who decades earlier used similar language to describe the foundation of MAP as a pivotal moment in the twentieth century when agricultural science scaled up from domestic industrialization to become a "device for power relationships between nations."[5] Historians of MAP were not the only ones to note its oversized influence. Reminiscing about the origins of MAP before a US Senate Committee in 1979, at which time MAP no longer existed, Norman Borlaug, by then already a Nobel Peace Prize laureate, made certain to underline that MAP "preceded all other foreign technical assistance programs in agriculture by at least 7 years" and that its establishment, at the request of the Mexican government, became a model for cooperative crop research.[6]

---

[2] For a solid overview of the foundation of CIMMYT and its placement within global concerns about food and poverty, as well as the need to strengthen national programs while building a global network, see Derek Byerlee, *The Birth of CIMMYT: Pioneering the Idea and Ideals of International Agricultural Research* (Mexico City: CIMMYT, 2016), https://repository.cimmyt.org/handle/10883/17705. Further histories of CIMMYT and its predecessors are cited below.

[3] In 2021 CIMMYT was the center with the second-highest expenditure, at $99 million, after Nigeria's International Institute of Tropical Agriculture (IITA). Since 2020, CIMMYT has had the third-highest dollar amount in active grants of all the centers, after the International Food Policy Research Institute (IFPRI) and IITA. For a breakdown of CGIAR's current research programs and the allocation of its budget, personnel, etc. by center, see www.cgiar.org/dashboards. For a comparison of the financials of all CGIAR centers, see www.cgiar.org/food-security-impact/finance-reports/dashboard/center-analysis. According to CGIAR statistics, CIMMYT has the second-largest workforce, after IITA; see www.cgiar.org/how-we-work/accountability/gender-diversity-and-inclusion/dashboards/cgiarworkforce. To further contextualize these patterns, see Selçuk Özgediz, *The CGIAR at 40: Institutional Evolution of the World's Premier Agricultural Research Network* (Washington, DC: CGIAR Fund, 2012).

[4] Derek Byerlee and John K. Lynam, "The Development of the International Center Model for Agricultural Research: A Prehistory of the CGIAR," *World Development* 135 (2020): 105080, at 1, 4; see also Margaret Carroll Boardman, "Sowing the Seeds of the Green Revolution: The Pivotal Role Mexico and International Non-profit Organizations Play in Making Biotechnology an Important Foreign Policy Issue for the 21st Century," *Mexico and the World* 4, no. 3 (1999): 1–34.

[5] John H. Perkins, *Geopolitics and the Green Revolution: Wheat, Genes, and the Cold War* (Oxford: Oxford University Press, 1997), p. 103.

[6] R. Norman Borlaug, "Statement before the United States Congress Senate Committee on Finance and United States Congress Senate Committee on Finance Subcommittee on International Trade," *North American Economic Interdependence II: Hearing before the*

Established in 1943 as an agricultural technical assistance agreement of the Mexican ministry of agriculture and livestock in partnership with the Rockefeller Foundation, the goals of MAP could be synthesized in a few key objectives: the training of Mexican scientists, increasing food production, and, equally important, stabilizing funding for the experiment stations that already existed in the country.[7] Beginning in 1950, the Rockefeller Foundation expanded beyond Mexico and began country-specific agriculture programs in Colombia (1950) and Chile (1955) in the Americas. These programs, described as "evolutionary extensions" of MAP, consisted of men who were part of that program moving "southward in successive stages, carrying with them materials, concepts, ideas and wisdom that they had acquired in helping to solve problems in agricultural production and human relations in Mexico."[8] But the influence of MAP was not confined to the Western hemisphere. In 1956 the Rockefeller Foundation signed an agreement with the government of India for a MAP-like program, and in 1960, in partnership with the Ford Foundation, it opened the International Rice Research Institute (IRRI) in the Philippines.[9] In addition to these MAP-influenced

Subcommittee on International Trade of the Committee on Finance, United States Senate, Ninety-Sixth Congress, First Session, October 1, 1979 (Washington, DC: US Government Printing Office, 1979), p. 76.

[7] Byerlee, *The Birth of CIMMYT.*

[8] E. C. Stakman, Richard Bradfield, and Paul C. Mangelsdorf, "Extending the Mexican Pattern: Action Programs in Colombia, Ecuador, and Chile," in Stakman, Bradfield, and Mangelsdorf, eds., *Campaigns against Hunger* (Cambridge, MA: Harvard University Press, 1967, reprint 2014), p. 216. On the Rockefeller Foundation's influence in agricultural research in Latin America beyond Mexico, see Timothy W. Lorek, "Imagining the Midwest in Latin America: US Advisors and the Envisioning of an Agricultural Middle Class in Colombia's Cauca Valley, 1943–1946," *The Historian* 75, no. 2 (2013): 283–305; Timothy W. Lorek, "Developing Paradise: Agricultural Science in the Conflicted Landscapes of Colombia's Cauca Valley, 1927–1967," Ph.D. dissertation, Yale University (2019); Hebe M. C. Vessuri, "Foreign Scientists, the Rockefeller Foundation and the Origins of Agricultural Science in Venezuela," *Minerva* (1994): 267–296; William San Martin, "Nitrogen Revolutions: Agricultural Expertise, Technology, and Policy in Cold War Chile," Ph.D. dissertation, University of California, Davis (2017); Chris J. Shepherd, "Imperial Science: The Rockefeller Foundation and Agricultural Science in Peru, 1940–1960," *Science as Culture* 14, no. 2 (2005): 113–137; Elta Smith, "Imaginaries of Development: The Rockefeller Foundation and Rice Research," *Science as Culture* 18, no. 4 (2009): 461–482.

[9] Tore C. Olsson, *Agrarian Crossings: Reformers and the Remaking of the US and Mexican Countryside* (Princeton, NJ: Princeton University Press, 2017), p. 155. Further resources on the Rockefeller Foundation in India include U. Lele and A. A. Goldsmith, "The Development of National Agricultural Research Capacity: India's Experience with the Rockefeller Foundation and Its Significance for Africa," *Economic Development and Cultural Change* 37 no. 2 (1989): 305–343; Marci Baranski, "Wide Adaptation of Green Revolution Wheat: International Roots and the Indian Context of a New Plant Breeding Ideal, 1960–1970," *Studies in History and Philosophy of Science* 50 (2015): 41–50;

programs, MAP supported cooperative programs for crop-testing far beyond Mexican fields. One of its most successful partnerships was the Central American Corn Improvement program, which focused on testing maize varieties.[10]

By 1960, with more than 800 Mexican scientists trained, the Rockefeller Foundation and the Office of Special Studies scheduled MAP's retirement. On January 1, 1961, the Office of Special Studies and the Mexican government's Institute for Agricultural Research were terminated. These two organizations merged into a newly formed research unit – the National Institute for Agricultural Research (INIA).[11] The Rockefeller Foundation had long anticipated that the Office of Special Studies, which oversaw MAP, would have an expiration date. As others noted, closing the Office for Special Studies allowed the Rockefeller Foundation to attain its objective of building "a strong wholly Mexican Agricultural Research Institution" while ensuring continuity of MAP's research agenda under INIA.[12] Since the research stations and installations of MAP had been built up from Mexican ones, the main transitional point was that of personnel. The staff of MAP transferred to INIA, while the few remaining foreign personnel would serve in an advisory capacity in the country or be reassigned.[13]

Marci Baranski, *The Globalization of Wheat: A Critical History of the Green Revolution* (Pittsburgh: University of Pittsburgh Press, 2022); Simi Mehta, Rattan Lal, and David Hansen, "US Land-Grant Universities in India: Assessing the Consequences of Agricultural Partnership, 1952–1972," *International Journal of Educational Development* 53 (2017): 58–70.

[10] Alejandro Fuentes, Carlos Salas, and Angel Salazar, "Origen e historia del Programa Cooperativo Centroamericano y del Caribe para el Mejoramiento de Cultivos Alimenticios y Producción Animal," *Agronomía Mesoamericana* 1 (2016): 93–96; Diana Alejandra Méndez Rojas, "Maize and the Green Revolution: Guatemala in the Global Context of Agricultural Research, 1954–1964," *Ciencia Nueva Revista de Historia y Política* 3, no. 1 (2019): 134–158; Rodolfo Araya Villalobos, *Programa Cooperativo Centroamericano para el Mejoramiento de Cultivos y Animales: 1954–2019* (Alajuela, Costa Rica: Universidad de Costa Rica, 2020), www.kerwa.ucr.ac.cr/handle/10669/81544.

[11] Bruce H. Jennings, *Foundations of International Agricultural Research: Science and Politics in Mexican Agriculture* (Boulder, CO: Westview Press, 1988), chapter 7.

[12] Ibid., p. 139.

[13] From its inception MAP worked both in concert and often at odds with Mexican national research programs. Researchers such as Karin Matchett have focused on the tension between the well-funded MAP and the Institute for Agricultural Research, which was often strapped for funds and equipment. As Matchett illustrates, these tensions led to a focus on different aspects of corn research and embracing certain methodologies over others, such as synthetic versus double-cross hybrid maize. See for example, Karin Matchett, "At Odds over Inbreeding: An Abandoned Attempt at Mexico/United States Collaboration to 'Improve' Mexican Corn, 1940–1950," *Journal of the History of Biology* 39, no. 2 (2006): 345–372.

At the time of its ending, MAP, though based in Mexico and focused on a handful of agricultural regions within the country, had for nearly two decades served as a blueprint for how to run agricultural research programs across the so-called developing world. The research model – using in-country field plots staffed by an internationally networked group of scientists, as well as training domestic scientists – could be successfully exported beyond the Americas. Yet its becoming a blueprint was not a given. Though there was knowledge-sharing among and between these networked programs, when MAP closed there was no effective institutional authority for global agricultural research as there would later be under CGIAR.

While the origins of MAP are widely and broadly covered by historians of the Green Revolution, less documented is the end of the program and the eventual emergence a few years later of what would become CIMMYT.[14] In fact, the timelines of where one program ends and the other begins are often entangled in these historical narratives.[15] Given the significant overlap of personnel, experiment stations, and research aims, as well as programming with the former MAP, in particular the Office of Special Studies, this confusion is not surprising.

CIMMYT had predecessor programs. For example, in 1958 the Rockefeller Foundation created the Inter-American Maize Improvement Program and the Wheat Improvement Program, but, as the historian Bruce Jennings describes, "the crop improvement programs in maize and wheat floundered. Part of this difficulty stemmed from the cooperative nature of these programs. They depended ... on the degree of cooperation arranged by host governments," which could be volatile.[16] In 1963, the president of Mexico, Adolfo López Mateos, and the president of the Rockefeller Foundation, J. George Harrar, created the International Corn and Wheat Research Institute with the idea "to

---

[14] On MAP and the origins of the Green Revolution, see Deborah Fitzgerald, "Exporting American Agriculture: The Rockefeller Foundation in Mexico, 1943–53," *Social Studies of Science* 16, no. 3 (1986): 457–483; David A. Sonnenfeld, "Mexico's 'Green Revolution,' 1940–1980: Towards an Environmental History," *Environmental History Review* 16, no. 4 (1992): 28–52; Perkins, *Geopolitics and the Green Revolution*; Olsson, *Agrarian Crossings*; Jonathan Harwood, "Whatever Happened to the Mexican Green Revolution?," *Agroecology and Sustainable Food Systems* 44, no. 9 (2020): 1243–1252; Jose Miguel Chavez Leyva, "Powerful Disruptions: Braceros, Campesinos, and the Green Revolution in Mexico, 1940–1965," *Agricultural History* 95 no. 3 (2021): 472–499; Baranski, *Globalization of Wheat*.

[15] CIMMYT's foundation is often mistakenly conflated with that of MAP. For example, the World Food Programme, in celebrating the fiftieth anniversary of Norman Borlaug's Nobel Prize, wrote that CIMMYT was started in 1943; see World Food Programme, "Mexico – CIMMYT," www.worldfoodprize.org/en/youth_programs/borlaugruan_in ternational_internship/international_internship_sites/mexico__CIMMYT.

[16] Jennings, *Foundations of International Agricultural Research*, p. 142.

fuse the two crop improvement programs into a single organization with an international mandate."[17]

The idea of this institute was first publicly mentioned in 1960 during a farewell dinner for remaining Rockefeller Foundation staff in Mexico. In addition to former MAP staff and cabinet members, there were several Mexican scientists in attendance who had trained via MAP. Listening to the long list of successes, the evening's host, President López Mateos, apparently remarked that he was "confused by this departure" because:

Just 2 months ago I visited Southeast Asia. Quite by chance, while I was in the Philippines, I was taken to the International Rice Research Institute, a magnificent organization. I was told that this was modelled after the Mexican agricultural program – the Rockefeller Foundation–Mexican government agricultural program – that we are saying goodbye to tonight. We know how much Mexico has benefited and since the model has been developed here, then I, as President of Mexico, strongly urge that my government and the two foundations [the Rockefeller and Ford Foundations] look for some way to establish an international center for maize and wheat improvement in Mexico, so that we can help other third world nations.[18]

It was clear that López Mateos positioned Mexico as both a model and a leader among developing countries.[19] The ambition to have Mexico as a key global player was neither farfetched nor unusual. As other accounts demonstrate, Mexican leaders, economists, and diplomats were not passive members of international organizations but for much of the twentieth century helped shape agendas, proposing new economic approaches and other interventions, including ambitious health care models.[20] Nor did the president's interest in crop research contradict his better-remembered campaign to accelerate Mexico's industrial development. Known as the Mexican

---

[17] Ibid.

[18] Reported in Borlaug, "Statement before the United States Congress Senate Committee," p. 78.

[19] Researchers such as Derek Byerlee trace the promotion of an international center for the tropics to an earlier period, 1950–51, especially when speaking about maize. According to Byerlee and John Lynam, then Rockefeller Foundation Vice President J. George Harrar embraced this idea, and it was first applied to the creation of IRRI in 1960. See Byerlee and Lynam, "The Development of the International Center Model." For a history of the much older project of the Tropical Plant Research Foundation, which Byerlee describes, see Stuart McCook, *States of Nature: Science, Agriculture, and Environment in the Spanish Caribbean, 1760–1940* (Austin: University of Texas Press, 2002).

[20] Christy Thornton, *Revolution in Development: Mexico and the Governance of the Global Economy* (Oakland: University of California Press, 2021). For an example of Mexico providing health models to the world, see Gabriela Soto Laveaga, "Poverty Alleviation from the Margins: Mexico's IMSS-COPLAMAR as a Challenge to Global Health and Economic Models, 1979–1989," *The Hispanic American Historical Review* 102, no. 4 (2022): 673–704.

Miracle, the period beginning in the mid 1950s through the early 1970s is often described as the golden age of the Mexican economy. Focused on industrial production, this was also a period of intense mechanization of the Mexican countryside. However, López Mateos' particular vision for Mexico as a maize and wheat global leader was not about *producing crops* but rather about *producing research* about such crops. Thus, the hopes were for Mexico, an agricultural country, to become a knowledge-production center for agriculture on a global scale. These were two intertwined but certainly distinct goals: food production and research production.

The International Corn and Wheat Research Institute created in 1963 by the government of Mexico and the Rockefeller Foundation quickly encountered difficulties, including disagreements over administration and allocation of resources, and an inability to attract funding. Resolving these issues without abandoning the idea of the institute required significant reconfiguration and resulted in the establishment of CIMMYT in 1966 as an international research institution independent of but in collaboration with Mexican governmental agencies.

Byerlee and Lyman trace the idea of a centralized, global research center focused on crop improvement not to that 1960 dinner but much earlier, to 1951.[21] More significantly, they signal the origins of a plan for a collaborative and networked crop-breeding model to ideas about efficiency espoused after World War I. Yet their research reveals that the invention of MAP and later CIMMYT, often attributed to US models, was nonetheless a "merger of the highly integrated international wheat program in partnership with the FAO [the United Nations Food and Agriculture Organization], a loose federation of country and regional maize programs, and associated basic research activities in Mexico."[22]

The 1960s vision of a Mexican institution modeled on a specific US–Mexico partnership but with a broader international focus morphed into something different when CIMMYT became a founding member of CGIAR in 1971. As part of CGIAR, CIMMYT, though still headquartered in Mexico, came to be perceived by both the public and scientists as part of a global network and not a national institution addressing domestic concerns. It would also, like the other centers, have a series of missions: the centralizing of functions for maximum efficacy (for instance, germplasm banks), close collaboration and sharing across institutes, and finally training "aimed to substitute for weaknesses in many developing national research systems."[23]

Put differently, once MAP was dissolved in 1961, scientists continued to travel to Mexico to conduct research and undergo training in

[21] Byerlee and Lynam, "The Development of the International Center Model."
[22] Ibid.   [23] Ibid., 2.

agricultural science, only now as part of a different program. This program had similar aims but additional funders: the Mexican government (supervised by INIA), the Rockefeller Foundation, the Ford Foundation, and FAO.[24] Indeed, at CIMMYT's founding on April 12, 1966, both the Mexican minister of agriculture, Juan Gil Preciado, and Rockefeller Foundation President J. George Harrar conveyed that the new center was in some ways a continuation of nearly twenty-three years of agricultural research between the Rockefeller Foundation and the Mexican government.[25] The center's goals were spelled out in that founding document: to conduct basic and applied research, distribute "superior" germplasm, train scientists, foster cooperation among scientists and breeders, and publish and distribute its findings.[26] Over the following years, these basic aims would be expanded and became, as community needs were considered, both more nuanced and specific (See Figure 4.1).

A Growing Center Reflects and Shapes the World, One Crop Germplasm at a Time

In 1960 MAP's International Wheat Program began to distribute "international trials" ("ensayos internacionales") of experimental lines of wheat.[27] Years later, in 1971, the same would be done for corn by CIMMYT, as Derek Byerlee and Greg Edmeades discuss in Chapter 9, this volume. What did these trials consist of? An international trial was composed of "identical experimental lines" shipped to research partners across the world, who planted these seeds following specific instructions and conditions and later compared them with local varieties.[28] But the process did not end there. All results were sent back to CIMMYT, where they were analyzed, discussed, later published, and broadly distributed. These international experimental aims were quite clear, as outlined in a CIMMYT publication of the time. In addition to the obvious ones, such as trying out new lines under vastly different climactic, pest, and disease conditions, was the important issue of standardization of the research. The international trials also served to train networked and partner scientists, as well as to obtain the germplasm needed to continue to make new crosses.[29]

---

[24] *Noticiero del CIMMYT* 1, no. 1 (July 1966): 1, 4.   [25] Ibid., 3.   [26] Ibid., 4–5.
[27] *Este es el CIMMYT*, Boletín de Información no. 8, March 1974, presentación 9, https://repository.cimmyt.org/handle/10883/19375.
[28] Ibid.
[29] By 1973 there were 1,429 international trials in 91 countries. On germplasm management, see Marianna Fenzi (Chapter 11) and Helen Anne Curry and Sabina Leonelli (Chapter 10), this volume.

Figure 4.1 Attendees of a 1968 international meeting held at CIMMYT, seated before a map of the research facilities in Mexico where its research programs initially were based. Rockefeller Archive Center, Rockefeller Foundation Photographs, CIMMYT Series 105, International Agricultural Meeting. Courtesy of Rockefeller Archive Center.

Mere months after CIMMYT's founding in 1966, *El Informador*, a newspaper based out of Guadalajara, reported on the center's research-driven mandate at its first meeting. It noted, clearly echoing the message and language of CIMMYT, that the "urgent need" to ensure an increased production of cereals using "modern technology" was a pressing, global one. The article went on to quote a "Rockefeller Foundation representative" as stating that the newly inaugurated CIMMYT would bring together research, experimentation, and training at the "highest levels" to increase maize and wheat yields.[30] It is worth pausing to explain that news of CIMMYT's mission was making it to the pages of a regional paper. Even if this article was a reprint from larger newspapers, as was the

---

[30] *El Informador* (September 20, 1966), no. 17, 406.

Figure 4.2 Wheat trainees inoculate plants at the CIMMYT research station in Toluca, Mexico, undated. CIMMYT repository.© CIMMYT.

practice, its inclusion suggests the broad appeal that this news of such a center had in other Mexican states.

With a vision of further training of young scientists, CIMMYT expanded its training program to include plant pathology, managing research stations, and wheat chemistry. Following the MAP model, foreign researchers travelled to Mexican research stations where both Mexican and international scientists were trained (Figure 4.2). They would return to their home countries with sample seeds and a core training in wheat and corn science.

Much of the initial focus of agricultural research centers was on the training of future scientists rather than the dissemination of germplasm directly to farmers.[31] Yet reports of famines in South Asia served as a catalyst to push for more extensive plant-breeding programs that could stretch from Mexico to farmers around the world. In hindsight,

---

[31] Recent research traces the networks of Latin American agronomists who, with grants from the Rockefeller Foundation, travelled to American land-grant colleges and other institutions to pursue postgraduate degrees in agricultural sciences. Diana Méndez's work, for instance, examines the dozens of Latin American agronomists who returned to become part of not just international organizations but also domestic research centers; see Diana Alejandra Méndez Rojas, "La agricultura como puente: Becarios guatemaltecos de la Fundación Rockefeller en México: Un viaje de ida y vuelta, 1949–1976," *Oficio Revista de Historia e Interdisciplina* 13 (2021): 49–70.

the inauguration of CIMMYT in spring 1966 seemed an auspicious time to launch an international agriculture research institution, given that at the time the spectre of hunger seemed to loom especially large across the ideologically divided Cold War world.[32] For example, a focus on famines happening in both India and Pakistan revealed that both countries had the lowest wheat yields since World War II.[33] Researchers believed that overpopulation and the depletion of resources would lead to more human hunger, increased violence, and political instability. In an ideologically separated world this meant potential communist insurrections which would, in turn, risk destabilizing Western societies, in particular the United States. Hence by zeroing in on global hunger, political instability could be averted by using science to increase yields that would, in turn, feed populations and create a more stable world. The globe, it seemed to Rockefeller and Ford Foundation personnel, was primed for an international organization rooted in agricultural science that could help improve crop yields – enough to stave off concerns of an overpopulated world.

In the fall of 1966 CIMMYT announced that it was broadening its scope via *Noticiero del CIMMYT*, or *CIMMYT News*, a bilingual publication available in seventy countries and devoted to detailing the latest scientific advances in wheat and corn research. Reporting on the fall meeting of CIMMYT's board, the *Noticiero* announced the board's apparent decision to fully concentrate CIMMYT's efforts on maize and wheat.[34] For the maize program, the center planned to establish projects "in plant breeding, agronomy, genetics and physiology as well as a broadened action for regional programs, such the Central American

---

[32] In influential histories of the Green Revolution, this is where CIMMYT enters the narrative as a reflection of global concerns about food security and overpopulation. See, for example, Nick Cullather, *The Hungry World: America's Cold War Battle against Poverty in Asia* (Cambridge, MA: Harvard University Press, 2010).

[33] Centro Internacional de Mejoramiento de Maíz y Trigo (CIMMYT), *La conmemoracion del 20 aniversario del CIMMYT* (Mexico: CIMMYT, 1987), https://repository.CIMMY T.org/handle/10883/3514.

[34] CIMMYT's board in 1966 featured a roster of leading Latin American ministers of agriculture and/or scientists, as well as key representatives of the Rockefeller and Ford Foundations and other international institutions: Juan Gil Preciado (chairman), secretary of agriculture, Mexico; J. G. Harrar (vice chairman), president of the Rockefeller Foundation; E. J. Wellhausen (secretary), director general, CIMMYT; Virgilio Barco, mayor of the City of Bogotà, Colombia; M. C. Chakrabandhu, director general of agriculture, ministry of agriculture, Thailand; Manuel Elgueta G., director, Institute of Agricultural Research, Chile; Emilio Gutierrez Roldan, National Seed Producing Agency, Mexico; Lowell S. Hardin, program officer for Latin America and the Caribbean, Ford Foundation; Carlos A. Krug, Brazil; Galo Plaza, Ecuador; Carlos P. Romulo, minister of education and president of the University of the Philippines; Nicolas Sanchez D., director, National Institute of Agricultural Research, Mexico; C. V. Subramaniam, India.

Cooperative program."[35] As for wheat, research projects would also be expanded to include cytogenetics, vital to understand the plant's cell biology and growth, and "enlargement of the activities of the milling and baking laboratory."[36] The latter was especially important to the work of wheat breeding. It was in these laboratories that wheat quality was tested. If a particular wheat variety did not yield flour that would easily rise when baked or did not pass a taste test, then that variety, regardless of rust resistance or other qualities valued in the field, would not be pursued as a successful strain. This work was considered so useful that the Rockefeller Foundation awarded a grant to the Mexican cereals chemist Evangelina Villegas to visit milling and baking laboratories in the United States and Canada[37] (Figure 4.3).

The growth of CIMMYT was programmed to be fast. At that same reunion it was proposed that by 1967, a year later, CIMMYT should have 121 technicians, and by 1970 the total would reach at least 189 (by 1973 there were 420 staff positions across the world).[38] Though headquartered in Mexico, the center's activities would be "multiplied through cooperative programs in many countries" where CIMMYT personnel would be based. The center would also offer trainees access to graduate education through an agreement with the nearby National School of Agriculture at Chapingo.[39] Researchers would continue to come, as they had under MAP, to Mexico.

[35] That the Central American Cooperative program was not explained in the publication makes it clear that this was a well-known initiative. Begun in 1954 with support from the Rockefeller Foundation, the Central American Cooperative Program for the Cultivation and Improvement of Food Cultivars (PCCMCA) was a network of agriculture and, initially, livestock programs. *CIMMYT News* 1, no. 4 (October 1966). For more on this program, see Fuentes, Salas, and Salazar, "Origen e historia"; Méndez Rojas, "Maize and the Green Revolution"; Araya Villalobos, *Programa Cooperativo Centroamericano*; Wainer Ignacio Coto Cedeño, "Semillas en disputa: Historias de vida y memorias del cambio tecnológico en la agricultura de la Papa en Costa Rica (1943–2015)," *Revista de Historia* 72 (2015): 75–100.

[36] *CIMMYT News* 1, no. 4 (October 1966).

[37] Gabriela Soto Laveaga, "When the Baker Is the Knowledge Maker: Evangelina Villegas and the Laboratories of the Green Revolution," presented at the Cain Conference, Philadelphia, PA, June 4, 2022 and "Worker Once Known: Thinking with Disposable, Discarded, Mislabeled, and Precariously Employed Laborers in History of Science," *Isis* 114, no. 4 (December 2023): 834–840. For more on Evangelina Villegas, see Diana Méndez Rojas, "Modernizar la agricultura, movilizar ideas: Trayectorias de los becarios en ciencias agrícolas de la Fundación Rockefeller en México, 1940–1980," Ph.D. dissertation, Instituto Mora (2022).

[38] *Este es el CIMMYT*, presentación 8.

[39] Mexico's National School of Agriculture, or Chapingo, as it is known locally, traces its origins to 1854, though this first iteration was quite dissimilar to today's sprawling campus, in its current location since 1923. Its long-cherished motto "Exploitation of the soil, not of man" reveals its postrevolutionary origins and the role of agronomists as agents of change in the nation.

Figure 4.3 The CIMMYT cereal scientist Evangelina Villegas (center) with other researchers and trainees, undated. CIMMYT repository. © CIMMYT.

In the early 1970s a key shift occurred when CIMMYT officials realized that it was often difficult for researchers from low-income regions and countries to travel to Mexico. It is uncertain how this realization came about, but to address the concern, the center launched a series of regional training programs. By the end of the decade there were four regionally based maize programs, four wheat ones, and, expanding beyond crop-centered research, four centers focused on regional economies.[40]

Regional centers also allowed for deeper understanding of how local farmers adopted new technologies and new seeds. Within Mexico, one such local model was the Puebla Project, which encompassed 47,000 families, mostly small-plot farmers, with whom CIMMYT researchers worked from 1967 to 1973. The aims of the project were, first, to increase technological transfer to smallholding farmers who relied on rainfed crops, especially maize, and, second, to train technicians from other regions. The lands of the Puebla Valley were selected because there was little irrigation infrastructure, as opposed to what could be found in CIMMYT's experiment station in Sonora. Also, locals reportedly seemed

---

[40] CIMMYT, *La conmemoracion del 20 aniversario*, p. 15.

eager to work with CIMMYT technicians.[41] With the Puebla Project, CIMMYT provided investment in maize for smallholding subsistence-level farmers. As a scholar noted, a new approach was needed to work with small farmers, especially since "enthusiasm was expressed for any attempt to bring the banking sector into closer contact with groups of producers who had traditionally remained outside their reach."[42]

The farmers' socioeconomic environment, which had not been an initial topic of interest for the architects of CIMMYT's goals, was becoming as important an area of focus as the crops these farmers planted. An additional shift was happening with a more region-centered, bottom-up understanding of agriculture. Yet despite the existence of the Puebla Project, which remained comparatively close to CIMMYT headquarters, CIMMYT was not yet reaching the most remote (often the poorest) farmers within Mexico or abroad. A sharper focus on these farmers would only come later in the century. Meanwhile the germplasm bank and wheat-breeding program, both core to the organization as it exists today, thrived in this era.[43]

At the center's one-year anniversary, in spring 1967, the president of the board of the Rockefeller Foundation, John D. Rockefeller III, and Rockefeller Foundation President J. George Harrar visited CIMMYT to learn about the ongoing "maize and wheat germplasm and cooperative research."[44] During their visit they discovered that Mexican wheat varieties planted in other countries already surpassed the surface area of wheat farming in Mexico, which demonstrated "the wide adaptability and acceptance of these varieties."[45] As the historian Marci Baranski shows, along with Harro Maat (Chapter 6, this volume) and others, the push for so-called wide adaptation was vital to the goals of international programs, for it allowed researchers to replicate findings from one location to another.[46] Though maize research was still important, by 1965 Norman Borlaug and Rockefeller Foundation scientists focused increasingly on wheat breeding and its purported adaptability to most soils. Wide adaptation would become a core tenet of CIMMYT's research agenda. Phrases such as "exchange of ideas," "fraternity with a common goal," and "a collective discussion," forged the sense of single and singular research community reinforced by the rush to try to feed the world's

[41] Michael Redclift, "Production Programs for Small Farmers: Plan Puebla as Myth and Reality," *Economic Development and Cultural Change* 31, no. 3 (1983): 551–570, at 555.
[42] Ibid., 553.
[43] CIMMYT, "CIMMYT Bread Wheat Breeding Program: Germplasm Movement and Planting Plans," 1970, https://repository.cimmyt.org/handle/10883/3911.
[44] *Noticiero del CIMMYT* 2, no. 6 (June 1967): 2, 4.   [45] Ibid., 4.
[46] Baranski, *Globalization of Wheat*.

hungry.[47] In the maize program, this communal sense of purpose was most visible in the speed with which CIMMYT's maize germplasm bank grew. The germplasm bank represented how agricultural research shifted from country-specific aims to global crop centers. For example, by 1974 the maize germplasm bank was already the largest in the world, with more than 12,000 samples from more than 47 countries.[48] With key breeding resources and connections to long-running training and breeding programs, CIMMYT symbolized Mexico's long-ascendant centrality to global maize and wheat research. Shortly thereafter CIMMYT expanded its aims once more.

CIMMYT on the Ground

As CIMMYT grew so did the scope of its programs. Here I focus on two examples of CIMMYT's vast projects, the wheat-breeding program and its training program, to showcase the deep local roots of global technology transfer.

Like the germplasm bank, the wheat-breeding program defined CIMMYT. From its foundation, the international breeding and testing nurseries attracted growing numbers of visiting scientists and trainees. The Bread Wheat Program operated in three Mexican locations: Ciudad Obregón in the arid, irrigated farming region of Sonora; Toluca in the central Mexican highlands; and at the CIMMYT headquarters at El Batán near Mexico City. (Today there are an additional two CIMMYT stations in tropical and subtropical settings: Agua Fría, Puebla, and Tlaltizapán, Morelo.)[49] The original locations – one at sea level near the Sonoran desert, the other two in rainy regions with high elevation – played a crucial role in experimentation and development of wheat lines with disease resistance.[50] But CIMMYT experimentation did not and

[47] *Este es el CIMMYT*, presentación 6/2.
[48] Ibid. As Helen Curry demonstrates, the collection and distribution of maize germplasm accelerated in the mid twentieth century, so CIMMYT was a significant (but not the sole) organization focused on germplasm exchange and collection. See Helen Anne Curry, *Endangered Maize: Industrial Agriculture and the Crisis of Extinction* (Oakland: University of California Press, 2022).
[49] Carolyn Cowan and Alfonso Cortés, "Experimental Stations in Mexico Improve Global Agriculture," CIMMYT blog, July 1, 2019, www.cimmyt.org/multimedia/experimental-stations-in-mexico-improve-global-agriculture.
[50] During the MAP era these were the stations that Borlaug used to develop shuttle breeding – the practice of shuttling seeds via truck from one region to the next to ensure that certain desired traits appeared in the next generation. See Liesel Vink, "Photo Essay: Mexico and the Launch of the Green Revolution," *RE:source* (November 5, 2019), https://resource.rockarch.org/story/photo-essay-mexico-and-the-launch-of-the-green-revolution/.

does not now remain limited to these five locations. Taking advantage of Mexico's extraordinary diversity of microclimates, wheat pathologists, for example, used nurseries across the country to screen for diseases. Meanwhile, breeders used seed multiplication plots to replicate stressors from across the globe. This research geography was and continues to be the lifeline of CIMMYT. It is in these spaces that researchers test new wheat and maize lines, examine the impact of pests and plant diseases, and host farmer workshops.

Genetic materials that survive these varied trials with natural and amplified stressors, such as heat tolerance or difficult tropical soil, have stronger viability in regions across the globe. In 1972 alone, nearly 5,500 crosses were made in bread wheat. But, from experience, less than 1 percent of these crosses would survive the center's "rigorous screening."[51] From generation two (F2) onward, experimental material was sent worldwide where plants' performance was observed for six generations in different conditions and in competition with local wheats. The most crucial aspect of CIMMYT's broad infrastructure, beyond its germplasm and experimental stations, was and continues to be the training of plant specialists. Between 1966 and 1988, the wheat improvement program served 471 trainees from 80 developing countries.[52] The trainee program, open mainly to researchers and extensionists from developing countries under the age of thirty, allowed participants to remain in Mexico from six to eighteen months. Some were later granted scholarships to pursue master's degrees, usually in Mexico. The range of trainees was broad, from government workers to postdoctoral fellows to visiting, well-established scholars. The benefits of this intergenerational mixing were significant, as program participants learned from each other. Similarly, the practice of working "shoulder to shoulder" engaged everyone in a hands-on approach.[53]

In addition to crop management, this hands-on practice consisted of "designing and managing field plots, choosing parental materials, making crosses ... scoring for tolerance and resistance to biotic and abiotic stresses, and selecting improved progeny."[54] The numbers of trainees in the first decade are telling. In 1966, there was a total of 22 scholars of all ranks (scholarship recipients, established scientists, temporary residents), but by 1973 there were 739.[55] The majority of these hailed from Latin America. It is important to recall that scholarships were also key for MAP.

[51] *Este es el CIMMYT*, presentación 7/2.
[52] R. L. Villareal and E. del Toro, "An Assessment of a Wheat Improvement Research Training Course for Developing Countries," *Journal of Natural Resources and Life Sciences Education* 22, no. 1 (1993): 38–43.
[53] Ibid.   [54] Ibid.   [55] *Este es el CIMMYT*, presentación 10.

The significant increase of twenty-two scholars in the first year of operation to a leap in hundreds of recipients mere years later is likely a reflection of the educational networks in place for two decades. In 1988, CIMMYT conducted follow-up questionnaires of 324 trainees to evaluate the program's effectiveness. The survey revealed that 74 percent of respondents worked for their government's research and extension services, and more than 50 percent continued to work with wheat.[56] Those who responded to the questionnaire hailed from forty-five countries across Asia, Latin America, North Africa, the Middle East, sub-Saharan Africa, and European countries. In other words, participants represented the global community of crop researchers. The 1988 survey revealed that the vast majority of trainees felt that they had gained something from participating in the program, including improved "plant breeding and plant pathology skills," and when returning they used "CIMMYT's methods in their training activity." The survey did reveal some discontent, and some trainees thought that the courses offered were too elementary, but these tended to be participants with either a doctorate or a master's degree.

CIMMYT 1975 – Thriving and Overextended

By 1974, the CGIAR network of international agricultural research centers focused on training and providing assistance to governments around the world. From the Philippines, to Nigeria, to Colombia and Peru, to India, to Kenya and Ethiopia, CGIAR leadership created a network of centers devoted to specific crops, livestock, and environments. In this larger circle of interconnected expertise, CIMMYT became the center focused on maize, wheat, barley, and triticale (the hybrid of wheat and rye), and, as such, a sort of scientific pilgrimage site for hundreds of researchers who regularly arrived in Mexico to study and exchange ideas. A decade after its founding, CIMMYT had become firmly established in both national and international agricultural research.[57] In other words, at this time CIMMYT envisioned itself more as a handmaiden to national projects, an additional research arm supporting domestic research.[58] In a similar vein, the larger CGIAR mission at this stage was to support national research programs and aid in pushing them to a higher level. Despite the global orientation of the CGIAR system, CIMMYT

---

[56] This survey is analyzed in Villareal and del Toro, "Assessment of a Wheat Improvement Research Training Course."

[57] At this time there were thirty-eight principal scientists, fourteen of whom were Mexican and nine American. *Este es el CIMMYT*, 6–1.

[58] Ibid., presentación 6/2.

publications continued to highlight the vast reach that the center maintained in Mexico. With maps and descriptions of the experimental stations in the country, CIMMYT materials emphasized the centralized coordination directed from El Batán, CIMMYT's headquarters. A detailed map of the central buildings – including dormitories for sixty scholars, baseball diamond, pool, basketball courts, and cafeteria – as well as laboratory and experimentation space, depicted this self-contained space as a sort of international scientific enclave.[59]

The built environment of CIMMYT was examined in a 1974 *New York Times* article that described El Batán as a "complex of modern buildings surrounded by 160 acres of experimental fields, three dozen agricultural scientists and scores of technicians, most from poor countries ... engaged in a major campaign to feed adequately the two billion people" who depended on wheat, corn, and similar crops to survive.[60] In this and dozens of other articles, it was the promise of science and how it could, if used appropriately, reduce hunger, which imbued CIMMYT with an aura of productive legitimacy. Two months later, after CIMMYT's participation in the World Food Conference in Rome (November 5–16, 1974), the center received more than twenty-five requests from individual countries seeking to increase food production to meet their populations' needs.[61] Despite these numbers, Haldore Hanson, CIMMYT's director general, turned down the majority of appeals. As he explained, the center's forty-five scientists were already overextended with consulting work and travel.[62] This high demand, however, brought about more changes in CIMMYT, especially in how it functioned on a global scale. As Hanson explained to the *New York Times*, CIMMYT would set up two-member

[59] Scholars have noted the importance of the built environment to convey messages about science and even agricultural experimentation. See Nikki Moore, "To Which Revolution? The National School of Agriculture and the Center for the Improvement of Corn and Wheat in Texcoco and El Batán, Mexico, 1924–1968," in Aggregate, ed., *Architecture in Development: Systems and the Emergence of the Global South* (London: Routledge, 2022), pp. 85–104.

[60] Boyce Aensberger, "Science Gives New Life to the Green Revolution," *New York Times* (September 3, 1974), www.nytimes.com/1974/09/03/archives/science-gives-new-life-to-the-greenrevolution-scientific-research.html.

[61] Victor K. McElheny, "Nations Demand Agricultural Aid," *New York Times* (August 3, 1975), www.nytimes.com/1975/08/03/archives/nations-demand-agricultural-aid-population-growth-spurs-appeals-to.html.

[62] Although the *New York Times* reported a total of forty-five scientists, a contemporary CIMMYT publication indicates that in February 1974 there were fifty-one international scientists at the Mexico campus and twenty-one scientists assigned to foreign posts. Yet these numbers do not fully reveal the extent of personnel. For example, in 1966 there were eight scientists in Mexico and twenty-five support staff. By 1974, there were 347 support staff working at CIMMYT. It is assumed that this support staff was different from fieldworkers, since at the time there were an additional ninety-five field workers. *Este es el CIMMYT*, presentación 8.

regional teams who would train scientists in their own countries. This effort to reach more farmers had begun earlier when CIMMYT joined CGIAR. This quintupled CIMMY's research budget from $9 million to more than $48 million in less than four years.[63] These funds were needed to push the use of high-yielding varieties and extend them to "small farms" across the world.[64] CGIAR funds also pushed for a reorientation of the organization taking place: production research that more accurately reflected farmers' needs.

A good example of a "typical" program (granted that all of these programs were unique to their locale) was CIMMYT's Regional Maize Program for Central America, Panama, and the Caribbean. A 1978 report on the program reveals the vast network of scientists, technicians, government workers, diplomats, farmers, and many intermediaries needed to make it function. The Regional Maize Program was sponsored by the Swiss government with the cooperation of fourteen countries. Modeled on the on-farm approach advocated by CIMMYT, it also included the core philosophy of the international organization: research at experiment stations, research and production of new technologies in farmers' fields, and demonstrations for technology transfer.[65] As part of this project, two maize scientists and an economist spent a total of 126 days consulting with ministers of agriculture and directors of national research institutions from nine governments (Costa Rica, El Salvador, Guatemala, Honduras, Nicaragua, Panama, Dominican Republic, Haiti, and Jamaica). Most meetings encouraged the "participation of local maize program leaders and technicians" with the stated aim that "technology generation" happening on farmers' fields would be more widely accepted.[66] Vital to this research was the establishment of maize nurseries and the running of experiments. In 1978, alone, a total of 164 experiments produced 9 new experimental varieties to be tested the following year. This research had to navigate differences from country to country and, indeed, between intra-country regions. For example, when it came to seed production, some countries had a "well-organized program" (El Salvador and Guatemala), while others were not developed. Pairing regionally specific characteristics (i.e., husk cover, propensity to ear rot, height of plants, leaf breadth) with desired crop yield was a scientific riddle that relied heavily on research conducted on farmers' fields.

[63] McElheny, "Nations Demand Agricultural Aid."   [64] Ibid.
[65] W. Villena and R. F. Soza, "1978 Annual Report: CIMMYT Regional Maize Program Central America, Panama and the Caribbean," September 1978, https://repository.cimmyt.org/handle/10883/3722.
[66] Ibid., 3.

The Regional Maize Program for Central America, Panama, and the Caribbean revealed that "production technology generated at experiment stations in the region often was not accepted by farmers."[67] This was due mainly to the fact that conditions at experiment stations were simply not replicable, and, crucially, the economic risk was not factored into the analysis of which varieties and technologies to propose to small-plot and medium-plot farmers. For instance, a factor that had not previously been examined was the difference between individuals and cooperative groups. The latter could afford the suggested herbicide while individual farmers found it difficult to even find it in local markets.[68] Working directly on farmers' plots expanded adoption of technology and led to subsequent yield increases.

Finally, in addition to in-country workshops, the 1978 Regional Maize Program introduced seven production program directors from El Salvador to Mexico's Poza Rica-Tuxpan, where the maize training program was located. Upon returning to their country these directors held a workshop to showcase what they had learned. This fruitful exchange allowed for knowledge and experience to ripple beyond national borders. This web of interconnected researchers, farmers, and state officials was anchored solely by its connection to CIMMYT. CIMMYT was becoming vital at all levels of regional agriculture development – local leaders, national bureaucrats, regional experts – reaching far beyond the research station.

The focus on "marginal zones," the buzz word in Mexican politics of the 1970s and 1980s, would trickle down into CIMMYT's lexicon and influence how the institution approached outreach.[69] Just as CIMMYT affected Mexico's framing of farmers' problems, so too did Mexico impact CIMMYT's framing of global problems. How, the organization asked, could agricultural advances and technology reach the most remote farmers, those who had not yet benefitted from CIMMYT's contributions? A greater focus on economic impact began to take shape.

A Global Center

In 1986, Mexican President Miguel de la Madrid presided over CIMMYT's twentieth anniversary. While some of the themes from the celebration echoed earlier research priorities, the international

---

[67] Ibid., 8.   [68] Ibid., 14.
[69] A "marginal zones" approach to Mexican socioeconomic problems in the countryside examined housing, health, nutrition, and education in Mexico's poorest sector. Providing for the *marginados* became a constant claim in the political speeches of President López-Portillo. CIMMYT, *La conmemoracion del 20 aniversario*, p. 17.

contributions and expansion of CIMMYT as part of the CGIAR network had emerged as the most critical. Of course, the context of the 1980s in Mexico proved quite different from the decade of CIMMYT's founding. By 1986, Mexico was in the midst of one of the worst economic crises in its history, and across the globe neoliberal reforms were on the rise. Further, widespread enthusiasm for the first generation of wheat seeds, which had promised to end world hunger in the 1950s and 1960s, gave way to growing critiques that pesticides, excessive fertilizer, and irrigation did more harm than good to small-plot farmers.

In this new era, CIMMYT's global impact was undeniable. More than 4,000 agricultural scientists from 125 countries had been trained at CIMMYT. The scholarship program had expanded to allow trainees to spend more time in Mexico. And by 1985 CIMMYT had the world's largest collections of wheat and maize germplasm, with more than 2 million seed packets sent on a yearly basis to nearly 120 countries.[70]

As a major global player facing economic crisis in Mexico, as well as criticism of its results, CIMMYT sought a more transparent accounting of the real costs to implement technological change. As CIMMYT Director General D. L. Winkelmann explained, "it is necessary to combine financial information with biological" research to assess how this knowledge made it on the ground.[71] In short, though still celebrating advances in plant breeding, there was once again a concerted effort to bring the farmer into greater focus, and to do this more successfully than in the past. This shift reflected a trend in international agricultural research, with a strong focus on more farmer-centered approaches to knowledge and technology development such as the frameworks of farming systems research, Farmer First, and Farmer-back-to-Farmer.

CIMMYT issued a series of publications reflecting this new interest. One example, *Gorras y Sombreros* – or *Baseball Caps and Sombreros* – focused on "paths of collaboration between technicians and peasants."[72] Taking the example of local knowledge transmission about farming with velvet beans, usually passed on from one generation to the next, CIMMYT organized a series of workshops that brought together state officials, nongovernmental organizations (NGOs), local leaders, and peasants from across southern Mexico and Central America to discuss velvet bean farming techniques as technologies worthy of study by an international organization. The velvet bean was introduced into the United States from Asia at the end of the nineteenth century.[73] From

---

[70] Ibid.   [71] Ibid., p. 17.

[72] D. Buckles, ed., *Gorras y sombreros: Caminos hacia la colaboración entre técnicos y campesinos* (Mexico: CIMMYT, 1993), https://repository.cimmyt.org/handle/10883/898.

[73] The history of the velvet bean is described in Buckles, ed., *Gorras y sombreros*, p. 4.

there it made its way to Central America via the United Fruit Company in the 1920s. The velvet bean had a specific appeal for plantation owners. Interspersed between corn stalks, mature beans could be used as forage for livestock, served as natural fertilizer for cotton or corn, and, if planted with oranges, worked as a natural weed deterrent. Indigenous farmers from southern Mexican states and Guatemala had been using velvet beans for decades. Despite its evident success in the fields, it was displaced by inorganic fertilizers and its use labeled "backward." By the early 1990s once-disdained practices were revisited, but there was little research on the bean's characteristics, what little knowledge existed was dispersed, and few, if any, controlled studies had been conducted, certainly not in experimental fields.

The push to bring farmers into conversation with scientists and extensionists became part of a growing trend that elevated local "practices" to the study of science. In the case of recuperating knowledge about the velvet bean, for instance, funds from several organizations were brought together to sustain a frank exchange of knowledge between professional researchers and farmers. Under the auspices of the Ford Foundation, twenty-four representatives from universities, NGOs, and both national and international agricultural programs from eight countries met with farmers. The group was not limited to the region but also included representatives from South America, West Africa, and the Philippines. While a significant focus of the workshop, which was held in Catemaco, Mexico, was research and new extension work on green fertilizers, the event kicked off with visits to two Indigenous communities experimenting with velvet beans in Soteapan and Mecayapan, Veracruz. These community visits served as the framing for the multiday event. Though just an example of a shift in CIMMYT's practices, the velvet bean meeting represented a growing determination to focus on farmers and the vigorous "exchange of technical knowledge."[74]

Questionnaires used to query farmers in this period reveal the level of local detail sought by CIMMYT experts. For instance, questions ranged from soil choice to tools used: How do you prepare your soil? In this cornfield, what was planted in the previous season? Why did you choose to use this lot and not another for experimental crops? Although completed questionnaires, if they have survived, remain hidden in the archival record, what is certain is that there was a concerted effort to tally the participation of local, small-plot farmers. Local farmers, for instance, "took control of experiments using simple and easy to understand practices." This ease could be translated as making the farmer feel

---

[74] Ibid., pp. iii–iv.

comfortable with experimentation by designing trials from previous farming experience. Furthermore, the design of the experiments was done collectively, with all participating farmers agreeing on what it was they sought to understand.[75] The push for openness and collective spirit was vital to give farmers a sense of control and equal footing with CIMMYT experts – although, as subsequent investigations into farmer participatory research illustrate, critically measuring "participation" is difficult, as is quantifying communication and other human-to-human interactions.[76]

By 1994 CIMMYT had only grown in its dominance as a producer of scientific agricultural knowledge. In that single year CIMMYT staff produced 410 publications on topics ranging from seed quality, to triticale improvement strategies, to disease resistance in Mexican landraces of maize, to networking for sustainable maize farming in Central America, to a traveling workshop on wheat-based sustainability in East Africa.[77] The institution's prominence was also evidenced in CIMMYT's global footprint. As reported by wheat scientist Sanjaya Rajaram, by 1994, 58 percent of the total bread wheat area in developing countries was planted by varieties directly or indirectly derived from CIMMYT germplasm.[78] In less than three decades, seeds developed in CIMMYT's experimental stations in Mexico had conquered the wheat fields of the world.

Tying together the themes of germplasm and focus on farmers, CIMMYT and the Mexican government launched a ten-year program known as Sustainable Modernization of Traditional Agriculture (MasAgro) in 2010 with the goal of reaching small-plot farmers, whose rainfed lands had previously been dismissed by agricultural research. In many ways, MasAgro, at least in writing, recalls how agricultural research that directly benefitted farmers was described in the early years of technical assistance – "to augment the productive capacity of small wheat and corn farmers" and guarantee "food security for the world's growing population."[79]

---

[75] Ibid., p. 65.
[76] Adrienne Martin and John Sherington, "Participatory Research Methods – Implementation, Effectiveness and Institutional Context," *Agricultural Systems* 55, no. 2 (1997): 195–216.
[77] *CIMMYT in 1994: Staff Publications* (Mexico: CIMMYT, 1994), https://repository.cimmyt.org/xmlui/handle/10883/19504.
[78] S. Rajaram and G. P. Hettel, eds., *Wheat Breeding at CIMMYT: Commemorating 50 Years of Research in Mexico for Global Wheat Improvement*, Wheat Special Report, No. 29 (Mexico: CIMMYT, 1995), p. 11.
[79] An agreement between Mexico's SAGARPA and CIMMYT that would run from October 2010 to December 2020 was to receive a total of $138 million. Secretaría de Agricultura, Ganadería, Desarrollo Rural, Pesca y Alimentación, Modernización Sustentable de la Agricultura Tradicional "MasAgro," Auditoría Financiera y de Cumplimiento: 13–0-08100–02-0300 DE-007.

Echoes of the original aims more than fifty years later reveal how decades of inputs and focused research projects have not always managed to transform agricultural fields. It is more difficult to change social context (poverty, or unequal land and water distribution, for example) than it is to create experimental plots.

Ten years prior, in 2003, the Mexican government had signed an agreement enabling CIMMYT to continue to function in the country as an international organization with a series of fiscal and judicial benefits reserved for international institutions in good standing and in acknowledgments of the important role that CIMMYT continued to play in the development of agricultural technology in the country. Crucially for its research mission, the agreement declared that seeds destined for CIMMYT research stations would continue to be exempt from Mexican law that prohibited the import of seeds. This latter point may seem obvious, given the nature of CIMMYT's research, but it also signaled the continued value placed on the ongoing work at the center and by its researchers. Fully 99 percent of CIMMYT's funding comes from external sources, but the Mexican government continues to provide about $300,000 yearly in addition to the lands and access to the nation's research stations.[80]

Conclusion

To understand the importance of CIMMYT today – and how we should best tell its history – we should first ask, was CIMMYT a technical assistance program? This depends on whom you ask. For example, in a 1979 hearing before the Subcommittee on International Trade of the Committee on Finance, John Pino, director of agricultural sciences at the Rockefeller Foundation, insisted that this was "no usual technical assistance program. We never, in fact, used that terminology."[81] Instead, those involved with CIMMYT preferred to focus on cooperative research and training programs as the goals. This is a vital distinction, because mid-twentieth-century technical assistance in practice, often associated with

---

[80] The sum of $300,000 is from a 2003 Mexican Senate discussion: "Discusión en la cámara de Senadores Acuerdo entre el gobierno de los Estados Unidos Mexicanos y el Centro Internacional de Mejoramiento de Maíz y Trigo relativo al establecimiento de la sede del centro en México, y de su protocolo adicional, 6 de noviembre de 2003."

[81] John A. Pino, "Statement before the United States Congress Senate Committee on Finance and United States Congress Senate Committee on Finance Subcommittee on International Trade," *North American Economic Interdependence II: Hearing before the Subcommittee on International Trade of the Committee on Finance, United States Senate, Ninety-Sixth Congress, First Session, October 1, 1979* (Washington, DC: US Government Printing Office, 1979), p. 73.

development aid, frequently disregarded local practices and knowledge. In its initial years, so did CIMMYT. As CIMMYT goals grew to incorporate socioeconomic impacts, its programs also sought to include more farmer participation. Farmers would not simply be passive recipients of information; rather, they became active participants and, as in the case of the velvet beans, vital designers of experiments. It was local farmers who understood the land at a deeper level, and it was farmers who, enmeshed in social networks and unspoken rules, could – and did – affect how science was conducted on the ground. Crucial, then, to the distinction of technical assistance versus an international research program was the role assumed by the Mexican government and Mexican research institutions. CIMMYT was conceived as initially a Mexican program, and in 1966 it embraced and reflected a Mexican nation which, like its president, was seeking to influence the globe, to become a leader in the so-called developing world.

When CIMMYT was incorporated into CGIAR a few years later in 1971, it joined the network of research institutions not as a recent creation but rather as an organization with a history that traces its origins decades earlier to the inauguration of MAP in 1943. These origins matter, for its aims and thus its research agenda reflect a divided world, a product of a post–World War II era, and the role that agricultural science can play in ending world hunger. Since that time, the organization's breadth and goals have modified to reflect the changing understanding that different actors – scientists, donors, NGOs – hold on food security, agricultural development, and agricultural research.

In 2020, CIMMYT's website and publications boasted that for more than fifty years it had used science to "make a difference," defining this as helping "tens of millions of farmers grow more nutritious, resilient and productive maize and wheat cropping systems, using methods that nourish the environment and combat climate change."[82] But at its origins, farmers themselves were not the focus of CIMMYT, and instead the driving engine for the organization was crop research, specifically for increased food production, yield, and ensuring a global food supply. In the new context of climate change and renewed calls to again increase crop yields, CIMMYT's historical adaptability will be put to the test.

[82] Statement quoted from CIMMYT's official website in 2020; see current version at www.cimmyt.org.

Part II

Science as Development

5   Solving "Second-Generation Development Problems"

ICRISAT and the Management of Groundnuts, Farmers, and Markets in the 1970s

Lucas M. Mueller

In March 1968, William S. Gaud, director of the United States Agency for International Development (USAID) proclaimed that a new revolution had taken place. In Pakistan, India, Turkey, and the Philippines, farmers brought in "record yields, harvests of unprecedented size" of wheat and rice.[1] Gaud attributed these harvests to a series of international agricultural interventions – new seeds, fertilizers, new attitudes among farmers, and new policies – that he described with the term "Green Revolution." Gaud juxtaposed the Green Revolution with the "violent Red Revolutions like that of the Soviets," which US leaders wished to forestall, and in his reflections likened it to the industrial revolution of the nineteenth century. To him, the Green Revolution could be just "as significant and as beneficial to mankind" as its industrial counterpart. He concluded his speech with a call to "to accelerate it, to spread it, and to make it permanent." However, when agricultural experts and representatives from the countries and institutions that had sponsored the research and extension of the Green Revolution met the next year for the first in a series of seven conferences, they were neither unequivocal in their assessment of the Green Revolution and its aftermath nor united about the next steps to be taken.

These seven conferences were convened by the Rockefeller Foundation at its estate, the Villa Serbelloni, in Bellagio on the shores of Lake Como (Figure 5.1). The discussions among donors and experts cast a different light on the aftermath of the Green Revolution. Historians have emphasized the specter of social disruption that haunted the officials, experts, and institutions behind the Green Revolution already by December 1968, when reports

Acknowledgments: I would like to thank Helen Anne Curry and Timothy W. Lorek, as well as the anonymous reviewers for their feedback.

[1] William S. Gaud, "The Green Revolution: Accomplishments and Apprehensions" (Address, The Society for International Development, Washington, DC, March 8, 1968), www.agbioworld.org/biotech-info/topics/borlaug/borlaug-green.html.

Figure 5.1 View of Villa Serbelloni, part of the Rockefeller Foundation property in Bellagio, Italy, where administrators gathered for successive meetings that gave rise to CGIAR, undated. Rockefeller Archive Center, Rockefeller Foundation photographs, series CMNS-2. Courtesy of Rockefeller Archive Center.

about civil unrest in Pakistan and India reached the United States.[2] This narrative intends to highlight a profound irony of the Green Revolution. Although agricultural interventions had the goal of containing social unrest in the developing countries by filling peasants' hungry stomachs and thereby closing their ears to the siren song of communism in the global Cold War, their actual fallout was further unrest. In contrast to this reading, I show in this chapter that the experts of the Green Revolution were at least as excited about new opportunities for interventions arising in the social fallout of their supposed triumph as they were haunted by these same patterns. Now that the stomachs had been filled and expectations had been raised, "second-generation development problems" beckoned to be tackled.

At the conferences, which were called Bellagio I to VII, experts discussed these "second-generation" (or sometimes also "later-generation") problems and possible solutions. One of the results was the foundation of a new research institute, the International Crop Research Institute for the

---

[2] Nick Cullather, *The Hungry World: America's Cold War Battle against Poverty in Asia* (Cambridge, MA: Harvard University Press, 2010), pp. 239–40.

Semi-Arid Tropics (ICRISAT) in 1972. Modeled on the international agricultural research institutes CIMMYT (the International Maize and Wheat Improvement Center) in Mexico and IRRI (the International Rice Research Institute) in the Philippines, at which scientists had developed the key interventions of the preceding decades, ICRISAT was one of the first institutes founded under the umbrella of the Consultative Group on International Agricultural Research (CGIAR). By analyzing the Bellagio conferences, the foundation of ICRISAT, and experts' discussions about expanding ICRISAT's mandate to breeding groundnuts, I show how experts reimagined agricultural development through the concept of second-generation development problems in the late 1960s and early 1970s. This reimagination was part and parcel of a more fundamental shift in international development towards a neoliberal international order. The idea of development as a great project of postcolonial states and international organizations slowly faltered, replaced by an imagination of an international order that was a space of free trade and competition between nations.[3] I argue that this reimagination entailed not a fundamental remaking of interventions but rather an expansion of existing strategies that experts imbued with new meanings to suit the changing order.

This chapter thus links the literature on the history of the Green Revolution and the emerging literature that seeks to understand the historical process of the "economization" of policy since the late 1960s. Historian Nick Cullather has argued that the concept of the Green Revolution gave "an artificial coherence to two decades of fragmented and often conflicting efforts to improve agriculture in the non-Western world" and became "a template for future action, in other words, a model."[4] However, it was not an unambiguous model. In this period, the coordinates of policymaking began to shift. Domestically, in the United States, think tanks and industrial organizations began to adopt and promote an "economic style of reasoning." Elizabeth Popp Berman has charted the rise of this style of reasoning in US policymaking since the late 1960s. While purporting neutrality, it carried implicit values, such as competition, choice, and efficiency, that eventually replaced others, such as equality, in US public policy.[5] For the domain of population politics, which had played a central role in agricultural policy, Michelle Murphy has shown how the introduction of "practices that differentially value and govern life in terms of their ability to foster the macroeconomy of the

---

[3] Sandrine Kott, *Organiser le monde: Une autre histoire de la Guerre Froide* (Paris: Seuil, 2021), p. 211.
[4] Cullather, *The Hungry World*, pp. 7, 233.
[5] Elizabeth Popp Berman, *Thinking like an Economist* (Princeton, NJ: Princeton University Press, 2022).

nation-state" shifted "racist accounts of differential human evolution into an economic rather than hereditary biological register."[6] This shift was not limited to the United States. Discussing the United Nations, Sandrine Kott has argued that the view of the world as a space of free trade and competition between nations replaced postwar internationalists' hopes of creating institutions that could successfully regulate the world with all its contradictions and instabilities.[7] However, the interventions and policies crafted for this new world order were not necessarily as new. Amy Offner has shown for the Americas that policymakers often repurposed mid-century strategies of the developmentalist states.[8] The successive Bellagio meetings provide an insight into the repurposing of agricultural development strategies along the lines of these broader patterns.

This chapter describes a process of reimagining development in international agricultural policy in the late 1960s and 1970s. The first part of the chapter examines the discussions of second-generation development problems and their relation to the Green Revolution among the group of international agricultural experts gathered at the Bellagio meetings. While the experts agreed in the aftermath of the Green Revolution that the job was incomplete, they were divided about what remained to be done. Was there still a problem of population growth outpacing food production? Or were there new problems that the success of the Green Revolution precipitated? While these interpretations initially implied different strategies, either scientific and technical research or socioeconomic reforms, the experts eventually converged on the former strategy – without necessarily converging on the objectives it targeted. The second part of the chapter explores that agreed-upon strategy by focusing on ICRISAT and its research program on groundnuts (also known as peanuts). Drawing on the example of ICRISAT's groundnut research, I show how agricultural experts deployed the same scientific-technical strategy of the Green Revolution, namely, that of developing new crop varieties and agricultural technologies, to address two distinct agendas. Some considered groundnut research at ICRISAT as a means of expanding the Green Revolution to previously underserved regions and populations in the semi-arid tropics. Others considered it as expanding the Green Revolution into the next stage of development, specifically towards empowering farmers to contribute to development by selling groundnuts for export. This chapter thus traces the changes to the political meaning of the Green Revolution –

---

[6] Michelle Murphy, *The Economization of Life* (Durham, NC: Duke University Press, 2017), p. 4.

[7] Kott, *Organiser le monde*.

[8] Amy C. Offner, *Sorting out the Mixed Economy: The Rise and Fall of Welfare and Developmental States in the Americas* (Princeton, NJ: Princeton University Press, 2019).

and the groundnut – within the CGIAR system, which provided a powerful bridge between different modes of development.

The World's Sorrows in Bellagio

In 1969, the Rockefeller Foundation convened all major players in agricultural development for a retreat in Italy. The directors of the Ford Foundation, the United Nations Food and Agriculture Organization (FAO), United Nations Development Programme (UNDP), World Bank, Organisation for Economic Co-operation and Development (OECD), Asian Development Bank, Inter-American Development Bank, Economic Commission for Africa, and the directors of the development agencies of France, Sweden, Japan, Canada, the United States, and the United Kingdom, as well as the Rockefeller Foundation's expert consultants, gathered in the Villa Serbelloni at Lake Como for a meeting that would become known as Bellagio I. The meeting was opened on April 23 by Will Myers, vice president of the Rockefeller Foundation. He invited the participants to deliberate on the "needs, potentialities, and priorities of programs designed to sustain and to expand the agricultural revolution."[9] This quotation – taken from published proceedings that were created after the end of the informal, off-the-record meeting at the request of the agency heads – reflected the participants' confidence in the agricultural interventions of the preceding decades. Over the next three days, they discussed the programs that had generated high-yielding varieties of wheat and rice, the technologies of intensified agriculture, and the capital flows and income transfers that surrounded the celebrated agricultural revolution. However, the participants were also confronted with a new set of problems that had emerged in the wake of this revolution.

Lowell S. Hardin, one of the expert consultants who attended the meeting, introduced these "later-generation development problems." Hardin was an agricultural economist on the faculty at Purdue University and a program officer of the Ford Foundation. He had participated in a science advisory committee to the US president, Lyndon B. Johnson, on the world food supply where, among other contributions, he had chaired a panel on "projected trends of trade in agricultural products."[10] More specific to CGIAR, Hardin co-authored the 1966 report on food production in the global tropics that led to the

---

[9] Rockefeller Foundation, "Agricultural Development: Proceedings of a Conference Sponsored by the Rockefeller Foundation, Bellagio, Italy, April 23–25, 1969 (Bellagio I)," [1969?], v, https://cgspace.cgiar.org/handle/10947/153.

[10] Panel on the World Food Supply, "The World Food Problem: A Report of the President's Science Advisory Committee," May 1967, vii, ix.

establishment of the International Center for Tropical Agriculture (CIAT), one of the four founding research centers that pre-dated CGIAR.[11] Hardin was thus a key architect of CGIAR behind the scenes, and his work reflects a steady and significant place for economic assessment within CGIAR institutional planning and development. At Bellagio I, Hardin explained that second-generation development problems centered on "those public and private decisions and actions necessary to promote continued economic growth – to achieve or maintain rates of output increase that appear to be within reach once major food deficits are reduced."[12] In other words, once food deficits were reduced – that is, the primary goal of the initial interventions of the Green Revolution had been achieved – a different set of policies would be necessary to continue agricultural and economic momentum. Hardin considered that achieving the first goal of reducing food shortages depended on developing science and technology to redefine "physical production limitations." However, the solutions for second-generation problems were not to be found in science and technology but instead in "resource allocation, marketing, international trade, diversification, distribution, and institutional matters."[13] In short, these were solutions that would require extensive and profound socioeconomic reforms – interventions that were the area of expertise of the (agricultural) economist.

The concept of second-generation development problems was discussed not just by Hardin and not just within the closed doors of Bellagio. One of the most prominent exponents of the Green Revolution, the plant pathologist and wheat breeder Norman Borlaug, who would be awarded the Nobel Peace Prize in 1970 "for having given a well-founded hope – the green revolution" and who was not at Bellagio I, adopted the term in a 1969 article to chide political leaders and economic planners for being ill-prepared to deal with these second-generation problems.[14] For Borlaug and his co-authors, the Green Revolution had closed the gap in food production and consumption, but it had also "injected a new rhythm of business activity into the formerly stagnant economies of these countries."[15] In addition to spending their income on agricultural inputs necessary to grow the new crops, farmers purchased consumer items, becoming active participants in an

---

[11] L. M. Roberts and Lowell S. Hardin, "A Proposal for Creating an International Institute for Agricultural Research and Training to Serve the Lowland Tropical Regions of the Americas," October 1966, https://hdl.handle.net/10568/72329.
[12] Rockefeller Foundation, "Agricultural Development," 44.   [13] Ibid.
[14] "Norman Borlaug – Facts," NobelPrize.org, 2023, www.nobelprize.org/prizes/peace/1970/borlaug/facts.
[15] Norman E. Borlaug et al., "A Green Revolution Yields a Golden Harvest," *Columbia Journal of World Business* 4, no. 5 (1969): 10.

emergent consumer economy. Borlaug warned that a potential source of unrest would emerge if people were denied participation in this economy, that is, if the Green Revolution was not maintained and expanded. (In contrast, they attributed the existing unrest of the late 1960s to students and labor leaders who were far removed from farmers.) Such a consumer economy, modeled on the contemporary United States, was seen to represent the highest stage of development in modernization theory, which guided much of postwar US development policy.[16] Hardin also shared this consumerist vision of the last stage of development, which consisted of "effectively widening the range of choice available to larger and larger numbers of people."[17]

At Bellagio I, Hardin and other participants did discuss the wider socioeconomic implications of the Green Revolution. Hardin emphasized that "technical production advances ... do have differential impacts," and that the unrest of people "left behind" could threaten political stability.[18] This prompted Hardin to ask whether "development assistance be limited essentially to the scientific-technological problems" and to propose a social-science think tank that could serve as a resource for individual sovereign nations to draw on in designing and implementing their own policies.[19] In *Foreign Affairs* in 1969, economist Clifton Wharton also considered the question of later-generation problems. He described how, in the wake of the Green Revolution, people migrated from rural areas to cities only to find employment opportunities in industry lacking, and observed that there were neither markets nor the infrastructure, such as storage units, to sell off excess harvests. However, Wharton saw also an opportunity in these developments, arguing "that the list of second-generation problems is a measure of what great opportunities exist for breaking the centuries-old chains of peasant poverty."[20]

Economists in Latin America, pan-African historians, and representatives from the "Third World" discussed proposals for radical reform to address persistent poverty and global inequality in the 1960s and 1970s. Economists in Chile developed different, partly conflicting versions of what would become known as dependency theory to understand the

---

[16] Nils Gilman, "Modernization Theory, the Highest Stage of American Intellectual History," in David C. Engerman et al., eds., *Staging Growth: Modernization, Development, and the Global Cold War* (Amherst: University of Massachusetts Press, 2003).
[17] Rockefeller Foundation, "Agricultural Development," 44.   [18] Ibid., 46.
[19] Ibid., 48.
[20] Clifton R. Wharton, "The Green Revolution: Cornucopia or Pandora's Box?," *Foreign Affairs* 47, no. 3 (1969): 464–476, at 475.

drivers of global inequality and divergent development.[21] Historian Walter Rodney built on such insights to understand "how Europe underdeveloped Africa" in his eponymous book.[22] Meanwhile, the countries of the "Third World" dominated the meetings of the United Nations Conference on Trade and Development (UNCTAD) and ultimately demanded a "New International Economic Order" that would include profound reforms of price stability and market access in international trade.[23] These were some of the issues that the participants in Bellagio I considered. However, as I will describe, their solutions were far from the radical reforms that the "Third World" proposed.

While the participants at Bellagio I discussed new, later-generation development problems, some cautioned against overemphasizing the excess production in specific local areas and thereby overlooking the vast and persistent deficiencies in available food supplies elsewhere. In fact, the specter of overpopulation had not disappeared. Myers warned that the increased harvests brought only temporary relief. By the end of the twentieth century, the world could be again "engulfed in a sea of famine," unless massive strides in the productivity and efficiency of their agricultural sector were made.[24] This echoed Borlaug, who also warned in 1969 that "the unrelenting increase in human numbers, with no relief in sight, continues to be the greatest unsolved multifaceted problem confronting mankind in its quest for a better standard of living for the world's masses."[25] In his concluding summary to Bellagio I, Myers emphasized the "vastly superior technologies of production" that were a "pervasive force in disrupting traditional agriculture and paving the way to its modernization and to great increases in agricultural production."[26]

At Bellagio I and beyond, the participants wavered between embarking on grand projects of economic development through agricultural exports and keeping the focus on extending the Green Revolution to "feed the world." Adekke Boerma, head of FAO, articulated the former spirit by stating that "in the development drama, agriculture is suddenly promoted from the neglected stepchild to the deus ex machina."[27] Agriculture – and agricultural research in particular – was not only the solution to overcoming a hungry world: it was now also envisioned as a potential driver of

---

[21] María Margarita Fajardo Hernández, *The World That Latin America Created: The United Nations Economic Commission for Latin America in the Development Era*, (Cambridge, Massachusetts: Harvard University Press, 2022).

[22] Walter Rodney, *How Europe Underdeveloped Africa* (1972, reprint London: Verso, 2018).

[23] Nils Gilman, "The New International Economic Order: A Reintroduction," *Humanity* 6, no. 1 (2015): 1–16.

[24] Rockefeller Foundation, "Agricultural Development," v.

[25] Borlaug et al., "A Green Revolution Yields a Golden Harvest," 19.

[26] Rockefeller Foundation, "Agricultural Development," 70.   [27] Ibid., 9.

economic development and growth. Elsewhere, Borlaug considered the two possibilities, the export potential of excess wheat production in Pakistan and the potential of growing additional crops during the winter season, such as oilseeds, pulses, and legumes, which could fill other nutritional needs. However, Borlaug cautioned that "little pertinent technology is available either within or outside Pakistan to increase yields of these winter pulses."[28] These kinds of crops would inspire the imagination of both the experts who were seeking to transform the Green Revolution into a driver of economic development and the experts who wanted to expand the Green Revolution to new frontiers of the hungry world.

The Crops and Centers of Later-Generation Development Problems

In early February 1970, the foundations convened a second meeting, Bellagio II, again on the shores of Lake Como, to discuss the next steps for agricultural development more concretely. Unlike Bellagio I, the attendees of this meeting were lower-level staff of development agencies. The participants were as excited as their predecessors about the vitality of the agricultural sector in many developing countries that would now reach traditional, even subsistence farms. They tabulated agricultural research needs, producing "a rough ranking of the adequacy of the technical knowledge available upon which to found the acceleration of agricultural modernization."[29] This exercise yielded the observation that "production technologies suited to harsher agricultural environments so that many more cultivators may participate in the harvest of development" were needed.[30] In short, the crops and areas that they considered in need of more "research-generated, superior technology" corresponded to the places where the fruits of the Green Revolution had not spread and discontent might threaten social stability. The openings that Bellagio I afforded – the discussions about agricultural development in a broader frame of global trade, prices, and markets – had already closed in Bellagio II, when participants centered on technical strategies to expand the Green Revolution to new regions and groups.

When the heads of assistance agencies met a few months later, in early April 1970, for Bellagio III, they discussed which new institutions could

---

[28] Borlaug et al., "A Green Revolution Yields a Golden Harvest," 16.
[29] Ford Foundation, "Accelerating Agricultural Modernization in Developing Nations: A Summary of Findings and Suggestions from Agriculturists from Development Assistance Agencies, Villa Serbelloni, Bellagio, Italy, February 3–6, 1970 (Bellagio II)," March 1970, 12, https://cgspace.cgiar.org/handle/10947/89.
[30] Ibid., 3.

be founded to advance research and development along these lines. One of the proposed institutions would be a "dry-land farming institute with concentration on sorghum and millets, and certain pulses (chickpeas, pigeon peas?)."[31] This proposal combined different needs that the participants of the previous meeting ranked highly, even as it registered some uncertainty about the specifics. The institute would expand the Green Revolution to new populations in the "drylands" by researching understudied legumes that would improve protein nutrition.

Drylands, uplands, or (semi-)arid regions and the populations that inhabited these climatic zones had long been a focus of colonial and postcolonial interventions (see also Courtney Fullilove, Chapter 1, this volume).[32] The Bellagio attendees considered that a new institute for unirrigated farming in drier regions should be situated in Asia, where the population pressure had seemed most urgent over the previous decades. Aid agency heads emphasized that the foundations would have to take the lead in ensuring that any new institute was well managed, reflecting the idea that agricultural research should take place in international institutions accountable to donors and not subject to national needs and desires. However, this did not necessarily reflect the realities on the ground, as Prakash Kumar (Chapter 2, this volume) shows: a new institute ultimately described as serving "semi-arid" regions was very much shaped by India's domestic and foreign policy priorities.

In addition to a proposal for "upland" crops, which were grown without access to wet or irrigated land, the participants of Bellagio III commissioned a report for research on food legumes. The agronomist and long-time Rockefeller Foundation employee Lewis M. Roberts wrote this report. He made the case for legumes based on the distinction between having not enough to eat, which the first period of the Green Revolution had addressed, and a lack of "vital nutritive elements," particularly protein to "produce sound growth and reasonable good health."[33] He wrote that "there is a growing awareness that the protein deficit problem is one of the most critical, complex aspects of the total food problem."[34] He thus

---

[31] Sterling Wortman, "Conference of Heads of Assistance Agencies, April 8–9, 1970," May 14, 1970, 2, https://hdl.handle.net/10947/415.

[32] Diana K. Davis, *Resurrecting the Granary of Rome: Environmental History and French Colonial Expansion in North Africa,* (Athens: Ohio University Press, 2007); Diana K. Davis, *The Arid Lands: History, Power, Knowledge:* (Cambridge, MA: MIT Press, 2016); Philipp Lehmann, *Desert Edens: Colonial Climate Engineering in the Age of Anxiety* (Princeton, NJ: Princeton University Press, 2022).

[33] Lewis M. Roberts, "The Food Legumes," November 1970, 130, https://hdl.handle.net/10947/1528.

[34] Ibid., 131.

recommended the "expansion and acceleration of research to increase production of certain of these high-protein crops." The focus on protein reflected a changing perception of malnutrition since the 1940s, when researchers in Africa found that malnourished children did not necessarily suffer only from a lack of calorie-rich food but also from a lack of protein.[35] Nutritional scientists in the United Kingdom, India, and elsewhere in academic, international, and industrial research institutes began to search for new sources of protein in plants and animals and for chemical processes that would synthesize protein. By the late 1960s, researchers and policymakers feared a full-blown global protein crisis. In 1968, the United Nations issued a report for "International Action to Avert the Impending Protein Crisis," and in 1971, the General Assembly adopted a resolution to address the problem.[36] International organizations such as the United Nations Children's Fund (UNICEF), FAO, and the World Health Organization (WHO) had worked hard to involve major food companies, including Unilever, Nestlé, and the Tata Group, in the research and marketing of protein rich foods.[37] This endeavor carried the promise of filling the protein gap and also creating new business opportunities in the developing world. This configuration of business and international programs around the promotion of infant formula over breastfeeding, which was part of this endeavor, would soon come under fire, ultimately giving rise to a consumer-based activism to challenge the global activities of Nestlé and other multinational companies.[38] At an international level, the protein question became a crucial arena of struggle over the moral and economic limitations of a market-based international order.

For the Bellagio meetings, Lewis Roberts considered different approaches for how international agricultural research could contribute to increasing availability of affordable protein, because animal or

---

[35] Jennifer Tappan, *The Riddle of Malnutrition: The Long Arc of Biomedical and Public Health Interventions in Uganda* (Athens: Ohio University Press, 2017).

[36] "International Action to Avert the Impending Protein Crisis," Economic and Social Council of the Advisory Committee on the Application of Science and Technology to Development (New York: United Nations, 1968); UN General Assembly, Resolution 2848 (26th Session), Protein Resources, A/RES/2848(XXVI), December 20, 1971, https://digitallibrary.un.org/record/192109.

[37] See further discussion of protein-deficiency concerns in Wilson Picado-Umaña, Chapter 8, this volume. Lucas M. Mueller, "Risk on the Negotiation Table: Malnutrition, Toxicity, and Postcolonial Development," in Angela N. H. Creager and Jean-Paul Gaudillière, eds., *Risk on the Table: Food Production, Health, and the Environment* (New York: Berghahn, 2021).

[38] Tehila Sasson, "Milking the Third World? Humanitarianism, Capitalism, and the Moral Economy of the Nestlé Boycott," *The American Historical Review* 121, no. 4 (October 3, 2016): 1196–1224.

synthetic proteins were too pricy for poor subsistence farmers and city slum dwellers in the developing countries. Groundnuts, which were produced in West and East African countries, were one of the possible "cheap" sources of protein. Roberts proposed to assign groundnuts to the existing International Institute of Tropical Agriculture (IITA) in Nigeria, where groundnuts were the most important export crop. Roberts emphasized the importance of breeding groundnuts and other legumes for improved quality, including quantitatively and qualitatively improved protein content, different amino acids, and the absence or reduced content of anti-metabolites and toxic factors. Such a research program would require widening the "narrow genetic base" of the food legumes through the collection of germplasm from cultivated and wild variants around the globe. The idea was to find inheritable traits that could be introduced to cultivated varieties, thereby producing food crops with the desired qualities. Roberts considered the timeframe of the project to be at least fifteen years. He thus emphasized that his recommendations should be accepted "only if the potential international supporting agencies are firmly committed to provide the financial backing that will be needed for a minimum period of 15 years."[39]

Roberts' proposal was discussed at the next meeting, Bellagio IV, this time held in New York, in December 1970.[40] At the same time, the International Bank for Reconstruction and Development (IBRD), UNDP, and FAO had initiated steps to bring together several existing and proposed agricultural research institutes under the umbrella of a new organization, CGIAR. Its constitutive meeting would take place just a month later, in mid January 1971, increasing the pressure to define the scope of CGIAR, its new institutes, and their research programs. The attendees of Bellagio IV thought that the proposed institute for upland crops would address sorghum and millet, which were considered staples for rural people in drier regions. While the institute would be established in Asia, it was to coordinate with the ongoing research on these crops in Africa.

The proposed upland crop institute would also accomplish some of the research on food legumes that Roberts had advocated. In spring 1971, a technical review panel of CGIAR, which included high-level members from the World Bank and foundations as well as lower-level participants from donor countries, met to discuss proposals on legume research, stating that "great benefits in nutrition would result from increased consumption

---

[39] Roberts, "The Food Legumes," 154.
[40] Nathan M. Koffsky, "Summary of Conference of Heads of Assistance Agencies, New York, December 3–4, 1970 (Bellagio IV)," https://hdl.handle.net/10947/1335.

of these crops. They are highly diverse and complex."[41] The participants pondered which institutions should study which legumes and proposed the following scheme: dry beans at the established CIAT, cowpeas at IITA, pigeon peas at the proposed "Upland" or IITA, chickpeas at "Upland" or CIMMYT, soybeans at CIAT or IITA, and groundnuts at IITA or African research organizations. At this meeting, the panel members were in consensus that research on soybeans and groundnuts was a low priority, because so much research was already being conducted on these species worldwide and because these were used and sold as cash crops, which were primarily exported. The study of legumes for nutrition had priority.

In October 1971, the Technical Advisory Committee (TAC) of CGIAR met for the first time to advise the newly formed CGIAR on the research program for its institutes. John Crawford, an economist and public servant from Australia, was the chairman. In an opening statement, Boerma of FAO echoed the discussions of the Bellagio conferences, highlighting the need to expand the promise of the Green Revolution to other regions of the world. R. D. Demuth, an observer from IBRD, added to this the need to apply the Green Revolution model to other crops and to livestock. IBRD considered research on food legumes as high-protein food sources, as well as research on rainfed crops, high-priority areas. The UNDP representative similarly emphasized the importance of edible proteins. Demuth also foregrounded the role of the TAC in advising CGIAR on priority areas for research and appropriate methodologies. He urged TAC members to make recommendations as soon as possible for financing in 1972.[42]

International research, according to the chairman, Crawford, was defined as: "research which, while located in a specific country, was of wider concern regionally and globally, independent of national interest or control, and free from political dictates of any one Government whilst retaining appropriate links with national research systems to ensure necessary testing of results and feed-back both of results and needs."[43] However, what was of wider concern was defined by international donors and foundations. The technical review committee members also considered French, British, and US research programs, finding that the regionalization within specific nation-states had been problematic. With this meeting, discussions about the research program of the institutes shifted to the TAC.

---

[41] CGIAR Technical Advisory Committee, "First Meeting of the Technical Advisory Committee, 29 June–2 July 1971: Summary Record," November 5, 1971, 5, https://hdl.handle.net/10947/1422.
[42] Ibid.   [43] Ibid., 3.

The TAC pursued the proposal for an upland crops institute that was put forth at Bellagio II and endorsed at Bellagio IV. Ralph W. Cummings of the Ford Foundation conducted a feasibility study. Hugh Doggett from the British Overseas Development Administration, John Comeau from the Canadian International Development Research Centre (IDRC), and L. Gauger of the Centre de Recherche Agronomique du Bambey, Senegal joined Cummings on field trips to determine the scope of the new institute. Their proposal was submitted on October 19, 1971 and called for a world center, ideally located in India, for the improvement of sorghum, millet, pigeon peas, chickpeas, and possibly additional crops such as groundnuts, and for the development of cropping patterns and farming in "the low rain fall, unirrigated, semi-arid tropics." The proposal followed the patterns and principles that had been developed with IRRI since 1960 and applied them to new areas. This included multidisciplinary research teams with links to regional programs, and an international board of "agricultural and scientific leaders" of the host country and other countries whose climatic and agricultural features fell into the domain of the institute.[44]

The institute was framed as an international institute whose "senior scientific staff should be drawn from among the best scientific talent available on an international basis," as the report stated.[45] The new institute was thus conceived as a domain with many diplomatic privileges. This included guarantees by the Indian government that people, scientific staff, and plants, especially seeds, were allowed to circulate in and out of the country as CGIAR needed. "Reasonable quarantine control" to avoid the introduction or export of pests and diseases was permitted but ideally through a quarantine unit directly associated with the institute. This legal framework would facilitate establishing an extensive germplasm collection with genetic material from around the globe in order to alleviate the problem of a narrow genetic base and breed crops with higher yields in greater quality.[46] Such collections would be pursued for groundnuts, as

[44] Ralph W. Cummings, L. Sauger, and Hugh Doggett, "Proposal for an International Crops Research Institute for the Semi-Arid Tropics (ICRISAT)," October 19, 1971, https://hdl.handle.net/10947/930.

[45] Ibid. Prakash Kumar (Chapter 2, this volume) highlights the political importance in India of this emphasis on the international nature of the institution.

[46] Helen Anne Curry, "From Working Collections to the World Germplasm Project: Agricultural Modernization and Genetic Conservation at the Rockefeller Foundation," *History and Philosophy of the Life Sciences* 39, no. 2 (2017): 5; Helen Anne Curry, *Endangered Maize: Industrial Agriculture and the Crisis of Extinction* (Oakland: University of California Press, 2022); Marianna Fenzi and Christophe Bonneuil, "From 'Genetic Resources' to 'Ecosystems Services': A Century of Science and Global Policies for Crop Diversity Conservation," *Culture, Agriculture, Food and Environment* 38, no. 2 (2016): 72–83.

they were for many crops in the CGIAR system (see Marianna Fenzi, Chapter 11, this volume). ICRISAT was established in 1972, and its funding structure was based on the new multilateral model. Rich nation-states, such as Australia, Belgium, Canada, the Federal Republic of Germany, the Netherlands, Norway, Saudi Arabia, Sweden, Switzerland, the United Kingdom, and the United States of America, contributed to ICRISAT's budget, as did the US foundations and also such international organizations as the European Economic Community (EEC), UNDP, the Asian Development Bank, and the World Bank.[47] ICRISAT thus was initially conceived as an expansion of the Green Revolution to new regions: the semi-arid, rainfed tropics and its populations.

Peanut Politics, or "Later-Generation Development Problems" in a Nutshell

Initially, ICRISAT focused on food crops of the semi-arid tropics. In 1973, the TAC charged a taskforce to develop a proposal for research on groundnuts. Adding groundnuts to CGIAR's research portfolio represented a departure from previous research endeavors. The peanut researchers A. H. Bunting, W. C. Gregory, J. C. Mauboussin, and J. G. Ryan were appointed to run the taskforce. Bunting, who held a faculty position in agricultural development at the University of Reading, UK, had worked on groundnuts in Tanganyika, Sudan, Nigeria, and other African colonies and countries. Mauboussin was from the Office for Overseas Scientific and Technological Research (ORSTOM), the French foreign-research organization, and had worked in Senegal, and James Ryan from Australia was an economist at ICRISAT. Walton Gregory from North Carolina State University was a peanut breeder and had collected wild forms in South America. The four men met in Hyderabad on March 20, 1974 and published their report later that year with the following conclusion:

[G]roundnut research at national stations in most countries (even in the United States) is not sufficiently extensive, penetrating, continuous or coordinated to allow progress at the rate which development programmes require. It would benefit very considerably from international cooperation, exchange of information, and training, and from the research in depth, and in new directions, which an international programme would provide. This is particularly the case in respect of genetic resources. As we explain later in this report, many thousands of cultivated varieties, and a remarkable wealth of wild species, offer prospects for genetic

---

[47] 10-AGD-377, "ICRISAT," Vol. III, FAO Archives.

improvement (including the control of some of the most important diseases) which can only be realised through the resources, scale of work, concentration in depth, continuity, and world-wide linkages of an international programme.[48]

Thus, they strongly recommended that groundnut research should be done at the international level. This scientific reasoning – especially the need for a collection of genetic material – justified international groundnut research, even though others considered that groundnuts were an export crop and thus outside the domain of international agricultural research. However, the authors countered this concern with the observation that "only by selling crops can farmers help to feed the nations as a whole." As they argued,

The possible counterargument that it [groundnut] is also an industrial and export crop, so that research for it should, therefore, in the first place be conducted by industry in cooperation with national governments seems to us to fail because there is, in fact, no such research (except in those parts of West Africa associated with France) and we know of no prospect of any. Moreover, by earning foreign exchange, groundnuts can help food production indirectly.[49]

Even though food production was the primary objective of CGIAR, groundnut research was still doable under this mandate, because it would indirectly lead to development through the acquisition of foreign exchange.

In their report the peanut experts also described the utility of groundnuts for nourishing developing nations. A kernel contained about 50 percent oil and 25 percent protein. The oil was used for food and cooking, as well as in the industrial production of margarine and soap. The protein could be used directly in human diets or for livestock projects that were also considered by CGIAR (see Rebekah Thompson and James Smith, Chapter 7, this volume). The protein-rich constituent of press-cake was an important component of feed for animals in Europe. Given this array of uses, including abroad and in industrial production, groundnuts were potentially "important contributions to the foreign exchange earnings of the semi-arid countries, which are so necessary to pay for the equipment and purchased inputs needed to expand food and other farm production."[50] For example, Senegal earned 50 percent of its foreign exchange in groundnuts, Nigeria 12 percent, and Sudan 8 percent. In short, the justification for the importance of groundnuts was primarily economic exchange, rather than food to feed the nations.

---

[48] A. H. Bunting, W. C. Gregory, J. C. Mauboussin, and J. G. Ryan, "A Proposal for Research on Groundnuts (Arachis) by the International Crop Research Institute for the Semi-Arid Tropics," ICRISAT, March 1974, 7, http://hdl.handle.net/10947/73.
[49] Ibid.   [50] Ibid., 4.

For the taskforce, the small yields in Asia and Africa in contrast to the United States were the central problem of groundnut agriculture. US farmers yielded 2,200 kilograms of groundnuts per hectare, while farmers in Asia harvested 830–840 kilograms per hectare and those in Africa 725. The United States remained the global standard for agricultural production, and its yields seemed to suggest that gains were possible elsewhere. The taskforce proposed several areas of research to close this gap, including the study of germplasm, protection against pests, viruses, fungal infections, improved production methods, and post-harvest technologies, as well as the creation of economic and social information about groundnuts. They primarily proposed to establish a world collection and register of wild and cultivated varieties and forms of *Arachis* – groundnuts – drawn from existing collections in India, the United States (especially the one at North Carolina State University), and elsewhere. Wild forms were considered particularly valuable for breeding varieties that were resistant to fungal diseases such as aflatoxin, and others that could only be controlled by costly and cumbersome procedures out of the reach of most small farmers. Ultimately, the proposal maintained that ICRISAT's focus should be on genetic studies with a duplication of existing collections starting in 1974 and sowing of known varieties in 1975 or 1976. These instructions were followed, and groundnut breeding started in 1976 with a focus on high yield, stability of yield, and resistance to disease and drought[51] (Figure 5.2).

ICRISAT was, however, not the only international institution concerned with groundnuts, which had gained a double importance – filling the protein gap and providing foreign exchange earnings.[52] In 1977, UNCTAD, which had been at the center of the efforts for the New International Economic Order, adopted resolution 93 (IV), an integrated program for commodities, including vegetable oils and oilseeds.[53] UNCTAD emphasized the political and economic international arrangements on vegetable oils and oilseeds, including:

improving the stability of the trade and income of individual developing countries; improving access to markets and the reliability of supplies; the diversification of production and expansion of processing in developing countries; improving the competitiveness of natural products; and improving market structures and the

---

[51] Research Projects, ICRISAT Files 1978, Governing Board, ICRISAT.
[52] Interdivisional Working Group, "Closing the Protein Gap," July 16, 1968, 12-ESN-516, FAO Archives, Rome.
[53] "Elements of Possible International Arrangements on Vegetable Oils and Oilseeds: Report Prepared Jointly by the UNCTAD and FAO Secretariats," United Nations, Geneva, June 3, 1977, https://digitallibrary.un.org/record/1639354.

Figure 5.2 Day laborers work in an experimental peanut field at ICRISAT's Hyderabad campus, 2016. Photo by Lucas M. Mueller.

marketing, distribution and transport systems for exports of raw materials and commodities from developing countries.[54]

This emphasis differed from ICRISAT's focus on producing groundnut varieties with specific characteristics. The UNCTAD report described the political economy of groundnuts much more extensively: "a substantial part of the total production of oilseeds enters world trade either as seed or in the form of oil and meal, exported by a large number of developing countries and some developed countries (especially the United States) partly to other developing countries but chiefly to western Europe and Japan."[55] The complexity of the trade, the differences between oilseeds, and the competition with synthetics and other agricultural products made it a tricky issue. Ultimately, however, the conference would only propose more research programs instead of addressing the political economic problems that were at the core of groundnut agriculture and international agricultural trade.

In subsequent years, groundnut research at ICRISAT continued to focus primarily on the technical dimensions of varieties. ICRISAT's

[54] Ibid., 17.   [55] Ibid., 4.

groundnut germplasm collection grew to include 11,641 accessions of cultivars and 115 in quarantine clearance, and new groundnut programs in Africa were established in the early 1980s. A decade after the beginning of its groundnut efforts, in May 1987, the legumes program at ICRISAT began to publish the *International Arachis Newsletter* in collaboration with the "Peanut Collaborative Research Support Program" headquartered in Georgia in the United States. The legumes program had been formed in 1986 by merging the ICRISAT pulses program (chickpea and pigeon pea) and the groundnut program. Its links with the Peanut Collaborative Research Support Program were many, showing the continued importance of the United States in this domain of international agricultural research.[56] The editors of *International Arachis Newsletter* introduced the key problems of groundnut research and the factors constraining yields, including diseases and pests, unreliable rainfall in the semi-arid tropics, recurring droughts, the lack of high-yielding adapted cultivars, poor agronomic practices, and the very limited use of fertilizers.[57] The biannual newsletters, whose title page signaled the global reach of ICRISAT's groundnut research (Figure 5.3), featured content with "current-awareness value to peer scientists" and were selected for news interest as well as scientific relevance. ICRISAT thus continued its focus on technical aspects of groundnut agriculture to scientifically address the problem of expanding the Green Revolution to new regions – and of making agriculture a driver of economic development more generally.

Conclusion

ICRISAT's history, and by extension the early history of CGIAR, provides insight not only into the expert discussions held in the aftermath of the self-proclaimed Green Revolution but also into broader changes of international development politics in the late 1960s and early 1970s. My account of the founding of ICRISAT and its unique groundnut research program suggests that there was no clear rupture between the tools and strategies of international research for agricultural development between those of the 1960s and those of the 1980s but instead a reworking of existing approaches and meanings towards ones that not only considered nutritional needs but also economized crop production with a view

---

[56] For example, R. W. Gibbons, the former leader of the groundnut program and now the head of the ICRISAT Sahelian center, sat on the board of the US-based Peanut Collaborative.

[57] These newsletters can be found at the Open Access Repository of the ICRISAT library: http://oar.icrisat.org.

Figure 5.3 The first issue of the *International Arachis Newsletter*, published in May 1987. The map on the cover identifies the main ICRISAT campus in Hyderabad and other ICRISAT locations as well as the hub of the USAID-funded Peanut Collaborative Research Program in Georgia and its international collaborators. By permission of ICRISAT.

towards global trade and markets. However, even as experts acknowledged the importance of markets, by continually emphasizing the need for research they precluded reforms of international agricultural markets and changes to the global economic order. They instead attempted to solve "second-generation development problems" through interventions grounded in scientific research and technical development that had become subsumed under the label of the Green Revolution.

6   Breeding Environments
WARDA and the Pursuit of Rice Productivity in West Africa

Harro Maat

In 2004 the World Food Prize was awarded to the rice breeders Yuan Longping from China and Monty Jones from Sierra Leone.[1] Yuan was lauded for applying the heterosis effect to rice, creating hybrid rice varieties that were widely grown in China from the mid 1970s. Jones, working for the West Africa Rice Development Association (WARDA), received the prize for breeding rice varieties from crossing African rice (*O. glaberrima*) and Asian rice (*O. sativa*). The interspecific varieties were considered a breakthrough for rice cultivation in Africa and therefore named New Rices for Africa, shortened to NERICA (Figure 6.1). From the moment NERICA varieties were released, they were heavily promoted as the best option for African rice farmers to increase their yields and were hailed as a marked success of WARDA. In the early 2000s the distribution of NERICA lines had just begun. The uptake by farmers and the effects on rice production in the various rice-growing regions of Africa were largely unclear. By the end of the decade, when more reports on NERICA's performance had appeared, it turned out that results were mixed at best.[2]

The excitement over NERICA and the combined award for Yuan Longping and Monty Jones seem emblematic of the history of WARDA since its inception in 1970. Granting the World Food Prize

Acknowledgments: My interest in the history of WARDA germinated more than ten years ago when I participated in a project on local knowledge of rice varieties in four West African countries. A big thanks to the researchers of that project, Alfred Mokuwa, Florent Okry, Béla Teeken, and Edwin Nuijten, for all the stimulating conversations. And I am very grateful for helpful comments on earlier versions of the chapter from Jonathan Harwood, Dominic Glover, Paul Richards, Yi-Tang Lin, and reviewers and editors of this volume.

[1] World Food Prize Foundation, "2004: Jones and Yuan," www.worldfoodprize.org/en/la ureates/20002009_laureates/2004_jones_and_yuan.

[2] A. Diagne, S. K. G. Midingoyi, and F. M. Kinkingninhoun-Medagbe, "Impact of NERICA Adoption on Rice Yield: Evidence from West Africa," in Keijiro Otsuka and Donald F. Larson, eds., *An African Green Revolution: Finding Ways to Boost Productivity on Small Farms* (Dordrecht: Springer, 2013), pp. 143–163.

Figure 6.1 A New Rices for Africa (NERICA) variety intended for use in lowland ecologies, one of several such varieties developed at AfricaRice in the 2010s. Photo by R. Raman, AfricaRice and reprinted by permission of AfricaRice.

to Yuan recognized his creation of hybrid rice and its contribution to the growth of rice production in China. The application of hybrid vigor or heterosis effect in rice requires a labor-intensive breeding and multiplication method. The technique results in F1 hybrid seeds that need replacement each year. The main advantage is that hybrids perform well with limited additional fertilizer. These features anticipated the limited production capacity for chemical fertilizer in China in the early 1970s, as well as the wide-ranging agricultural research and extension system embedded in rural communes.[3] The NERICA varieties were also a technical achievement in that they were based on crossbreeding two species, *glaberrima*, a rice species native to West Africa, and the *sativa* or Asian rice species. The main complicating factor for this breeding strategy is overcoming high sterility levels in the offspring, which was achieved by back-crossing interspecific lines with *sativa* lines. Linkages with the many rice farmers in West Africa, however, were poorly developed. The new varieties were tested at the WARDA farm, a set of experimental plots near the research station, rather than on actual farms in the region.

---

[3] Sigrid Schmalzer, *Red Revolution, Green Revolution: Scientific Farming in Socialist China* (Chicago: University of Chicago Press, 2016).

The work by Jones and his team thus was a technical achievement without the kind of effects on rice production in Africa that had been seen with hybrid rice in China. The criticism WARDA received for the triumphant claims over NERICA was acknowledged in later years and taken as an incentive for further testing of NERICA varieties in different African countries.[4] The fact that such further testing happened after the launch of the NERICA lines, and not before, suggests that the technical challenge of achieving an interspecific hybrid was prioritized over questions about what kinds of rice varieties were needed and how these were best distributed to African rice farmers. Moreover, the *sativa* varieties selected for backcrossing made the NERICA varieties fertilizer-responsive, like Asian improved varieties. As various studies have pointed out, agricultural improvements in Africa are typically framed as an African version of the Green Revolution in Asia, a framing in which technical similarities are considered capable of overcoming ecological, social, and economic differences.[5] Such framings turned the NERICA lines into evidence that WARDA was a rice-breeding institute comparable to the International Rice Research Institute (IRRI), famed for its contribution of the "miracle rice" IR8 to the Green Revolution in Asia (Figure 6.2), and that similar effects on rice production would follow from WARDA's breeding program.

The question of whether and how WARDA's trajectory compared with that of IRRI is indeed central in most historical accounts of the organization, which was renamed the Africa Rice Center in 2009 after widening its scope and membership to other African countries.[6] A key feature of IRRI is that it operated as a centralized research institute, concentrating scientific and technical expertise at a single research location. A major assumption of the centralized model was the isolation of research and plant-breeding techniques from diverse and locally specific environments. The aim was to develop "breakthrough" rice varieties that would

[4] S. Orr, J. Sumberg, O. Erenstein, and A. Oswald, "Funding International Agricultural Research and the Need to Be Noticed: A Case Study of NERICA Rice," *Outlook on Agriculture* 37, no. 3 (2008): 159–168; E. Tollens, M. Demont, A. Sié, M. Diagne, K. Saito, and M. Wopereis, "From WARDA to AfricaRice: An Overview of Rice Research for Development Activities Conducted in Partnership in Africa," in M. Wopereis, ed., *Realizing Africa's Rice Promise* (Boston, MA: CABI, 2013), pp. 1–23.
[5] E. H. P. Frankema, "Africa and the Green Revolution: A Global Historical Perspective," *NJAS Wageningen Journal of Life Sciences* 70–71 (2014): 17–24; Otsuka and Larson, eds., *An African Green Revolution*.
[6] John R. Walsh, *Wide Crossing: The West Africa Rice Development Association in Transition, 1985–2000*, SOAS Studies in Development Geography (Aldershot: Ashgate, 2001); Derek Byerlee and John K. Lynam, "The Development of the International Center Model for Agricultural Research: Prehistory of the CGIAR," *World Development* 135 (2020): 105080; Tollens et al., "From WARDA to AfricaRice."

Figure 6.2 IRRI's semidwarf IR-8 rice variety, the standard against which later rice-breeding efforts would be measured. Rockefeller Archive Center, Rockefeller Foundation photographs, series 242D. Courtesy of Rockefeller Archive Center.

have "wide adaptability," suggesting the variety would be transferable across regions with largely similar conditions. The association of centers with research eminence created expectations that IRRI delivered on in 1966 with the launch of IR8.[7] Awards and prizes subsequently conferred stardom to plant breeders and confirmed the status of research institutes as "centers of excellence."[8]

Existing historical accounts of WARDA take this model as the leading principle for understanding how the institute emerged and, in the 1990s,

---

[7] M. R. Baranski, "Wide Adaptation of Green Revolution Wheat: International Roots and the Indian Context of a New Plant Breeding Ideal, 1960–1970," *Studies in History and Philosophy of Biological and Biomedical Sciences* 50 (2015): 41–50; Nick Cullather, *The Hungry World: America's Cold War Battle against Poverty in Asia* (Cambridge, MA: Harvard University Press, 2010); Jonathan Harwood, "Coming to Terms with Tropical Ecology: Technology Transfer during the Early Green Revolution," *International Journal of Agricultural Sustainability* 19, nos. 3–4 (2021): 1–14; Randolph Barker, Robert W. Herdt, and Beth Rose, *The Rice Economy of Asia* (Washington, DC: Resources for the Future, 1985).

[8] The most prominent example is the Nobel Peace Prize for Norman Borlaug, often referred to as the "father of the Green Revolution." IRRI breeders Henry Beachell and Gurdev Khush received the World Food Prize in 1996; see World Food Prize Foundation, "1996: Beachell and Khush," www.worldfoodprize.org/en/laureates/19871999_laureates/1996_beachell_and_khush.

was ultimately turned into a centralized rice research institute ready to produce transformative rice varieties in the African context. Delays in WARDA's development are typically explained as resulting from an unclear research mandate at its inception in combination with limited budgets and shortages of trained staff. As WARDA's main historian, John Walsh, put it: "in the fundamental matter of research strategy WARDA had gotten off on the wrong foot."[9] However, a broader historical examination of research strategy, one that considers the breeding environment, including the ecology, economy, and social-political features of rice farming, puts WARDA's history in a different light. As I demonstrate in this chapter, WARDA initially focused on rice-farming environments defined in the colonial period. The colonial focus on rice was closely linked to exports to Europe and related commercial interests. Moreover, the colonial policies excluded a major environment where West African farmers grew rice, namely the forested humid uplands zone.

WARDA shifted its focus to the humid uplands in the early 1990s, with decisive consequences. As argued by experts within and outside CGIAR (Consultative Group on International Agricultural Research), the humid uplands required a different research strategy, one in which variation in farm types was the starting point. The CGIAR Technical Advisory Committee (TAC) urged WARDA to engage with such an approach, known as farming systems research. Experts also argued that a research agenda for the humid uplands required a decentralized breeding strategy. Although these consequences of its changed research target were acknowledged in the official documents, WARDA chose a different route by further centralizing research and concentrating on the technical ambition of breeding interspecific hybrids. Here I present the history of rice research in West Africa between the 1930s and 1990s from a perspective that considers rice farming and rice breeding as coproduced by ecological and social environments. In the conclusion, I reflect on how this historical perspective sheds a different light on the controversial launch of NERICA varieties as a breakthrough in rice improvement.

Colonial Rice Environments in West Africa, 1930–60

From the early decades of the twentieth century colonial policies were framed as "civilizing missions" that aimed to improve the living standards of people in colonized territories by investing in the local

---

[9] Walsh, *Wide Crossing*, p. 12.

economy.[10] Colonial investments in agriculture chiefly focused on crops that supplied European industries and consumers. For the French and British territories in West Africa, the main products were cotton, coffee, cocoa, palm oil, and timber. A key problem for the colonial enterprise, including the production of these agricultural exports, was labor. West Africa had been a major area of enslavement, and colonizing powers sought new mechanisms to secure human labor after the abolition of slavery. Colonial administrations and private companies used the disguise of taxation and dodgy contracts to force African people's labor on plantations or enforce their delivery of specified quantities of agricultural produce. From about the 1930s colonial powers introduced settlement schemes to boost agricultural production in Africa.[11] Large areas of land in low-population areas were prepared as new production sites where relocated families were given plots to produce certain crops, usually a combination of food crops for local consumption and crops for export. The settlement schemes provided all the facilities needed to farm the land, and scheme managers promised prosperity to settler farmers if they produced the prescribed crops in sufficient quantities.

One of these schemes, the Office du Niger, was built along the Niger River in French Sudan (present-day Mali). The history of the irrigated land settlement program of the Office du Niger, vividly documented by Monika van Beusekom, shows a gradual shift in the main crop from cotton to rice.[12] French colonial officials had pointed out the potential of rice cultivation along the Niger River when planning the scheme, which had initially focused on providing cotton for the French textile industry. The major grain crops in the region were millet and sorghum, but farmers were growing rice on the riverbanks, using the seasonal flooding of the river. Colonial authorities anticipated further demand for rice in other colonized areas, for example Senegal, where the French had invested primarily in groundnuts.[13] The Senegalese groundnut schemes supplied

---

[10] In the words of a French colonial report from the early 1920s: "The indigenous populations are incapable of developing their country alone ... their negligence too often leads them to cultivate land insufficient for obtaining the produce necessary for satisfying their yearly needs." Quoted in Alice L. Conklin, *A Mission to Civilize: The Republican Idea of Empire in France and West Africa, 1895–1930* (Stanford, CA: Stanford University Press, 1997), p. 237.

[11] For an overview, see Christophe Bonneuil, "Development as Experiment: Science and State Building in Late Colonial and Postcolonial Africa, 1930–1970," *Osiris* 14 (2001): 258–281.

[12] Monica M. Van Beusekom, *Negotiating Development: African Farmers and Colonial Experts at the Office du Niger, 1920–1960* (Portsmouth, NH: Heinemann, 2002).

[13] Christophe Bonneuil, "Penetrating the Natives: Peanut Breeding, Peasants and the Colonial State in Senegal (1900–1950)," *Science, Technology and Society* 4, no. 2 (1999): 273–302.

the French oil-seed industry and increased the local demand for food. The growing rice imports from Asia to Senegal were another incentive to stimulate rice in the West African region, which the Office du Niger did by building irrigation infrastructure to facilitate a more permanent water supply to a larger area.

The Office du Niger irrigation and settlement scheme exemplifies the overall transformation of agriculture in West Africa set in motion by colonial agricultural policies. Colonial powers shared an optimism about science and technology, expecting lush harvests and quick returns on investments in roads, irrigation infrastructure, machinery, and mineral fertilizers. Researchers and technicians played a leading role in the African settlement schemes.[14] Although illustrating colonial policies in general, the schemes were also in many ways specific to the semi-arid Sahel region. In particular, the scale of the schemes and their dependency on irrigation infrastructure made them costly and necessitated a substantial layer of managerial and technical staff. After 1945 the French colonial authorities continued the investments in irrigated rice, although new schemes were substantially smaller in size.

The irrigated river schemes provided a major impetus for rice research. Two research stations were established to serve these schemes. The first was created in 1927 in Diafarabé (Jafarabe) in the Mopti region and attached to the Office du Niger in 1930.[15] Because the Office du Niger scheme opened up large stretches of land for which rice was a new crop, a principal task of the Diafarabé station was testing rice varieties that would perform well under irrigated conditions. Researchers also tried to find out more about these farming environments, for example studying the different soil types and soil fertility levels. The second research station was established in the 1940s at Richard Toll, attached to one of the smaller irrigated rice schemes along the Senegal River. These stations were later included in WARDA, together with two further stations located in Ivory Coast and Sierra Leone.

These additional two stations had different origins. The French colonial policy of stimulating export crops and food crop production in Ivory Coast focused on two regions. A forest region, covering roughly half the country northward from the coast, featured export crops that were perennials, mainly cocoa and coffee. Timber was also an export product. Producing these exports formed a major share of the economy, resulting in an increasing inflow of migrants. Although people grew food crops, including rice, in the forest zone, these types of farming were considered

[14] Bonneuil, "Development as Experiment."
[15] Van Beusekom, *Negotiating Development*, p. 18.

detrimental for the forest.[16] Growing food crops was instead stimulated in the drier savanna area further north, in combination with cotton production. The rural migration triggered by economic policies mixed ethnic groups as well as cropping patterns across the region. Whereas rice cultivation in Ivory Coast had been limited to the southwest forest region, halfway through the twentieth century rice was grown by smallholder farms across the colony, resulting in a wide variety of rice-farming methods and preferences for particular rice varieties.[17] From 1955, the French colonial administration also introduced small irrigation schemes in the savanna zone, together with further intensification of cotton cultivation.[18] To support these schemes, a rice research station in Ivory Coast was located in the southern savanna area at Bouaké, which was a major center for the cotton industry thanks to its location on the railway line running north from Abidjan. Rice research at the Bouaké station focused on selection of varieties for the dry upland rice farms.

One other rice research station eventually included in WARDA was created by the British colonial administration in Sierra Leone. Policies in the British West African territories were broadly similar to those in the countries controlled by France. The colonial economy of Sierra Leone had a somewhat different history, in that after the 1930s mining became increasingly important and by the 1960s, when it was an independent nation, completely dominated the country's exports.[19] Investments in food crop production therefore were less directly connected to areas developed for export crops than in the settlement and irrigation schemes elsewhere. However, rice was considered a potential export crop, and indeed Sierra Leone exported rice in the 1930s and 1950s when farmers responded to favorable global price fluctuations.[20] As colonial experts quickly found out, rice yields in lowland areas along the rivers and inland swamps were higher than in the humid uplands. Concerns over the negative effects of slash-and-burn farming in the forest zone further

[16] J. Fairhead, and M. Leach, *Misreading the African Landscape: Society and Ecology in a Forest-Savanna Mosaic* (Cambridge: Cambridge University Press, 1996).

[17] There are at least eight distinct rice-cropping systems in Ivory Coast. See Laurence Becker and Roger Diallo, "The Cultural Diffusion of Rice Cropping in Côte d'Ivoire," *Geographical Review* 86, no. 4 (1996): 505–528.

[18] Ibid.; Thomas J. Bassett, "The Development of Cotton in Northern Ivory Coast, 1910–1965," *The Journal of African History* 29, no. 2 (1988): 267–284.

[19] A. B. Zack-Williams, "Merchant Capital and Underdevelopment in Sierra Leone," *Review of African Political Economy* 9, no. 25 (1982): 74–82.

[20] David Moore-Sieray, "The Evolution of Colonial Agricultural Policy in Sierra Leone, with Special Reference to Swamp Rice Cultivation, 1908–1939," Ph.D. dissertation (The School of Oriental & African Studies, University of London, 1988), p. 177; Paul Richards, *Coping with Hunger: Hazard and Experiment in an African Rice-Farming System*, London Research Series in Geography No. 11 (London: Allen & Unwin, 1986).

motivated a focus on lowland areas. Colonial experts considered the rice farms in the Great Scarcies River area, located in the southwest bordering Guinea, to have the most potential. There, the nearness of the ocean caused tidal flows and extended mangrove forests, coastal conditions that farmers used to their advantage, resulting in one of the most productive rice areas of the country. Colonial officials thought that yields could be further increased with water control and other introduced techniques, making the Scarcies region a model for the rest of the country.[21] A rice research station was operational from 1934 in Rokupr, one of the nodes in the trade network for rice from the Scarcies region.

In sum, the rice research stations in the French and British colonial contexts of West Africa focused on farming environments that contributed, or were complementary, to an export-oriented economy. In the upper Sahel region, the river-based flooded farms were turned into much larger irrigation schemes where rice became the dominant food crop. The stations at Diafarabé and Richard Toll addressed the demands of farming rice in these conditions. Meanwhile, in Ivory Coast and Sierra Leone, two countries with substantial forest zones in which rice was an important crop, colonial officers largely ignored rice cultivation in the forested humid uplands. Intensification of rice cultivation in Ivory Coast concentrated on the northern dry savanna area, where rice complemented cotton growing. In Sierra Leone the emphasis was on tidal flooding in the coastal area and other lowlands that allowed for inundated rice cultivation.

Despite the restricted focus of rice research on dry uplands and inundated lowlands, research activities had effects beyond these target environments. The search for rice varieties that performed best in the Sahelian schemes, coastal areas, or dry savanna regions triggered a lively exchange of rice varieties on a global scale. Colonial experts were well networked and often travelled between different colonial territories, bringing rice varieties themselves or making requests to have varieties sent over. Varietal improvement during the first half of the twentieth century consisted primarily of sorting, recording, and testing the many different rice types. Named varieties were often so-called landraces, groups of morphologically similar types, which breeders further split up into "pure" lines. The starting point for breeders' selection work was the varieties that farmers held in their fields. An example is the variety Demerara Creole, originating from British Guiana in South America. In the early twentieth century, sugar

---

[21] Michael Johnny, John Karimu, and Paul Richards, "Upland and Swamp Rice Farming Systems in Sierra Leone: The Social Context of Technological Change," *Africa: Journal of the International African Institute* 51, no. 2 (1981): 596–620.

cultivation in British Guiana relied on indentured laborers from India, many of whom continued as smallholder rice farmers in the coastal zone. Colonial agronomists identified Demerara Creole as a landrace introduced to British Guiana by these Indian plantation workers.[22] The first descriptions of the variety in British Guiana date from the 1900s, and soon thereafter it was introduced to Sierra Leone.[23] From there it spread along the coast and further inland, also passing the border with French Guinea, where it was observed by the French botanist Roland Portères.[24] Many other varieties from within and outside the West African region were collected for further selection, followed by intentional and unintentional distribution.[25] Whereas colonial rice breeders primarily looked for varieties for targeted rice zones, the varieties also circulated more widely via informal distribution channels, reaching all rice areas, including the forested rice zones.

Colonial experts were aware of the wider effects of their work. The emergence of research stations and the often long-term involvement of experts in colonial programs implied that local agricultural practices were observed, if not closely studied.[26] Various colonial experts in West Africa persistently reported on the importance of local farming practices and the value of farmers' knowledge and skills. For example, farmers in the Office du Niger cultivated a variety of crops, including rice, on their own fields, within and outside the scheme, mostly producing better-quality and higher yields. The scheme's main agronomist, Pierre Viguier, concluded in the late 1940s that the colonial policy had failed. What West Africa needed, he argued, was not more foreign technology but "the application of *African* formulas, inspired no doubt by more evolved techniques, but thought out by Africans, adapted to the needs, the means, the aspirations

---

[22] Harro Maat and Tinde van Andel, "The History of the Rice Gene Pool in Suriname: Circulations of Rice and People from the Eighteenth Century until Late Twentieth Century," *Historia Agraria* 75 (2018): 69–91.

[23] Moore-Sierray, "Evolution of Colonial Agricultural Policy," p. 65.

[24] Portères mentions the following synonyms of Demerara Creole used in French Guinea: Dixie, Dissi, DC, Dixie-Kabak, Dixie I, Dixie II, Dissi Kouyé. See Roland Portères, "Les variérés de riz de l'Île du Kabak (Guinée Française)," *Journal d'Agriculture Tropicale et de Botanique Appliquée* 4, no. 5 (1957): 185–211, at 209.

[25] At the Bouaké station, about 4,000 rice varieties were assembled and tested for the savanna conditions up until 1975, with less than 25 percent considered fit for cultivation. See Michel Jacquot, "Varietal Improvement Programme for Pluvial Rice in Francophone Africa," in I. W. Buddenhagen and G. J. Persley, eds., *Rice in Africa: Proceedings of a Conference Held at the International Institute of Tropical Agriculture Ibadan, Nigeria, 7–11 March 1977* (London: Academic Press, 1978), pp. 117–129.

[26] Joseph Morgan Hodge, *Triumph of the Expert: Agrarian Doctrines of Development and the Legacies of British Colonialism* (Athens: Ohio University Press, 2007); Helen Tilley, *Africa as a Living Laboratory: Empire, Development, and the Problem of Scientific Knowledge, 1870–1950* (Chicago: University of Chicago Press, 2011).

of African peoples."[27] Colonial administrations, however, seemed fully convinced about the superiority of foreign farming technology. In the 1950s the introduction of tractor plowing and mechanization of other tasks in 4,500 hectares of the Office du Niger scheme was to overcome "indiscipline, labour bottlenecks, or incompetence on the part of the farmers."[28] By the late 1950s the results showed an overall higher yield in the nonmechanized parts of the scheme. Similar observations were made in colonial Sierra Leone. Experts like the soil scientist H. W. Dougall had noticed that farmers used higher fields at the margins of the mangrove swamps "to establish their rice nurseries, cassava and sweet potato beds and occasional banana groves. It is doubtful if the system could be profitably improved upon."[29] The colonial administration in Sierra Leone nevertheless continued with the introduction of mechanization and improved irrigation infrastructure in the lowland areas. These technologies appeared very costly, and once authorities gave up support for maintenance and other operational costs, farmers quickly reverted to their "African formulas."

Dependent and Independent West African Rice Environments in the 1960s

Between the end of colonial domination around 1960 and the establishment of WARDA in 1971, two major developments affected the course of rice research in West Africa. The independent West African nations started to transform colonial policies into national plans to stimulate the economy and agriculture. Few countries managed to do so without foreign support, and therefore former colonial powers and new international donors entered the scene. A second development was the strategy of the Rockefeller and Ford Foundations, backed by the US government, to boost food production in Asia and produce what has since been known as the Green Revolution. The strategy focused on plant breeding, resulting in varieties of wheat and rice that allowed for double cropping and overall high yields in areas with proper irrigation facilities and availability of mineral fertilizer. The philanthropies and US aid agencies pushed other international actors and Western allies to help apply the same strategy in other areas, including Africa.

Although the economic challenges for the independent West African states were broadly similar, there were major differences among country-level policies. For example, Ghana, independent in 1957, profited from

---

[27] Quoted in Van Beusekom, *Negotiating Development*, p. 130. Emphasis in the original.
[28] Ibid., p. 171.   [29] Richards, *Coping with Hunger*, p. 13.

the high price of cocoa in the 1950s, providing the government with revenues for investment in the economy. Rice was a relatively recent food crop in Ghana, and by the mid 1960s the government aimed for an expansion of rice production, with the help of Chinese experts and funds.[30] In Sierra Leone, where rice was the predominant food crop, the government had few financial reserves and fully relied on foreign support for rice improvement, which was provided by Taiwan. In 1968, when elections in Sierra Leone created a regime change, the new government opted for support from China.[31] The presence of China and Taiwan in West Africa makes clear that assistance programs for food production in Africa, as in Asia, were affected by the Cold War.[32]

The new geopolitical constellation of the Cold War was also entangled with ongoing colonial dependencies. France and Britain held a firm overall grip on the economic interests in their former West African territories, mainly by securing commercial exploitation of lucrative export crops. France also coordinated agricultural research activities for tropical agronomy and food crops through a new umbrella organization created in 1960, the Institut de Recherches Agronomiques Tropicales et des Cultures Vivrières (IRAT).[33] Although new independent governments were formally in charge of agricultural research, IRAT's headquarters were in France, and the budgets, staffing, and research agendas of the stations in West Africa were controlled from Paris.[34] Circumstances differed in former British colonies. Although independent governments had taken over all responsibilities for agricultural research, many British colonial experts continued to advise and support these countries, switching jobs from colonial service to national and international aid

---

[30] Kojo Amanor, "South–South Cooperation and Agribusiness Contestations in Irrigated Rice: China, Brazil and Ghana," in James E. Sumberg, ed., *Agronomy for Development: The Politics of Knowledge in Agricultural Research* (London: Routledge, 2017), 32–43; Deborah Bräutigam, *Chinese Aid and African Development: Exporting Green Revolution* (Basingstoke: Macmillan, 1998).

[31] Richards, *Coping with Hunger*; Zachary D. Poppel, "Quick Rice: International Development and the Green Revolution in Sierra Leone, 1960–1976," in C. Helstosky, ed., *The Routledge History of Food* (London: Routledge, 2014), pp. 332–351.

[32] Cullather, *The Hungry World*, p. 252.

[33] IRAT was included in the Centre de Cooperation Internationale en Recherche Agronomique pour le Développement (CIRAD) in 1984. Benoit Daviron and Janine Sarraut-Woods, "History of Public Organizations and Associations Specializing in a Single Agricultural Commodity and Related to Francophone Africa," in Estelle Biénabe, Alain Rival, and Denis Loeillet, eds., *Sustainable Development and Tropical Agri-Chains* (Dordrecht: Springer, 2017), pp. 29–40.

[34] The French aid system in Africa from the 1960s "signalled an intensification (rather than a loosening) of colonial ties: the golden age of French science and medicine in Africa began after independence." Quoted in G. Lachenal, "At Home in the Postcolony: Ecology, Empire and Domesticity at the Lamto Field Station, Ivory Coast," *Social Studies of Science* 46, no. 6 (2016): 877–893, at 879.

agencies.[35] The British centralized research structure established in 1953, in which the Rokupr research station had coordinated rice research across all British West African countries, was dismantled. Rokupr became the national research center for Sierra Leone in 1962, integrated with the agricultural research station and college at Njala.[36] In countries like Nigeria, Liberia, and Ghana, rice research was similarly taken up in national agricultural research agendas. As I describe below, the continuation of a centralized research organization by France would play a major role in the establishment of WARDA.

Towards the end of the colonial period, French and British researchers had started crossbreeding experiments to further fit rice varieties to West African environments. The breeding programs had resulted in various new releases for the target regions but were also distributed to other areas through local trading networks. Continued plant-breeding experiments at the research stations after independence included crossings between *sativa* and *glaberrima* rice varieties. Halfway through the 1960s such interspecific crosses were made in Sierra Leone and Nigeria. IRAT researchers did the same somewhat later.[37] None of these experiments resulted in stable lines that were released for distribution. Among other challenges, the decolonization of research in former British colonies quickly reduced the research capacity. By the end of the 1960s there were four full-time rice breeders, one each in Sierra Leone and Liberia and two in Nigeria.[38] Local resources compensated for new limitations. At Rokupr, for example, the experimental fields were tended by women who grew their own crops at the edges of the rice fields and helped in sorting seeds of the different varieties.[39] The linkages with Njala University, in the eastern part of the country, created the option for training new rice experts. However, most government and donor money was spent on improvement projects. In sum, by the time US donors planned to expand international agricultural research to West Africa in the late 1960s, there was a rather strong French-run research infrastructure in former French territories, and a relatively weak research infrastructure in former British colonies.

The proposal for a new international rice research institute in West Africa emerged alongside the preparations for the International Institute

[35] Hodge, *Triumph of the Expert.*
[36] Moore-Sierray, "Evolution of Colonial Agricultural Policy," p. 268.
[37] S. S. Virmani, J. O. Olufowote, and A. O. Abifarin, "Rice Improvement in Tropical Anglophone Africa," in I. W. Buddenhagen and G. J. Persley, eds., *Rice in Africa: Proceedings of a Conference Held at the International Institute of Tropical Agriculture, Ibadan, Nigeria, 7–11 March 1977* (London: Academic Press, 1978), pp. 101–116; Jacquot, "Varietal Improvement."
[38] Virmani et al., "Rice Improvement."   [39] Poppel, "Quick Rice."

of Tropical Agriculture (IITA), which had been initiated by the Rockefeller Foundation officials George Harrar and Will Myers in 1962. Harrar and Myers proposed that the Nigerian government set up IITA on the campus of the University of Ibadan. The new research center would focus on various tropical crops – except rice, as it was initially assumed that IRRI would provide the research input for rice breeding.[40] However, the importance of rice as a food crop in West Africa and the distinctive rice ecologies of Africa invited calls for further institutional development. Two years after the opening of IITA in 1967, the American donors, joined by the United Nations Food and Agriculture Organization (FAO) and the United Nations Development Programme (UNDP), made arrangements for a central rice research center in West Africa, a move resisted by France as it continued to coordinate several rice research stations in the region.[41] The negotiations among these actors for the future of rice research in West Africa were chaired by UNDP officer Paul Marc Henry, who had previously worked in the region for the French government.

There were three organizational models on the table. In the first, the new rice center would primarily disseminate research results from IITA, IRRI, and the French rice stations across the region. A second model was to make it a coordinating institute for the existing rice research stations. A third option was to create a central research institute with the four existing stations as satellites. This last model was preferred by the representative of the Rockefeller Foundation, Will Myers, and IRRI's director, Robert Chandler. The UNDP delegation proposed a variant that reinstated the coordinating role for the Rokupr station in Sierra Leone.[42] Final agreement was reached at a 1970 meeting in Rome hosted by FAO. The outcome followed the second model of establishing a coordinating institute for the existing research stations. The model suggested a centralized research institute, yet without its having a firm lead in research activities. In contrast to the somewhat obfuscated research mandate of the overarching institute, the task of the four research stations was spelled out in clear terms.

Research activities at the newly designated WARDA stations focused on irrigated and deep-flooded (or floating) rice in Senegal and Mali, mangrove rice in Rokupr, and upland rice in Bouaké. This division implied a rather straightforward continuation of the colonial-era research agendas, and in many ways it was a continuation. However, colonial dependencies had become entangled with the newly emerging aid

[40] Byerlee and Lynam, "The Development of the International Center Model."
[41] Ibid., 11.   [42] Ibid., 57.

relationships, Cold War politics, and diverging national policies across the independent states. Countries like Ghana and Nigeria, for example, started to invest in rice production under irrigated conditions that were very different from the irrigated rice zones in the Sahel region. The Rokupr station in Sierra Leone never exclusively focused on mangrove rice. It had had a much broader mandate in the 1950s and then turned into a national research station in the 1960s, at which point it focused on all rice environments in the country. The agendas of the four research stations were further challenged after WARDA became operational.

Redefining Rice Farming Environments, 1970–2000

From 1970 a growing number of international donors became involved in rice improvement projects in West Africa. This was further intensified by the integration of WARDA into CGIAR in subsequent years. One of the challenges for WARDA was to broaden the scope of the four research institutes it coordinated. As explained in the previous section, the stations had built up expertise for specific rice-farming environments within institutional contexts defined by the British and French colonial empires. In the new setting of WARDA, the colonial distinctions were formally gone, and country borders were to be exceeded. For example, Rokupr had little experience with research on mangrove rice other than in Sierra Leone. However, mangrove rice environments stretched out along the coast of West Africa, crossing various national and linguistic borders. Likewise, irrigated rice schemes existed in West African countries other than Senegal and Mali. Throughout the first three decades of WARDA's existence, the focus of the four stations on distinct rice-farming environments gradually changed. As I show here, there were two major reasons for the shifts in focus: first, changes within the farming environments themselves and, second, increasing criticism about the exclusion of the upland humid forest zone from research agendas.

Connections between WARDA and IRRI implied that the improved rice varieties introduced in Asia would become available for African farmers. Direct transfers, however, were hardly an option. By the early 1970s most varieties released by IRRI focused on Asian rice environments, where irrigation and fertilizer were available and two or even three rice crops per year were possible.[43] The irrigation schemes in Senegal and Mali would in principle match these conditions. However, the Sahelian

[43] Robert S. Anderson, Edwin Levy, and Barrie M. Morrison, *Rice Science and Development Politics: Research Strategies and IRRI's Technologies Confront Asian Diversity, 1950–1980* (Oxford: Clarendon Press, 1991).

climate includes a relatively cold winter period, requiring further selection of cold-resistant varieties in order to achieve double cropping.[44] The farming environments of the Sahelian irrigation schemes not only differed from Asian irrigated rice schemes but also from other irrigated rice areas in West Africa. Compounding these concerns, the operational conditions of irrigated rice schemes in the Sahel changed substantially during the last decades of the twentieth century.

In Mali, the legacy of the Office du Niger had created disparities among farmers in terms of entitlement to land and irrigation water. Subsequent governments tried to solve inequities through state-imposed farmer cooperatives and village advisory councils. These social structures were often caught in a crossfire from ongoing government involvement and complaints from farmers.[45] The introduction of farmer-operated motor pumps aimed to reduce the dependency on national managerial bodies. Pump-based village schemes, applied in Mali and Senegal from the end of the 1970s, were largely successful in decentralizing rice production but did bring up new problems. Researchers noted, for example, that support mainly went to male farmers, marginalizing women who often had a major share in agricultural work. In one of the Senegalese schemes a Dutch-funded program developed new pump-irrigated farmland in collaboration with women's groups. The women opted for vegetable crops rather than rice, as these better served their household food security concerns and relied on their knowledge about local food markets.[46] The women's decision to prioritize their own food security concerns rather than government demands for rice was an example followed by many rice farmers when, in the 1990s, the winds of neoliberal policy reforms stopped the supply of cheap fuel, maintenance for the motor pumps, credit for fertilizers, and subsidies on rice. Farmers' interest in growing rice dropped significantly, as did acreage of rice cultivation, and many village schemes were abandoned.[47] These changes dampened the high

---

[44] CGIAR Technical Advisory Committee, *Report of TAC Quinquennial Review Mission to the West African Rice Development Association* (Rome: FAO, 1979), p. 38, https://hdl.ha ndle.net/10568/118482.

[45] R. James Bingen, *Food Production and Rural Development in the Sahel: Lesson from Mali's Operation Riz-Segou* (Boulder, CO: Westview Press, 1985).

[46] H. Maat and P. P. Mollinga, "Water bij de uien: Technologische en andere ontwikkelingen op het Ile à Morphil, Senegal," *Kennis en Methode* 18 (1994): 40–63; G. Diemers and F. P. Huibers, *Gestion paysanne de l'irrigation dans la vallée du Fleuve Senegal: Implications pour la conception des amenagements hydro-agricoles: Rapport de fin de projet, gestion de l'eau de l'adrao* (Wageningen: Wageningen University, 1991).

[47] The older farming techniques on the riverbanks made use of the occasional floods, but this option was blocked once the Manatali dam became operational. The only other option then was to revert to drought-resistant crops, such as sorghum, millet, and maize. See William G. Moseley, Judith Carney, Laurence Becker, and Susan Hanson,

expectations about the potential of irrigated rice schemes in West Africa. Although irrigated rice continued as one of WARDA's targets, its prominence was overtaken by attention to a different environment: rice farming in the forested humid uplands.

Soon after the establishment of WARDA, agricultural experts working in the region pointed out that rice farming in the humid upland zone deserved more attention. Despite the overall consensus that yield increases had to come from irrigated rice, the size of the humid uplands, in terms of acreage and number of farm households, could no longer be ignored. In its first quinquennial review of WARDA in 1979, CGIAR's TAC noted that some WARDA officers considered the focus on dry upland rice by the station in Bouaké inadequate for the humid uplands.[48] The committee's advice was to investigate and discuss the issue. In the next quinquennial review, conducted in 1984, the committee repeated the advice in stronger terms, noting that humid uplands are "a badly neglected area of rice research and development which deserves increased attention from WARDA, national rice research programs, and IRRI and IITA."[49] The importance was repeated in subsequent years. By the early 1990s the continuum of rice ecologies in the humid uplands was the main focus for WARDA's research. Of the four farming environments that had been set as WARDA's research targets in 1970, only irrigated rice remained as a priority.[50] The focus on humid uplands was a major shift in WARDA's strategy that had further consequences for its research agenda (Figures 6.3 and 6.4).

A key feature of the humid upland zone is that it contains a variety of soils with different water levels. The topographical sequence implies a continuum from higher parts with low water levels, to a middle zone with saturated soil conditions, to a lower swamp area with standing water

"Neoliberal Policy, Rural Livelihoods, and Urban Food Security in West Africa: A Comparative Study of the Gambia, Côte d'Ivoire, and Mali," *Proceedings of the National Academy of Sciences of the United States of America* 107, no. 13 (2010): 5774–5779; A. Adams, "The Senegal River: Flood Management and the Future of the Valley," Issue Paper, International Institute for Environment and Development, January 2000.

[48] CGIAR Technical Advisory Committee, *Report of the TAC Quinquennial Review Mission*, p. 46.

[49] CGIAR Technical Advisory Committee, *Report of the Second External Program Review of the West Africa Rice Development Association (WARDA)* (Rome: FAO, 1985), p. 69, https://hdl.handle.net/10568/118618.

[50] The exclusivity of the Sahel region for irrigated rice was questioned by the external review committee in 2001, recommending an "expansion of the Irrigated Rice Programme so as to address effectively irrigated systems beyond the Sahel with emphasis on breeding for the humid and sub-humid zone." CGIAR Technical Advisory Committee, *Report of the Fourth External Programme and Management Review of the West Africa Rice Development Association (WARDA)* (Rome: FAO, 2001), p. 28.

Figure 6.3 Rice demonstration plots featuring "Upland Germplasm" and "Lowland Germplasm" (the latter including NERICA lines) that were associated with a WARDA collaboration in Liberia funded by Japan, 2009. Photo by R. Raman, AfricaRice and reprinted by permission of AfricaRice.

for most of the rainy season. The variation in ecological conditions and water levels makes it difficult to breed or select a single rice variety suitable on all types of farmland. The anthropologist Paul Richards, who had studied rice farms in the humid zone of Sierra Leone since 1977, concluded that selecting and experimenting with a broad set of different rice varieties was common practice among farmers.[51] Richards reported in detail about farmers' preferences and their selection of rice varieties that addressed different soil and water conditions, resistance to weeds or pest damage, growth duration in anticipation of labor peaks during harvest time, and integration with other crops. Richards' studies found some resonance in reports produced within CGIAR. For example, the focus on ecological variation and integration with crops other than rice was a research area with overlapping interests at WARDA and IITA. In the 1980s IITA had initiated an "agro-ecological characterization of rice-growing environments" in West Africa.[52] In a report of this effort

---

[51] P. Richards, *Indigenous Agricultural Revolution: Ecology and Food Production in West Africa* (London: Hutchinson, 1985); Richards, *Coping with Hunger*.
[52] P. N. Windmeijer and W. Andriesse, *Inland Valleys in West Africa: An Agro-ecological Characterization of Rice-growing Environments*, No. 52, International Institute for Land Reclamation and Improvement, 1993; W. Andriesse and L. O. Fresco, "A Characterization of Rice-Growing Environments in West Africa," *Agriculture, Ecosystems and Environment* 33, no. 4 (1991): 377–395.

Figure 6.4 Two rice researchers at an Africa Rice Center upland rice-breeding site on the Danyi Plateau, Togo, in 2007. Photo by Harro Maat.

published in the early 1990s, the researchers argued for a further integration of the work of WARDA and IITA in analyzing farming systems and called for improved varieties based on regional variation.

The message was taken on board by the next external program review committee, which wrote in its 1993 assessment that "WARDA needs a farming systems approach to research with a strong ecological focus so as to be of short- and long-term environmental, social and economic benefit."[53] What the studies and reports hardly mentioned was that a farming systems approach had implications for WARDA's approach to rice breeding. A "strong ecological focus" in West Africa would require a variety of projects embedded in nation- and region-specific research and extension facilities. Rice breeding then would follow from region-specific characteristics and farmers' needs. As Richards explained, farmers'

---

[53] CGIAR Technical Advisory Committee, *Report of the Third External Programme and Management Review of the West Africa Rice Development Association (WARDA)* (Rome: FAO, 1993), p. 23, https://hdl.handle.net/10947/1579.

selection strategies were likely to be seen as irrational in the eyes of
breeders and extension agents, who would "replace a profusion of uncertain and unstable variants of local land races with a much smaller range of
fixed and reliable seed types."[54] The breeder's perspective would be
valuable, Richards and colleagues argued, but only if based on
a continuous dialogue between farmers and breeders to find the common
ground in what "wide varietal adaptability" means from both
perspectives.[55] One such breeding approach, participatory varietal selection (PVS), gained some leverage in national rice-breeding projects in
West Africa. An evaluation study of PVS projects over the late 1990s
showed that national researchers were positive about the approach and
most concerned by the limited capacity of their national research organizations to expand the work. The study also pointed out that financial
support from WARDA was very modest.[56] In the 1990s, the actual rice-
breeding agenda of WARDA had fully focused on the interspecific crossing experiments.

Conclusion

The evidence presented in this chapter provides a very different history
of WARDA from what can be found in the limited historiography of the
institute. Existing accounts perceive rice research primarily as on-
station research and breeding capacity. Such capacity was, by the time
WARDA was created, scattered over different regional and national
stations. WARDA's initial mandate gave the institute a coordinating
rather than a leading role in rice research and breeding. In the received
view, the alleged lack of control of WARDA's headquarters staff over the
research agenda was resolved with the centralization of research at the
Bouaké station in the 1990s, with a clear research focus on interspecific
crossing. In contrast, WARDA's history as presented here explored rice
research from an environmental perspective, examining linkages
between farming environments and rice research. This shows a trend
in which WARDA continued to focus on the rice-farming areas defined
in the colonial period, addressing European commercial interests rather
than the concerns of West African rice farmers. The 1990s, was the

---

[54] Richards, *Coping with Hunger*, p. 145.
[55] S. S. Monde and P. Richards, "Rice Biodiversity Conservation and Plant Improvement
in Sierra Leone," in A. Putter, ed., *Safeguarding the Genetic Base of Africa's Traditional
Crops* (Rome: CTA/IPGRI, 1994), pp. 83–90.
[56] Nina Lilja and Olaf Erenstein, "Institutional Process Impacts of Participatory Rice
Improvement Research and Gender Analysis in West Africa," Participatory Research
and Gender Analysis (PRGA), Working Document No. 20, PRGA-Centro Internacional
de Agricultura Tropical (CIAT), Cali, Colombia, 2002.

decade in which the inclusion of the humid uplands implied that WARDA finally covered the variety of rice-growing environments in West Africa, acknowledging the importance of studying the diversity of farming systems and addressing them with such methods as PVS. This version of WARDA's history provides a puzzling contradiction with existing historical accounts.

One possible explanation for the contradicting trends in WARDA's history is the limited overall research budget that, together with increasing regional political instability in the 1990s, enforced a decision to concentrate resources on a restricted rice-breeding program. CGIAR faced shrinking budgets in the 1990s, and a civil war in Liberia prompted WARDA to move its headquarters to Ivory Coast. However, the challenging circumstances of the 1990s do not explain why centralization of research was not implemented in earlier decades, when pretty much all the options were there and known to CGIAR officials. Moreover, improved rice varieties were produced by various national and regional breeding stations. From the 1960s to the 1990s, this added up to almost 200 releases of improved varieties.[57] As noted above, the early breeding activities also included interspecific crossing experiments with *glaberrima* and *sativa* rice. In other words, centralization of research and breeding was not a necessary condition for developing the scientific or technical knowledge to produce interspecific crossing lines or improved varieties more generally. And as shown in the account of the negotiations leading to WARDA's establishment in the late 1960s, the US donors preferred a central breeding institute but did not block other options. Moreover, by the 1990s, as centralization took hold, there was much more evidence that diverse and decentralized research would be a better fit for West African conditions.

An alternative explanation considers the contradictions in the history of WARDA as the effect of a delayed decolonization of rice research in West Africa. The research stations set up in countries formerly under French rule formed the backbone of WARDA's research. During the colonial period, Britain and France largely applied the same policies. France, however, continued its control over research facilities in its former colonial territories until the 1990s, mainly to secure uninterrupted access to export crops.[58] The historiography from the colonial period further makes clear that experts were aware that, in the selected environments, rice cultivation methods aiming for high yields required high investments. Moreover, they acknowledged that the farming practices of West African

---

[57] Tollens et al., "From WARDA to AfricaRice."
[58] Daviron and Sarraut-Woods, "History of Public Organizations," pp. 29–40.

rice farmers were highly effective, given prevailing conditions, covering a wider variety of farming environments. Nevertheless, the research organization established for WARDA in 1970 largely continued the research agenda set in the colonial period.

This may justify a conclusion that colonial interests controlled rice research at WARDA, in particular through the continued influence of French officials. However, official documents of WARDA only provide indirect evidence for this. And even in the unlikely scenario that French influences largely determined the course of WARDA's research, CGIAR donors had various options to invest in a complementary rice research agenda. Rokupr, for example, had earlier dealt with a regional mandate, and in the 1960s countries such as Ghana and Nigeria had also initiated rice research. Another route to diversify rice research was through IITA, which indeed did conduct complementary rice research, including rice breeding. Moreover, CGIAR donors other than France were also active in countries like Senegal, suggesting that the focus on irrigated rice environments was considered an attractive option for multiple actors at least until the 1990s. This would compromise a conclusion that attributes WARDA's research agenda and the limited and late attention to humid uplands to French intentions alone.

CGIAR donors likely considered irrigated rice the best option until the moment protectionist policies were dismantled. There is a striking synchrony between the economic reforms that ended the economic viability of large, irrigated rice environments and the promotion of a new model focusing on the humid uplands. This explanation still does not account for the decision to ignore the expert advice to focus on farming systems research and diversification of the breeding strategy – leaving as a final explanation the allure of a centralized research model, mentioned in the introduction.

The summary version of WARDA's history would then be that there was sympathy, but never strong conviction, for a diversification of WARDA's research agenda. The need to strengthen decentralized, national research capacities in West Africa was acknowledged but never seen as a CGIAR task, other than by showing that top-notch scientific research leads to superior crops under favorable conditions. The example of IRRI was to be followed, and, when the opportunity came in the 1990s, it was pushed through with fervor. Such an explanation would have to discard the contradicting developments and statements presented in this chapter. Moreover, it would have to discard evidence from the history of IRRI, showing that by the early 1970s the centralized research and breeding model, including the notion of wide adaptability, was put up for debate, leading to a gradual change of IRRI's research agenda.[59] IRRI breeders themselves advised against a centralized research agenda in Africa, stating at a conference in Nigeria in 1977 that "rice varieties should be tailor-made for specific locations, conditions and systems. So-called widely adapted varieties are probably nothing more than a reflection of a past void in local research capability."[60] Statements like this are difficult to make chime with the course of rice research at WARDA after the 1990s. Perhaps donors and CGIAR decision-makers have had a stronger conviction regarding the powers of research excellence and science-based technologies than CGIAR researchers and other experts.

The dichotomy between the actual research work and representations of science by policymakers also played a prominent role in the life and work of Yuan Longping. As Sigrid Schmalzer has shown, Yuan's work was embedded in wider networks of breeders, agronomists, and rice farmers across China. His role in hybrid rice breeding was largely unknown by the public until after 1976, when post-Maoist reforms started to kick in and the history of agricultural development was rewritten along the lines of research excellence and individual prestige rather than team effort.[61] Having opened this chapter with the two winners of the 2004 World Food Prize, I should end with the caution that although the prize winners no doubt have laudable merits as individual researchers, the prizes as such are poor emblems of the research traditions and history of the science of which they were a part.

[59] Harwood, "Coming to Terms." Jonathan Harwood, "Could the Adverse Consequences of the Green Revolution Have Been Foreseen? How Experts Responded to Unwelcome Evidence," *Agroecology and Sustainable Food Systems* 44, no. 4 (2020): 509–535.

[60] W. R. Coffman, G. S. Khush, and H. E. Kauffman, "Genetic Evaluation and Utilization Programme of the International Rice Research institute (IRRI)," in I. W. Buddenhagen and G. J. Persley, eds., *Rice in Africa: Proceedings of a Conference Held at the International Institute of Tropical Agriculture, Ibadan, Nigeria, 7–11 March 1977* (London: Academic Press, 1978), pp. 137–146, at 137.

[61] Schmalzer, *Red Revolution, Green Revolution*.

7 Reconsidering "Excellence"

Natural and Social Science Approaches to Livestock Research at ILRI

Rebekah Thompson and James Smith

What we now know as ILRI – the International Livestock Research Institute – had a bifurcated beginning, born out of two institutions that were launched to tackle the problem of unproductive African livestock in quite different ways. The International Laboratory for Research on Animal Diseases (ILRAD) focused on medical solutions: "[to] serve as a world center for the improvement of animal production by developing means of conquering major animal diseases, particularly those associated with pathogenic protozoa which seriously limit animal industries in many parts of the world."[1] Meanwhile, the International Livestock Centre for Africa (ILCA) was to develop applied solutions for livestock systems: "research programs designed to solve the basic production and socioeconomic problems that are serving as constraints to livestock development."[2] The history of these two institutions, one focused on the micro and the other on the macro, and their subsequent merger, raises a number of questions about the notion of "excellence" as it relates to science policy, particularly in an African context. It raises questions about what types of knowledge are valued, what knowledge is valued for, and ultimately who values that knowledge. It speaks to the history of the institutionalization of veterinary science in and for Africa, as well as to broader challenges within the Consultative Group on International Agricultural Research (CGIAR) as it continually seeks to reinvent itself in the face of political, economic, scientific, and organizational challenges.

In this chapter, we show how the establishment of ILRAD and ILCA, as two research centers with two fundamentally different research agendas, influenced the ways in which human–livestock relationships, diseases

---

[1] W. Pritchard, A. Robertson, and R. Sachs, "Proposal for an International Laboratory for Research on Animal Diseases," Report Commissioned by the Rockefeller Foundation and Consultative Group on International Agricultural Research (CGIAR), 1972.

[2] G. H. Beck et al., "An International African Livestock Centre: Task Force Report," 1971, ILCA Library, accession number 35311.

in livestock, and research excellence were conceptualized by CGIAR in sub-Saharan Africa. After a brief discussion of the notion of scientific research excellence, followed by historical introductions to the institutions at the heart of our analysis, we draw in the second half of this chapter on two contemporary case studies – one examining the development of transgenic, trypanosome-resistant cattle, and the other exploring the establishment of CGIAR Research Programs (CRPs) and the outcomes of an agricultural research for development (AR4D) program – to show how the legacy of ILRI's predecessors has continued to shape, influence, and define the trajectory of its projects. We conclude that it is important to recognize how institutions and funding bodies conceptualize excellence, as this shapes the way in which knowledge is produced and how research impact is ultimately perceived.

Natural Science, Social Science, and Centers of Research Excellence

In her account of international medicine, the historian Deborah Neill traces the emergence of the new field of "tropical medicine" in the late nineteenth and early twentieth centuries.[3] As Neill highlights, this field was driven by transnational collaboration borne out of European colonialism and new scientific networks. Tropical medicine was one key backdrop for the establishment of livestock research in Africa. A second was the pursuit of agricultural research as international aid. As other contributions to this volume describe, the perceived successes of crop development and dissemination at the International Maize and Wheat Improvement Center (CIMMYT) and the International Rice Research Institute (IRRI) had led by the late 1960s and early 1970s to what John McKelvey, an entomologist and associate director for agricultural programs at the Rockefeller Foundation, called "institute fever": a growing investment in international institutes as tools to drive modernization and development, using science.[4] Thus ILRAD and ILCA, as with other earlier CGIAR institutions, were established in order to produce scientific solutions to address agricultural issues, and ILRI inherited that legacy.

Since its founding, ILRI, like many other CGIAR centers, has presented itself as a center of research excellence. Yet, as researchers have shown, what excellence is and how it is defined remains contested. Excellence carries significant weight in terms of recognition, policy,

[3] D. Neill, *Networks in Tropical Medicine: Internationalism, Colonialism, and the Rise of a Medical Specialty, 1890–1930* (Stanford: Stanford University Press, 2012).
[4] J. J. McKelvey, *Reflections: Living and Traveling in the 20th Century* (Brookfield, NY: Worden Press, 2000).

funding, prioritization, and practice. Its framing can be influenced more by politics and policy – donor priorities, for example – than any impartial assessment of quality, and indeed assessment of quality is itself often subjective.[5] Furthermore, assessment based on supposedly objective measures introduces other biases, for example, privileging outcomes that can be counted.[6] As Lucas M. Mueller likewise chronicles in Chapter 5, this volume, there are strong associations between investment in scientific excellence and economic development in its broadest sense, in terms of both wealth producing the best science and scientific investment producing economic growth.[7] This introduces a spatial element into understandings of excellence that maps onto political economic geographies and draws from existing narratives of institutional excellence. The latter often revolve around perceptions of the primacy of certain disciplines, for example the natural sciences over the social sciences, or basic over applied sciences.[8]

Scientific excellence is incredibly complicated: it is contested in multiple ways; it is subjective; it is hierarchized and creates its own hierarchies; and no matter how good the science may be, its outcomes are uncertain. The notion of scientific excellence has nonetheless led to decades-long intense interest in finding institutional mechanisms to concentrate and harness international scientific activity, build research capacity, and drive innovation both globally and specifically in Africa.[9] In many respects, CGIAR and its institutes exemplify this interest.

The pursuit of scientific excellence has had implications for CGIAR research. Within CGIAR centers, scientific solutions have historically been presented as the ultimate answer to agricultural problems. This aligned with a dominant conceptualization of science as global in reach and therefore, to varying degrees, unconcerned with local realities. Agricultural problems were subsequently framed as technical issues that

---

[5] D. Sridhar, "Who Sets the Global Health Research Agenda? The Challenge of Multi-Bi Financing," *PLoS Med* 9, no. 9 (2012): e1001312; K. H. Hove, "Does the Type of Funding Influence Research Results – and Do Researchers Influence Funders?" *Prometheus* 36, no. 2 (2020): 153–172.

[6] D. W. Aksnes, L. Langfeltd, and P. Wouters, "Citations, Citation Indicators, and Research Quality: An Overview of Basic Concepts and Theories," *Sage Open* 9, no. 1 (2019): 1–17.

[7] D. King, "The Scientific Impact of Nations," *Nature* 430 (2004): 311–316.

[8] R. Tijssen, "Re-valuing Research Excellence: From Excellentism to Responsible Assessment," in E. Kraemer-Mbula, R. Tijssen, M. Wallace, and R. McLean, eds., *Transforming Research Excellence: New Ideas from the Global South* (Cape Town: African Minds, 2020), pp. 59–78.

[9] T. Hellstrom, "Centres of Excellence and Capacity Building: From Strategy to Impact," *Science and Public Policy* 45, no. 4 (2018): 543–552; R. Tijssen and E. Kraemer-Mbula, "Research Excellence in Africa: Policies, Perceptions, and Performance," *Science and Public Policy* 45, no. 3 (2017): 392–403.

could be dealt with in isolation, for instance in a laboratory or field trial, and solutions were perceived to be easily disseminated, often through a relatively apolitical process of diffusion. This understanding of science as globally applicable has had significant repercussions, particularly for social scientists, whose findings and solutions are almost always tailored to specific, bounded contexts. Records show that social scientists were late to join agricultural research programs, and that, when they did, their work was often perceived to be of less importance than contributions from other scientific disciplines.[10] As the Dutch sociologist D. B. W. M. van Dusseldorp noted in 1977, for every thousand natural scientists working in agricultural research, there was fewer than one social scientist.[11]

These tensions, framed and mediated by dominant perspectives of scientific excellence, are encapsulated in the history of ILRI and its precursors. We now turn to the institutional history of ILRI with a view to illustrating how, at least in part, ILRI has had to manufacture and negotiate the complex contours of "scientific excellence" and the demand for scientific solutions as it sought to fulfill its important and ambitious mandate.

The International Laboratory for Research on Animal Diseases

Historians of veterinary medicine have shown the close links between the establishment of veterinary systems and colonial expansion in sub-Saharan Africa.[12] This is evident when examining diseases of cattle, such as trypanosomiasis, which was perceived as threatening to the stability of colonial rule.[13] As historians have described, trypanosomiasis, which is caused by a parasite and spread by the tsetse fly, was troubling for colonial authorities as it caused serious illness and death in both humans and cattle. The pervasiveness of trypanosomiasis across much of sub-Saharan Africa prompted imperial governments to invest substantial sums of money in parasitology and tropical medicine in attempts to

---

[10] D. E. Horton, *Social Scientists in Agricultural Research: Lessons from the Mantaro Valley Project, Peru* (Ottawa: IDRC, 1984).

[11] D. B. W. M. van Dusseldorp, "Some Thoughts on the Role of Social Sciences in the Agricultural Research Centres in Developing Countries," *Netherlands Journal of Agricultural Science* 25, no. 4 (1977): 213–228.

[12] W. Mwatwara and S. Swart, "'If Our Cattle Die, We Eat Them but These White People Bury and Burn Them!' African Livestock Regimes, Veterinary Knowledge and the Emergence of a Colonial Order in Southern Rhodesia, c. 1860–1902," *Kronos* 41, no. 1 (2015): 112–141.

[13] The same was true for East Coast fever (ECF – see below in this chapter). See T. T. Dolan, "Dogmas and Misunderstandings in East Coast Fever," *Tropical Medicine & International Health* 4, no. 9 (1999): A3–A11.

control the prevalence and spread of the disease.[14] Thus, parasitology was to a large extent spurred on by colonialism, as parasitic diseases risked the spread and profitability of colonial investment. It nevertheless remained relatively isolated as a field of study and efforts to control trypanosomiasis in the colonial period were ultimately unsuccessful. This left an enduring problem for researchers to solve.

By the early 1970s – as philanthropies, international organizations, and aid agencies formalized the system that would become CGIAR, and experts gathered at sites like Bellagio to determine its portfolio of institutions and research programs (see Lucas M. Mueller, Chapter 5, this volume) – parasitology appeared on the brink of profound change. Within and beyond the field, there was a belief that the benefits of recent biological research, especially molecular biology, could make an important contribution to parasitology and the control of parasite-borne diseases. To the experts organizing CGIAR, it appeared that a research center focused on animal diseases would be a potential opportunity to bring the benefits of modern parasitology to those living in developing countries.

At the successive Bellagio meetings, participants debated what the exact function and focus of a livestock disease research center – soon to be known as ILRAD – would be. Ultimately the decision was taken that the center's initial emphasis would be on haemoprotozoan diseases – commonly known as blood parasites – and immunological aspects of African animal diseases. As the entomologist and early proponent of an international center on animal diseases, John McKelvey, described:

> to focus sharply on one, possibly two, diseases [East Coast fever, or ECF, and African animal trypanosomiasis, or AAT], and on one problem, immunization techniques, to combat the diseases would afford greater chance of success than to range widely over many problems of cattle production in Africa. The Rockefeller Foundation successes in the medical sciences, combating yellow fever, for example, and in the agricultural sciences, maize and wheat improvement, reinforced this belief.[15]

The focus of ILRAD was therefore on parasitic diseases that have well-known causes, a tight focus intended to guarantee success. McKelvey's nod to the Rockefeller Foundation's prior public health successes points to other anticipated payoffs of this focus. Protozoa also affect many people in the developing world, and thus the suggestion was that with

---

[14] Maryinez Lyons, *The Colonial Disease: A Social History of Sleeping Sickness in Northern Zaire, 1900–1940* (Cambridge: Cambridge University Press, 1992); I. Maudlin, "African Trypanosomiasis," *Annals of Tropical Medicine and Parasitology* 100, no. 8 (2006): 679–701.

[15] J. M. McIntire and D. Grace, *The Impact of the International Livestock Research Institute* (Nairobi: CABI International, 2020), p. 13.

the right sort of investment a considerable improvement could be made for human lives as well as livestock. The challenge would prove formidable.

Immunology as a key to combating cattle diseases was the livestock equivalent of the "isolable problem" of raising cereal yields that had been identified in the first international agricultural research centers as a means to combat rural poverty and underdevelopment. A vaccine would be a quick, transformative solution. As a later ILRAD annual report stated, "Vaccines are a more sustainable way of controlling disease than vector control using insecticides or parasite control using drug treatments, which have contaminative, drug residue or drug resistance side effects."[16] There was sustained confidence both within ILRAD and CGIAR more generally about what the institution could achieve. In 1971, planners imagined that a vaccine for ECF, a disease caused by a protozoan parasite (*Theileria pava spp.*) and typically spread by a tick bite, could be commercially available relatively quickly: "one half or perhaps three fourths of the research towards vaccine production has been accomplished but to complete the final stages of this research will probably require five to ten years."[17] After the ECF vaccine was complete, ILRAD researchers would focus on AAT. ECF vaccine development was seen as a "short-term program" and trypanosomiasis research as a "long-term problem."[18]

To meet its research goals, ILRAD brought together an elite group of international scientists to focus on the development of molecular tools and novel vaccines. As other researchers have noted, administrators within CGIAR believed that the best method for producing scientific solutions to problems was to give research centers and the scientists working within them independence and flexibility.[19] Thus, ILRAD functioned as an independent research center, with scientists in theory shaping its research independently from CGIAR influence. This allowed ILRAD to remain an "island of excellence," with its ambitions to produce excellent applied sciences for the benefit of developing country livestock systems.[20]

---

[16] *ILRAD 1988: Annual Report of the International Laboratory for Research on Animal Diseases* (Nairobi: ILRAD, 1989), p. 1, https://hdl.handle.net/10568/49681.

[17] CGIAR Technical Advisory Committee, "East Coast Fever and Related Diseases: A Technical Conference" (Rome, Italy, 1971), March 8, 1971, 285–286, https://hdl.handle.net/10947/486.

[18] Ibid., 286.

[19] J. Chataway, J. Smith, and D. Wield, "Shaping Scientific Excellence in Agricultural Research," *International Journal of Biotechnology* 9, no. 2 (2007): 171–187.

[20] Ibid.

Both ECF and AAT were – perhaps ambitiously in retrospect – perceived by CGIAR administrators and ILRAD staff as diseases that could be controlled through new molecular techniques and would be synergistic in terms of the skills required, even if ECF was the "short-term program" and trypanosomiasis the "long-term problem." Earlier research on the causative protozoan agents of the two diseases (trypanosomes and theileria) had shown that while it was possible to immunize livestock against reinfection with the specific strain used in the vaccine, this did not confer immunity to other strains of the parasites.[21] This meant that ILRAD would have a more ambitious mandate from the other early CGIAR centers, which were primarily established to conduct translational research – for example by adopting established breeding techniques to create new rice or wheat varieties. ILRAD, too, had a translational mandate, but it also had substantive fundamental research to undertake, namely establishing the nature of immunity against the parasites in question and the mechanisms causing the failure of earlier immunization efforts.

Scientists who worked at ILRAD have suggested that in terms of African development goals, AAT and ECF may not have been the most appropriate diseases for the institution to have focused its attention on.[22] The reasons for this were twofold: one, there was little evidence that these were the top two diseases of concern for the majority of East African livestock farmers; and two, AAT and ECF proved to be much more difficult to develop vaccines for than other diseases – particularly those caused by bacteria or viruses.[23] Moreover, while ILRAD's scientists were focused on research programs that were both original in concept and highly experimental in method, something was missing. As a 1972 taskforce organized by CGIAR and led by the Australian agricultural scientist Derek Tribe reported:

> The primary cause of the disappointing growth in animal productivity in tropical Africa has been the failure to integrate the biological, economic and sociological components of research and development programmes ... Technical answers are available to many of the specific problems facing livestock development in Africa. The major constraint lies rather in the difficulty of introducing change into existing socio-economic systems, combined with inexperience in adapting technologies to suit local conditions.[24]

[21] McIntire and Grace, *Impact of the International Livestock Research Institute*, p. 14.
[22] B. D. Perry, "The Control of East Coast Fever of Cattle by Live Parasite Vaccination: A Science-to-Impact Narrative," *One Health* 2 (2016): 103–114; ILRI, *Strategic Planning Process* (Nairobi: ILRI, 1999), p. 99.
[23] ILRI, *Strategic Planning Process*, p. 99.
[24] D. Tribe et al., "Animal Production and Research in Tropical Africa," Report of the Task Force commissioned by the African Livestock Sub-Committee of the Consultative Group on International Agricultural Research (CGIAR), 1972, ILCA, accession number 00129.

In other words, even where potentially valuable tools and knowledge were available, they were not in use. This observation, and others like it, led to the establishment of ILCA in 1974.

The International Livestock Centre for Africa

Situated in Addis Ababa, Ethiopia, the International Livestock Centre for Africa (ILCA) was founded in the belief that existing solutions to Africa's livestock problems were not being applied because of a significant lack of research on, and knowledge about, local livestock systems.[25] CGIAR administrators saw an opportunity to bridge that knowledge gap. As initially imagined, the function of the new center would be "to assemble a multi-disciplinary team of scientists to develop research programs designed to solve the basic production and socio-economic problems that are serving as constraints to livestock development."[26]

In its early years, ILCA staff conducted systems surveys that described the major agro-ecological zones of sub-Saharan Africa and their production systems. This approach involved scientists working in interdisciplinary teams to study livestock production systems holistically, identifying and testing possible innovations, and defining high-priority areas for more intensive research.

The organization and operation of ILCA was significantly influenced by "systems thinking," which developed throughout the 1970s in response to the perceived failure of conventional scientific methods in addressing agricultural issues, particularly in developing countries.[27] Systems thinking moved researchers outside of the confines of the laboratory to consider the ways in which components of complex systems interact and influence one another. The formation of ILCA was, therefore, an acknowledgment that in order to develop sustainable and long-term solutions for unproductive livestock systems, a more comprehensive approach was needed. As the 1972 taskforce led by Tribe stated:

The first task of the interdisciplinary team would be to gain a basic appreciation of the major livestock production systems of Africa, by the study of all available

---

[25] *Improving Livestock Production in Africa: Evolution of ILCA's Programme 1974–94* (Addis Ababa: ILCA, 1994), https://hdl.handle.net/10568/5456.
[26] Ibid., p. 1.
[27] D. Gibbon, "Systems Thinking, Interdisciplinarity and Farmer Participation: Essential Ingredients in Working for More Sustainable Organic Farming Systems," in *Proceedings of the UK Organic Research 2002 Conference* (Aberystwyth: Organic Centre Wales, Institute of Rural Studies, University of Wales, 2002), pp. 105–108.

literature, a review of ongoing research programmes, and widespread travel and survey. From the base the team will then be expected to devise its own programme of studies.[28]

Following this vision, the newly created ILCA established a network of sites in tropical Africa to monitor livestock production systems. This network approach to "systems thinking" similarly impacted the establishment of the Africa Rice Center, as Harro Maat describes in Chapter 6, this volume. ILCA's zonal research teams measured the productivity of cattle, sheep, and goats. The first baseline surveys "diagnosed" general factors constraining animal production in the various zones.[29] These included low dry-season feed quality, inadequate water supplies, and competition between people and calves for limited milk supplies in arid pastoral systems. The surveys also focused on animal diseases and animal mortality, poor feed quality, the availability of animal draught power, and inefficient water conservation and utilization. The initial activities carried out by ILCA researchers, such as literature analysis and field surveys, were not bounded by common delineations, such as language and region. This early work pursued a "problem analysis" as a basis for developing interventions at the farm level, undertake more intensive studies, and assess systems-level production alternatives.

Nevertheless, ILCA's first quinquennial institutional review, completed in 1981, strongly suggested the institute should move away from systems description and place more emphasis on component research.[30] These analyses would build on the identification of constraints to livestock productivity up until that point by exploring options for overcoming these constraints. This was especially important, as early research had shown that, somewhat contrary to prior proclamations, introduced technologies generally did not offer any great advantages over traditional methods, given the economic and ecological constraints facing many African producers.[31] ILCA teams thus focused their attention on designing and researching possible improvements. These included the use of crossbred cows for dairying and cattle for traction, incorporating legumes into the cropping system, making better use of Indigenous feeds, alley cropping, establishment of "fodder banks" of leguminous pasture for dry-season grazing, selective harvesting and handling of crop residues to improve livestock nutrition and soil management, and improving the drainage of soils prone to waterlogging.

---

[28] Tribe et al., "Animal Production and Research in Tropical Africa."
[29] G. Gryssels, J. McIntire, and F. Anderson, "Research with a Farming Systems Perspective at ILCA," *ILCA Bulletin* no. 25 (1986): 17–22.
[30] Ibid.   [31] Ibid.

In some respects, ILCA was the antithesis of ILRAD. ILCA grew out of systems thinking recognizing Indigenous knowledge systems and the importance of local context. ILRAD had been founded on faith in cutting-edge science and universalizable technologies. Yet ILCA, like ILRAD, struggled to produce solutions. As William Pritchard, a renowned leader in tropical veterinary medicine, later observed, many of the challenges that ILCA faced stemmed from its adoption of systems thinking.[32] The systems approach was conceptual rather than organizational; it suggested an approach to research as opposed to a method for developing interventions. It did not necessarily lead to solutions to problems.

Moreover, while the research conducted within ILCA was intended to reflect and build upon real farming systems, the performance of the center was still measured against conventional scientific criteria. Accounts of ILCA staff expressing frustration that the reality of smallholder farming systems affected the operation and outcomes of their trials suggest that research ambitions and development objectives did not always align.[33] Consequently, with an underlying expectation of precision knowledge and scientific productivity, as opposed to systems understanding and on-farm benefits, the contribution of social scientists in ILCA's multidisciplinary teams was eventually limited to economists only.[34]

Thus, while one might simplistically characterize the research targets of ILRAD as "upstream" and ILCA as "downstream," one could equally argue that both were high-concept approaches. In either case, both institutions and approaches struggled to gain currency and momentum within the core business of CGIAR.

"Isolable Problems" versus "Systems Thinking"

In 1987, an external review of ILCA commissioned by the CGIAR's Technical Advisory Committee (TAC) recommended that the institute further narrow its focus to avoid spreading its resources too thinly over a broad spectrum of activities.[35] ILCA's original mandate stood, but it was asked to work more closely with and to strengthen the capacity of national agricultural research systems. In addition, ILCA was advised to focus its work on six narrowly defined "thrusts." These were three

[32] P. Gardiner, Interview, ILRAD, Nairobi, February 24, 1991. Interview transcript shared with James Smith.

[33] A. Waters-Bayer and W. Bayer, "Driving Livestock Development through Multi-disciplinary Systems Research: An Impact Narrative," ILRI Research Brief, 2014.

[34] Ibid.

[35] CGIAR Technical Advisory Committee, "Report of the Second External Program Review of the International Livestock Centre for Africa (ILCA)," December 1987, 4, https://hdl.handle.net/10947/1275.

"commodity thrusts" (cattle, milk, and meat; small ruminant meat and milk; and animal traction, with an aim to increase production and outputs), and three "strategic thrusts" (animal feed resources; trypanotolerance; and livestock policy and resource use).[36] With one exception, the thrusts were focused on animal production, health, nutrition, and genetics, all of which could be measured scientifically.

Meanwhile, at ILRAD change was also on the horizon. Although there had been some progress in ILRAD's research and training programs in its first decade, there was a sense that its short- and medium-term priorities needed revision. As was true at ILCA, much of ILRAD's early success was about basic rather than applied research, as evidenced by the centrality of the yearly tally of academic publications to successive annual reviews. Mapping this research productivity onto disease control priorities was very much secondary, a reflection of the institute's initial upstream focus. Furthermore, although research on ECF moved at a faster pace than trypanosomiasis research in the 1980s, as expected, ECF vaccine development nevertheless lagged behind the earlier, rather ambitious, timescales. It was only in 1989 that a review group recommended the establishment of a project area on vaccine formulation.[37] By comparison, trypanosomiasis research remained at a much earlier stage.

There was, therefore, a noted lack of progress in ILRAD's vaccine program, alongside a lack of strategic direction within ILCA's applied research that the instantiation of research "thrusts" attempted to correct. At the same time, Africa's food needs were rapidly growing and financial possibilities were shifting – with donor priorities focusing increasingly on the environment and funding in general constrained by global recession.[38] External assessments of animal agriculture in sub-Saharan Africa (conducted by the Winrock International Institute for Agricultural Development and supported by many of CGIAR's major donors) argued for a sharper focus of activities.[39] This layering of concerns, both within and beyond CGIAR livestock centers, began to point towards their closer collaboration. Similar suggestions were made at the 1992 CGIAR annual meeting in Washington, DC, where it was recommended that ILCA and

[36] S. Watanabe, "ILCA's Strategy for Improving the Output of Livestock in Sub-Saharan Africa based on Six Research Thrusts, " *Tropical Agriculture Research Series* 25 (1992): 92–103.

[37] ILRAD, *Annual Scientific Report 1989* (Nairobi: ILRAD, 1990), p. 2, https://hdl.handle .net/10568/91143.

[38] D. Byerlee, "The Search for a New Paradigm for the Development of National Agricultural Research Systems," *World Development* 26, no. 6 (1998): 1049–1055.

[39] Winrock International, *Assessment of Animal Agriculture in Sub-Saharan Africa* (Morrilton, AR: Winrock International Institute for Agricultural Development, 1992), https://hdl.handle.net/10947/186.

ILRAD work towards "closer cooperation through joint program committees and cross board membership" and that "joint funding opportunities should be explored."[40] A further external review published in January 1993 suggested that ILCA focus primarily on applied research on crop-livestock farming systems and build collaborative research networks and livestock research capacity in Africa's national agricultural research systems.[41] The days of ILCA as a center focused on broadly surveying African livestock production systems and setting research agendas appeared to be numbered.

Meanwhile, ILRAD was being asked to respond to similar externalities. Its 1993 medium-term plan signaled the coming change when it noted "Depending upon the levels of funding obtained, ILRAD and ILCA foresee increased collaboration, utilizing the complementary expertise and approaches of both institutes, in areas of mutual concern."[42]

When CGIAR had established ILRAD and ILCA in the early 1970s, those involved considered it likely that the two centers ultimately would come together as a unified research institute. Indeed, the gestation of the centers had included discussions about whether there should be two centers in the first place. It was not until two decades later, however, in the early 1990s, that the merger was set in motion. In May 1993 CGIAR took the decision to unify the centers, setting up a committee "to identify priority activities for international livestock research, which would be managed through a single institution and be constrained by the current proportion of CGIAR resources allocated to livestock."[43]

ILRI was established in September 1994. It had a huge task, as its remit would no longer be limited to Africa but encompass global needs. The institute came into existence during a period of flux and resource constraints within CGIAR and the broader donor community, and it had to deal with the realities of merging two fundamentally different entities. The merger was fraught with difficulty. With differing mandates, research cultures, and disciplinary representation, staff later commented that inviting ILRAD employees to support ILRI was "like asking turkeys to

[40] "CGIAR International Centers Week, Washington, DC, October 26–30, 1992: Summary of Proceedings and Decisions," January 1993, 9, https://hdl.handle.net/1094 7/280.

[41] ILCA, "Report of the Third External Programme and Management Review of International Livestock Centre for Africa (ILCA)," January 1993, https://hdl.handle.ne t/10947/1571.

[42] ILRAD, *ILRAD 1994–1998 Medium-Term Plan for Research on Livestock Diseases* (Nairobi: ILRAD, 1993), p. 19: https://cgspace.cgiar.org/handle/10568/49797.

[43] ILRI, *A Global Livestock Research Institute* (Nairobi: ILRI, 1995), pp. 2–3.

vote for Christmas."[44] A 2000 external report commissioned by the CGIAR TAC stated that it recognized the continued difficulties in unifying ILCA and ILRAD, with the "two centers [maintaining] widely different cultures."[45] In 2001, the ILRI annual review attempted to reframe these challenges as a reason for institutional pride:

> It is no mean achievement to have successfully made the transition despite the dramatic external and internal changes that ILRI had to withstand. From 1995 to 2001, when drastic falls in funding for international agricultural research and the change from dependency on unrestricted grants to reliance on project funding severely taxed the morale and programmatic integrity of all CGIAR centres, ILRI handled in addition the evolution out of two centres that could hardly have been more different in goals, culture and modes of operation.[46]

The merger, although partially demanded by cutbacks in CGIAR's budget, was perhaps premature. It was not ideal to merge primarily through financial exigency rather than strategic choice. ILRI was largely unable to exploit its new comparative advantages as it might have hoped, hampered by the sheer complexity of its vaccine-based research agenda and by the organizational challenge of effectively drawing together the existing scientific and systems-thinking approaches.

In addition to working within the organizational legacies of ILRI's two constituent institutions, ILRI administrators and staff, like others in the CGIAR system, had to work through the much longer legacies of conducting scientific research for development. As the historian Deborah Fitzgerald observes, "While some have argued that the technologies exported to developing countries are inappropriate, one might extend the argument by locating the inappropriateness in the institutional structures and ideologies from which these technologies have emerged."[47]

Building on this observation, we offer two case studies below to demonstrate how the legacies of ILCA and ILRAD, as two separate research institutions, continue to have repercussions for more recent ILRI projects. In the first case study, we discuss the development of transgenic, trypanosome-resistant cattle. We suggest that the roots of this work reflect the science-led values and notions of excellence as defined within

---

[44] O. Nielsen, "The Consultative Group on International Agricultural Research (CGIAR): The International Livestock Research Institute (ILRI)," *The Canadian Veterinary Journal* 40, no. 9 (1999): 642–644, at 642.

[45] CGIAR Technical Advisory Committee and CGIAR Secretariat, "Report of the First External Programme and Management Review of the International Livestock Research Institute (ILRI)," June 2000, 8, https://hdl.handle.net/10947/552.

[46] ILRI, *ILRI Annual Report 2001: The Poor and Livestock Mapping: Targeting Research for Development Impact* (Nairobi: ILRI, 2002), p. 3, https://hdl.handle.net/10568/49691.

[47] Deborah Fitzgerald, "Exporting American Agriculture: The Rockefeller Foundation in Mexico, 1943–1953," *Social Studies of Science* 16, no. 3 (1996): 457–483.

ILRAD. In the second case study, based on ethnographic research conducted in 2015, we present research on the pig value chain in Uganda. Here we highlight the similarities between the project and the systems-led research undertaken by ILCA. These case studies, in turn, demonstrate the ways in which the historical lineages of ILCA and ILRAD, and the prior concepts of research excellence and science for development on which those institutions were premised, have continued to influence and affect ILRI research after the merger in 1994.

Transgenic, Trypanosome-Resistant Cattle

Despite a long history of scientific attempts to control the different species and subspecies of *Trypanosoma* parasites, trypanosomiasis remains a major challenge for many countries in sub-Saharan Africa. As we described above, scientists at ILRAD attempted to develop a vaccine to control trypanosomiasis; however, no product ultimately came to fruition during the institute's independent existence – nor have they since. Thus, the current techniques used to prevent the spread of the disease predominantly focus on the tsetse-fly vector, with control programs involving methods such as the release of sterile male tsetse flies and the continued and increasingly innovative use of insecticides to limit tsetse populations.[48] For example, in multiple African countries, tsetse control programs have deployed "Tiny Targets," small, blue pieces of insecticide-impregnated cloth that attract and kill tsetse.[49]

In 2013, responding to the continued failures to eliminate trypanosomiasis, two scientists, Jayne Raper of City University of New York and Steve Kemp of ILRI in Nairobi, proposed genetic modification to produce cattle with 100 percent resistance to all species of trypanosomes.[50] The project, known as the Mzima project, received initial funding from the US National Science Foundation and was presented as a means through which to reshape livestock systems in Africa.[51] The subsequent

---

[48] A. M. Abd-Alla, M. Bergoin, A. G. Parker et al., "Improving Sterile Insect Technique (SIT) for Tsetse Flies through Research on Their Symbionts and Pathogens," *Journal of Invertebrate Pathology* 112 (2013): S2–S10.

[49] J. B. Rayaisse, F. Courtin, M. H. Mahamat, M. Chérif, W. Yoni, N. M. O. Gadjibet, M. Peka, P. Solano, S. J. Torr, and A. P. M Shaw, "Delivering 'Tiny Targets' in a Remote Region of Southern Chad: A Cost Analysis of Tsetse Control in the Mandoul Sleeping Sickness Focus," *Parasites & Vectors* 13, no. 1 (2020): 1–16.

[50] M. Yu, C. Muteti, M. Ogugo, W. A. Ritchie, J. Raper, and S. Kemp, "Cloning of the African Indigenous Cattle Breed Kenyan Boran," *Animal Genetics* 47, no. 4 (2016): 510–511.

[51] ILRI, "Mzima Cow Project: A Transgenics Approach to Introducing Resistance to Trypanosomiasis Translating Genetic Research to Adoption and Social Value," poster, March 12, 2018, https://hdl.handle.net/10568/91998.

research involved a number of international partners based in the United Kingdom (the Roslin Institute at the University of Edinburgh), United States (City University of New York and Michigan State University), and Kenya (ILRI).

ILRI was tasked with developing the technologies, skills, and infrastructure required to undertake the majority of the work in Kenya. ILRI's production of Tumaini ("Hope" in Swahili), the first cloned bull in Africa, was subsequently described by ILRI scientists as the first step towards producing trypanosome-resistant cattle, as the cloned bull opened up "the possibility of making genetically modified Kenyan Boran (see below) with foreign genes or desired traits."[52] The next step was subsequently to produce Boran cloned cattle with modified genes that would naturally confer resistance to trypanosomiasis. Scientists planned to achieve this by inserting baboon genes into cow genomes. Baboon genes were selected because, as Jayne Raper had established, they contain trypanosome lytic factors (TLFs) – a serum that has the ability to kill both animal and human infective trypanosomes.[53] In relation to previous ILRAD and ILRI research, the attempted development of the trypanosome-resistant cow represented a significant shift. Earlier scientific work on trypanosomiasis focused on the pathogen or the vector, whereas the Mzima project focused on the cow itself.

Some cattle breeds – notably the West African N'Dama – possess some natural resistance to trypanosomiasis. However, N'Dama were considered too small and unproductive to provide a solution for controlling the disease. Instead, the scientists working on the project selected the much larger Boran cattle, an Indigenous breed of East African zebu (*Bos indicus*) reared almost exclusively in Kenya. In its 2015 corporate report, ILRI stated that the final step of this long-term project was to introduce trypanosome-resistant cattle to breeding schemes across Africa.[54] The report set out that this research subsequently offered "a reliable, self-sustaining and cost-effective way of protecting tens of millions of African cattle against disease."[55] Yet the Mzima project is underpinned by an understanding that its transformed cow can be developed and integrated

---

[52] M. Yu et al., "Cloning of the African Indigenous Cattle Breed Kenyan Boran," *Animal Genetics* 47 no. 4 (2016): 510–511.

[53] R. Thomson, P. Molina-Portela, H. Mott, M. Carrington, and J. Raper, "Hydrodynamic Gene Delivery of Baboon Trypanosome Lytic Factor Eliminates Both Animal and Human-Infective African Trypanosomes," *Proceedings of the National Academy of Sciences* 106, no. 46 (2009): 19509–19514.

[54] ILRI, *Corporate Report 2014–2015* (Nairobi: ILRI, 2015), p. 45, https://hdl.handle.net/10568/68631.

[55] Ibid.

into a range of different contexts.[56] This assumption does not take into consideration that Boran cattle, although deemed more suitable by ILRI scientists, require different levels of care from other types and that this may be at odds with the ways in which people currently live with their cattle. As trypanosomiasis is not confined to one area but instead affects farmers across a number of countries in sub-Saharan Africa, the cow selected for development into a disease-resisting technology matters. For instance, despite both being zebu cattle breeds, the predominant breed of cattle reared in Sudan, Kenana cattle are reared for dairy, while Boran cattle are reared for beef.[57] A milking cow and a beef-producing cow have different roles in farmers' lives and may be valued differently. This, however, was not taken into account by scientists working on the Mzima project. As a 2017 report from a workshop on the Mzima Cow Strategy highlighted, "the exact effect of the transgene on milk and meat production is currently not known and must be carefully assessed in impact and safety studies."[58] Thus, while ILRI may well produce a trypanosome-resistant cow, scientists' seemingly singular pursuit of a biotechnological achievement at the expense of other considerations means that if it comes to fruition the final cow may not be suitable for every farming context. In short, the production of Tumaini raises questions about how scientific outputs translate into society, and whose benefit they serve. By proposing the development of trypanosome-resistant cattle, ILRI scientists conceptualized the cow as a technology that could be developed in the laboratory and integrated into farmers' lives. Yet a cow is not just ground-breaking science: it is also an animal that is understood in diverse ways across different contexts.

There are important continuities between the Mzima project and the ILRAD research projects of the 1970s and 1980s. Specifically, a group of international scientists were brought together to develop a magic-bullet technology that could be scaled up and integrated into existing livestock systems. As in the case of ILRAD vaccine development, the objective was to produce research excellence in the form of a cutting-edge scientific solution, without sufficiently exploring the applicability and acceptability of the output beyond the laboratory. The insurmountability of technical

[56] See M. Green, "Dairying as Development: Caring for 'Modern' Cows in Tanzania," *Human Organization* 76, no. 2 (2017): 109–120.

[57] O. Mwai, O. Hanotte, Y. J. Kwon, and S. Cho, "African Indigenous Cattle: Unique Genetic Resources in a Rapidly Changing World," *Asian-Australasian Journal of Animal Sciences* 28, no. 7 (2015): 911–921, at 911.

[58] C. Canales, N. Manson, and B. Jones, *Mzima Cow Strategy & Theory of Change: Translating from Genetic Research in Africa to Adoption and Social Value: Workshop Report*, Genetics for Africa – Strategies and Opportunities (Nairobi: ILRI, 2017), https://sti4d.com/wp-content/uploads/2022/06/report-mzima-workshop.pdf.

hurdles may be another continuity. Although ILRI staff were able to clone Tumaini, they were unable to develop a transgenic cow as planned. Thus, despite initial optimism about the benefits that novel scientific techniques could bring to disease control efforts, the problem of trypanosomiasis endures.

Smallholder Pig Value Chains in Uganda

In 2010, CGIAR's Funder Forum indicated that its research was not sufficiently translating into development outcomes.[59] In response, and in recognition that research alone was not generating impact, CGIAR introduced cross-institutional CRPs in 2011.[60] The CRPs were intended to act as a mechanism for funding AR4D programs, bringing together experts from across the (then) fifteen CGIAR centers to design and implement large-scale interventions. Prior to the introduction of the CRPs, ILRI had already begun to recognize that nontechnical innovations were needed in its livestock development programs, particularly to translate research into development impact.[61] As a result, ILRI administrators had already reincorporated a multidisciplinary systems approach, earlier adopted by ILCA, and social scientists had started to be reintroduced into ILRI's research teams.

CGIAR designated each center to lead a CRP, with ILRI leading on the CRP Livestock and Fish. ILRI subsequently created nine country-based hubs, with research in each hub focused on a single species or commodity. These research sites were to "serve as laboratories for characterizing and assessing smallholder value chains."[62] When ILRI staff were asked about the CRP Livestock and Fish, many spoke positively about its introduction and how it would affect the impact of their research.[63] As a veterinary epidemiologist in Nairobi asserted, "Old ILRI was all about writing papers, conducting research, developing careers that way. Science is now being used in a new way, to create development impact."[64]

---

[59] CGIAR, *A Strategy and Results Framework for the CGIAR* (Washington, DC: CGIAR, 2011), p. 3, www.iwmi.cgiar.org/About_IWMI/PDF/CGIAR_SRF_2011.pdf.
[60] Ibid.   [61] Waters-Bayer and Bayer, "Driving Livestock Development."
[62] D. Baker, A. Speedy, and J. Hambrey, *Report of the CGIAR Research Program on Livestock and Fish Commissioned External Evaluation of the Program's Value Chain Approach* (Nairobi: ILRI, 2014), p. viii.
[63] For her Ph.D., Rebekah Thompson undertook thirteen months of ethnographic research in Uganda (January–December 2015 and October–November 2017). For her Master's research, Thompson spent one month (April–May 2014) at the ILRI office in Nairobi, Kenya and one week at the ILRI office in Addis Ababa, Ethiopia (May 2014), studying the history of ILRI and AR4D programs.
[64] Thompson interview with ILRI staff, ILRI office Nairobi, Kenya, May 2014.

Figure 7.1. A Camborough pig on a farm in Mukono, about thirty-two kilometers east of Kampala, Uganda, 2015. The introduced breed is prized for being fast-growing and producing large litters, among other qualities. Photo by Rebekah Thompson.

In line with the expectations of the CRP Livestock and Fish, ILRI established several new offices and multidisciplinary teams, one of which was located in Kampala, Uganda. In Uganda, the first project funded under the CRP focused on pigs and aimed to catalyze the smallholder pig value chain – that is, the steps followed by small-scale farmers, traders, slaughterhouse workers, and butchers to raise, sell, and profit from pigs and pork products. This project, which was funded by the European Commission/International Fund for Agricultural Development and Irish Aid, focused its research activities on the pig value chain in three districts of Uganda (Mukono, Kamuli, and Masaka). In Uganda, pig farming has rapidly increased since the 1960s in concert with a rising demand for pork, particularly in urban areas (Figure 7.1). It is now recorded as having the highest per capita consumption of pork in East Africa.[65] With this social and economic backdrop, the main objectives for the project were to identify constraints and opportunities along the smallholder pig value chain, from

[65] K. Roesel, F. Ejobi, M. Dione et al., "Knowledge, Attitudes and Practices of Pork Consumers in Uganda," *Global Food Security* 20 (2019): 26–36.

farm to fork, and to design and test "best bet interventions."[66] Despite the importance of pigs for people's livelihoods in Uganda, until the introduction of the CRPs, pigs had not been central to research conducted by ILRI, ILCA, or ILRAD, with ILCA going as far as excluding pigs from all research projects during its years of operation (1974–94).[67] Thus, although pigs had been a major source of income and nutrition for people in Uganda, it was not until ILRI as a research institute began to perceive the pig value chain as a potential means of generating "impact" that the significance of pigs as objects of research in Uganda was recognized. This focus on pigs as potentially generating the right kind of research outputs simultaneously transformed pigs from a livestock animal into a research object and a development target.

When one of the authors visited the ILRI office in Kampala in 2015, it immediately became clear that all the work conducted through the office was on pigs. The walls were lined with posters of pigs and corresponding statistics from ILRI's research activities. As one of the staff members enthusiastically commented, "We speak pigs, we eat pigs, everything is pigs."[68] The research being carried out by ILRI staff at the site ranged from work on pig husbandry practices to pork consumption habits to trading patterns to slaughterhouse processes.

The multidisciplinary team in Uganda was composed largely of international staff. These staff members were expected to build relationships with a range of local partners, who in turn would translate the research conducted by ILRI into observable development outcomes. Straightforward on paper, establishing partnerships was difficult in practice. As an ILRI-employed capacity development consultant emphasized, "People are capturing certain knowledge but there is no situation in which to apply the knowledge. Then the knowledge goes to waste. Basically, the knowledge is not being applied."[69] Echoing this claim, ILRI staff often described local partnerships as unequal or "on and off," with this profoundly affecting the outcomes that they could expect, especially within the limited time periods dictated by funding bodies.

Funding was another major issue, and often discussed as a key reason why research was conducted with certain partners. Funding also explained why ILRI outputs regularly followed a similar format.

---

[66] W. O. Ochola, "Report of the Value Chain Assessment & Best Bet Interventions Identification, Workshop, Kampala, April 9–10, 2013," April 30, 2013, https://hdl.han dle.net/10568/29031.

[67] R. Blench, "A History of Pigs in Africa," in R. M. Blench and K. MacDonald, eds., *Origins and Development of African Livestock: Archaeology, Genetics, Linguistics and Ethnography* (Oxford: Routledge, 2000), 355–367.

[68] Thompson Interview with ILRI staff, ILRI office Kampala, Uganda, February 2015.

[69] Thompson Interview with ILRI staff, ILRI office Kampala, Uganda, February 2015.

Funding partners in Europe dictated the form of outputs, typically publications and project reports. These outputs did not always meet expectations, even when they took the "right" form. While the Uganda team did successfully publish numerous papers and reports from different scientific studies conducted along the pig value chain, external evaluators maintained that the outputs from the CRP Livestock and Fish were lacking overall in terms of "high-quality" peer-reviewed publications and that it should be producing more "excellent rather than good or acceptable grey literature."[70]

In order to translate the content of their publications and reports into observable development impact on the ground, ILRI staff often held training sessions. In these sessions, relevant stakeholders were educated on topics such as biosecurity or food hygiene measures. Participants were also often provided with products that ILRI staff deemed they should find useful, including information sheets, bleach, fly nets, or "tippy taps" (a low-cost, hands-free device for handwashing). These interventions, in line with the broader objectives of the CRP Livestock and Fish, were also easily transferable into other contexts outside of Uganda.[71] However, when we visited pig farms and pork butchers in Uganda, it became evident that ILRI interventions were not consistently generating sustainable impact for people working along the pig value chain (Figures 7.2 and 7.3). Pork butchers, for example, described fly nets as obscuring customers' vision of the meat, and many butchers further explained that, as they were unable to read, they could not understand the text on food safety information sheets. The form that ILRI interventions were forced to take reduced the complexity of the pig supply chain and the relationships within it to a single workshop session or a training manual. As a result, people were often trained on "best practices" as defined by international organizations such as the United Nations Food and Agriculture Organization (FAO) or World Health Organization (WHO). These interventions, in turn, were used as evidence that ILRI research was generating development impact along the pig value chain.

To summarize: in Uganda, a research team comprised almost entirely of international staff adopted a transdisciplinary, systems approach in an attempt to transform the smallholder pig value chain. Yet the outputs generated did not consistently translate back into meaningful impact for stakeholders working with pigs. This sketch of their work shows how certain forms of outputs – most commonly the publications and reports

---

[70] CGIAR-IEA, *Evaluation of CGIAR Research Program on Livestock and Fish* (Rome: CGIAR, 2016), p. 12, https://iaes.cgiar.org//sites/default/files/pdf/LF-EVAL-Report-Volume-1_FINAL-1.pdf.
[71] Ibid., p. 2.

Figure 7.2 Pork products for sale in Mukono, Uganda, 2015. Photo by Rebekah Thompson.

Figure 7.3 Transporting pigs by bike in Uganda, 2017. Photo by Rebekah Thompson.

that transcended local contexts – were still perceived within CGIAR and by external funding bodies as markers of excellence. As a result, actual development outcomes were reduced to little more than a series of quick fixes that could be used as evidence of impact when reporting back to funding bodies.

The example of the smallholder pig value chain project in Uganda indicates how some earlier criticisms of ILCA continue to apply to contemporary ILRI projects. More specifically, "high-quality" peer-reviewed publications and globally transferable solutions still ultimately govern what is perceived to be excellence in terms of project outcomes. In practice, this means that ILRI's AR4D continues to be geared towards generating excellence in terms of solutions that can be inserted into any livestock system, rather than sustainable development that is tailored to local contexts.

Conclusion

In this chapter, we illustrated how the formation of ILRAD and ILCA carved two distinct trajectories – one focused on generating scientific solutions for veterinary medicine, the other on livestock systems research. Although distinct in their mandates, these two institutions merged in 1994, creating ILRI. We have shown that despite this amalgamation, there are continuities stretching from the distinct historical lineages of ILRAD and ILCA to recent projects carried out by ILRI. More specifically, we have demonstrated how research excellence within ILRAD and ILCA was shaped by the continued privileging of certain notions of science and the production of scientific solutions for agricultural problems within international institutions. ILRAD was set an almost impossible technical task, but the promise of advances in immunology gave it significant momentum in pursuit of its vaccines. By contrast, ILCA's systems approach was scientifically more feasible, yet its researchers struggled to gain recognition for the quality of their work. Ultimately the limits of both approaches were exposed, and this in turn rationalized the merger. ILRI continues to undertake important work, but the specter of "excellence" – what it is and who defines it – has continued to loom large.

Reflecting on this history, we contend that despite a renewed focus on generating impact from AR4D programs, as an institution ILRI has continued to strive for recognition as a global center of scientific excellence, shaped by notions of excellence ascribed by global institutions and ideas about cutting-edge science that can be abstracted from specific farming contexts and therefore adopted by researchers and policymakers around the world. Yet, we argue that by framing excellence in terms of scientific solutions, ILRI staff have had little option but to overlook the complexities of livestock supply chains and the nuances of human–animal relationships in specific contexts. This limitation is entrenched further as ILRI is forced to look globally for its funding. As Derek Byerlee and Greg Edmeades (Chapter 9, this volume) similarly conclude with relation to CIMMYT, it is in some ways easier to attract funding by igniting donor interest in the biotechnological possibility of Tumaini than in worthy farmer-facing breeding programs for poultry and livestock that ILRI also leads, such as the African Dairy Genetic Gains program.

As we move into the next fifty years of CGIAR research, there is a pressing need to examine how excellence is conceptualized by CGIAR institutions, funding bodies, and researchers, and to recognize the implications that this conceptualization has for the ways research is conducted and the outputs it generates.

8 The Protein Factor

CIAT's Bean Improvement Research in Central America

Wilson Picado-Umaña

Beans are the most popular legume for human consumption and historically have been valued by the poorest populations around the world as a source of vegetable protein. Accounting for more than 30 million hectares, the legume is one of the most cultivated crops in Africa, Asia, and Latin America. Despite its significance for human nutrition, studies on the history of bean improvement are less known among scholars, compared with studies of such grains as wheat, maize, and rice.[1] Regarding plant breeding in legumes, several studies focus on the research developed from within the Consultative Group on International Agricultural Research (CGIAR).[2] Yet few studies examine the research developed by the International Center for Tropical Agriculture (CIAT) based in Colombia, and fewer still focus on CIAT's specific impact in Central America.

This chapter analyzes the bean-breeding programs developed by CIAT in Central America between 1970 and 1990. The region is an ideal laboratory for studying plant breeding in beans during the Green Revolution. On the one hand, Central America became a Cold War hot spot from 1960 to 1990, as the United States waged counterinsurgent campaigns to stymie the spread of communism in this region. At the same time, Central America experienced rising rates of malnutrition, particularly in rural areas. Amidst this so-called "protein gap," world organizations and scientific institutions conducted numerous surveys and field research to identify the nutritional deficiencies in the region's rural

---

[1] Alan L. Olmstead and Paul W. Rhode, *Creating Abundance: Biological and American Agricultural Development* (New York: Cambridge University Press, 2008); Dana G. Dalrymple, *Development and Spread of High-Yielding Rice Varieties in Developing Countries* (Washington, DC: Agency for International Development, 1986); Jack R. Kloppenburg, *First the Seed: The Political Economy of Plant Biotechnology, 1492–2000*, 2nd edn. (Madison: University of Wisconsin Press, 2004).

[2] Thomas M. Arndt et al., *Resource Allocation and Productivity in National and International Agricultural Research* (Minneapolis: University of Minnesota Press, 1977); Vernon Ruttan, *Agricultural Research Policy and Development* (Rome: FAO, 1987).

population, as well as its causes and potential solutions. What role did the bean varietal improvement process play at this juncture of Cold War and nutritional crises? What were the results?

The first part of this chapter addresses CIAT's origins, its organizational framework, and the research and training programs it established. The second part examines the center's endeavors in approaching the "protein crisis" in the so-called Third World through the creation of a bean research program. The third part delves into the development of a bean-breeding program aimed at enhancing nutritional conditions among rural populations in Central America. The last part argues that the obstacles in advancing a bean monoculture and the consequences of civil war on peasant agriculture hindered the development of a Green Revolution in beans in the region.

CIAT's Research Programs

CIAT's precursor was the Colombian Agricultural Program (CAP), which was established by a 1949 agreement signed between the Rockefeller Foundation and Colombia's government. During World War II, the foundation funded research at the agriculture and veterinary programs of the Universidad Nacional (in Medellín and Bogotá, respectively), as well as by the Tropical Agriculture School in Cali (which would soon relocate to Palmira and join the Universidad Nacional system with Rockefeller Foundation support). At each campus, the foundation invested in building facilities and equipment. At the same time, the Rockefeller Foundation offered training scholarships for Colombian students and faculty at universities in the United States and through the Mexican Agricultural Program (MAP).[3]

The Colombian program was built from the blueprint of MAP. It came under the initial direction of Lewis M. Roberts and Joseph A. Ruppert, prominent US scientists who specialized respectively in maize and wheat in Mexico. In Colombia, Roberts led the program at the Tulio Ospina experiment station in Medellín, while Ruppert operated out of the new Tibaitatá experiment station outside Bogotá, which possessed favorable ecological conditions for wheat production.[4] The founding of Tibaitatá, which replaced the older Picota station in 1951, was due in part to Edwin J. Wellhausen, a veteran of the Mexican program, who modeled the new

[3] Rockefeller Foundation, *Annual Report 1947* (New York: Rockefeller Foundation, 1947), pp. 166–167. The history of CIAT is further detailed by Timothy W. Lorek, Chapter 3, this volume.

[4] E. C. Stakman, Richard Bradfield, and Paul C. Mangelsdorf, eds., *Campaigns against Hunger* (Cambridge: Harvard University Press, 1967), p. 219.

station on Mexican experimental stations.[5] Both stations, in Medellín and Bogotá, experimented with new varieties, using genetic material brought from Mexico. In 1955, once the research projects were fully consolidated, the Rockefeller Foundation advised in the creation of a new Colombian government agency, the Ministry of Agriculture's Department of Agricultural Research, or Departamento de Investigación Agrícola, which emulated the coordinated approach of the Office of Special Studies in Mexico.

Following the precedent for centralizing coordination of agricultural research in Colombia, represented by the Department of Agricultural Research, the Rockefeller Foundation then supported the establishment of the Colombian Agricultural Institute (ICA) in 1962. This was part of a broader land reform project funded in part by the Alliance for Progress, a program initiated by US President John F. Kennedy to foster economic ties between the United States and Latin America. After the establishment of ICA, the Rockefeller Foundation gradually withdrew from direct involvement in Colombian domestic agricultural research and partnered with the Ford and Kellogg Foundations to redirect their Colombian assets into a new international center. CIAT was created in 1967, the same year the International Institute of Tropical Agriculture (IITA) opened in Nigeria. Both institutions were funded by the United States Agency for International Development (USAID), the Rockefeller and Ford Foundations, and their respective state governments. Both centers' first goal was the development of agricultural research in tropical environments. Thus, research prioritized crops grown by peasants for local consumption, such as cassava and legumes, over global cereal crops, such as wheat and maize (Figure 8.1).

CIAT's particular framework stemmed from the organizational models of other CGIAR institutes. Following the experience of the International Rice Research Institute (IRRI) in the Philippines, CIAT built a brand-new campus in Palmira, Colombia, adjacent to the older ICA facilities that once formed part of the CAP.[6] It organized its research projects and scientific teams following the framework developed by the International Maize and Wheat Improvement Center (CIMMYT) in Mexico.[7] Furthermore, CIAT pursued livestock management research and prioritized support for smallholders over larger farmers. CIAT also worked in close contact with national research programs in Latin America in order to develop training programs and scientific networks.

[5] Ibid., p. 220.
[6] CIAT, *Annual Report 1969* (Cali, Colombia: CIAT, 1969), p. 8, https://hdl.handle.net/10568/61840.
[7] John Lynam and Derek Byerlee, *Siempre pioneros: CIAT: 50 años contribuyendo a la sostenibilidad alimentaria futura* (Cali, Colombia: CIAT, 2017), p. 19.

Figure 8.1 Beans featured among the objects of research and breeding at the Rockefeller Foundation's agricultural program in Colombia. Here a small group considers beans growing in the greenhouse, ca. 1954. CIMMYT repository. © CIMMYT.

Between its creation and the 1980s, CIAT developed six broad programs encompassing plant-breeding research and training. The rice program, launched in 1967, took advantage of an alliance with IRRI in the Philippines and ICA in Colombia.[8] On the one hand, cooperation with IRRI turned CIAT into a "genetic bridge," allowing the transcontinental exchange of rice varieties and the introduction of the high-yielding variety IR8 germplasm to Latin America. Following IRRI's experience, the program developed high-yield varieties suited for irrigated rice, and created an intensive agrochemical package for pest control and plant disease.[9] On the other hand, the association with ICA enabled experimentation and field testing of Asian rice varieties in Colombia and

---

[8] CIAT, *Annual Report 1969*, p. 27.
[9] CIAT, *Seminar on Rice Policies in Latin America* (Cali, Colombia: CIAT, 1971), https://hdl.handle.net/10568/56374.

facilitated the development of lines adapted to Latin America's tropical agricultures. Initially established in Colombia, the pasture program pursued live-stock management systems that focused on herd improvement and ani-mal health. After 1975, the program spread across other Latin American countries, focusing on grass improvement. Scientists at CIAT deemed depleted soils, droughts, and seasonal water availability changes as causes for low productivity of local grass varieties. Grass improvement required the creation of a germplasm collection (samples of varieties and popula-tions thought to be useful in breeding) and the development of science-based knowledge on local pastures and grasslands. Thus, CIAT built the International Tropical Pastures Network, enabling germplasm exchange and research between the center and programs in other Latin American countries. Emphasis on increasing meat and dairy production on the acidic and infertile soils of the Latin American tropics reserved, in turn, more fertile lands for export-led agriculture, as Timothy W. Lorek chron-icles in Chapter 3 of this volume.

The cassava program had no direct ties to other programs previously developed by the Rockefeller Foundation.[10] Therefore, CIAT designed the program's framework and built an international scientific network. The program started by assessing the disadvantages stemming from the crop's particular traits. For example, cassava improvement did not favor the development of varieties with general adaptability, as was the case with wheat or rice. Second, the crop was primarily grown by peasants in a slope farming system characterized by the low use of chemical inputs and the lack of irrigation. To overcome both limitations, the cassava program took over the creation of a germplasm collection, as well as the development of a technological package and agricultural practices that would "adapt the crop to the environment, instead of adapting the envir-onment to the crop." Furthermore, the program established a global scientific network, allowing the exchange of knowledge with programs and institutes in Asia and Africa, particularly Thailand and Nigeria.[11]

Varietal improvement of beans to increase yields per hectare became one of CIAT's most complex tasks. For instance, Latin American

[10] CIAT, *A Proposal for the Improvement and Development of Cassava, A Tropical Root Crop* (Cali, Colombia: CIAT, 1971), https://hdl.handle.net/10568/72110.
[11] CIAT, *Informe anual del programa de yuca 1979* (Cali, Colombia: CIAT, 1980), pp. 99–101, https://hdl.handle.net/10568/77963; E. R. Terry and Reginald MacIntyre, eds., *The International Exchange and Testing of Cassava Germ Plasm in Africa* (Ottawa: IDRC, 1975), https://hdl.handle.net/10568/77936; CIAT, *CIAT 1984: Reseña de los logros principales durante el período 1977–1983* (Cali, Colombia: CIAT: 1984), p. 37, https://hdl.handle.net/10568/70299; CIAT, *CIAT Annual Report 1987* (Cali, Colombia: CIAT, 1987), p. 37, https://hdl.handle.net/10568/65942.

peasants grew a wide diversity of bean varieties in plots with other crops using local slope farming systems that enabled cultivation on sloping land while conserving water and soil. Thus, the bean program developed varietal selection processes through close coordination with national programs to take advantage of local experts' knowledge, producing varieties suited to each country's different agro-ecological and social environments. Accordingly, CIAT established a training program that offered internships and short courses in Colombia and other participant countries. In the 1980s, the successful results that CIAT achieved in Latin America spurred the launching of similar programs in Africa.

CIAT's genetic resources program managed the germplasm collections created by these breeding efforts. The bean program benefitted from the seed samples collected by the US Department of Agriculture in Mexico during the 1960s and from the collection gathered by the Rockefeller Foundation's CAP in Colombia.[12] Possessing more than 9,000 samples, or accessions, in 1980, the pasture program obtained its genetic material from the Commonwealth Scientific and Industrial Research Organization (CSIRO) in Australia and the University of Florida in the United States.[13] It also acquired fodder legume accessions identified by CIAT's scientists in Brazil, Venezuela, and Southeast Asia. Furthermore, CIAT co-managed the cassava collection with IITA in Nigeria. By and large, the center became the most important global distributor of bean, cassava, and grass-breeding materials.

The training program supported the specialization of government officials in specific agricultural research fields, as well as the distribution and adoption of breeding materials and new varieties.[14] In the pursuit of both goals, CIAT published booklets and pamphlets, facilitating knowledge exchange with scientists from many Latin American countries. The program also organized field trips to Mexico, Colombia, and Central America, as well as visits to North American universities and Asian institutes, to become acquainted with various training programs. Throughout the 1970s, CIAT increasingly moved training activities from its facilities in Colombia to partner countries and provided education for scientists from the cassava, beans, rice, and grass programs. CIAT supported these programs directly and received funding from such entities as the Kellogg Foundation, USAID, the Inter-American Development Bank, and others.[15] The exchange of knowledge that resulted from this program set the foundation for CIAT's global network.

[12] Lynam and Byerlee, *Siempre pioneros*, p. 110.   [13] Ibid., p. 111.
[14] CIAT, *Training and Conferences Report* (Cali, Colombia: CIAT, 1984), https://hdl.handle.net/10568/69052.
[15] Lynam and Byerlee, *Siempre pioneros*, p. 25.

CIAT and the Third World's "Protein Crisis"

During the 1950s and 1960s, the World Health Organization (WHO), the United Nations Children's Fund (UNICEF), and the United Nations Food and Agriculture Organization (FAO) acknowledged protein consumption deficiencies as one of the most significant problems affecting the Third World's population, marking what became known as the "protein crisis."[16] Multiple studies carried out during the 1930s in North America and Africa showed the impact of low protein intake on children's physical development and early child mortality. These findings led world organizations to establish food programs and scientific commissions, such as the United Nations Protein Advisory Group in 1955, and to hold symposiums and conferences in Mexico (1960), Italy (1963), and the United States (1960, 1964, and 1965) to address protein consumption in Third World countries. In 1964, FAO's survey "Protein: At the Heart of the World Food Problem" pushed the topic to the fore of its agenda. A few years later, the United Nations issued its 1968 "International Action to Avert the Impending Protein Crisis," declaring the "protein gap" a global-level emergency.[17] That same year, *Life* magazine published an image of malnourished children in the Biafra War in West Africa on its cover, bringing protein deficiency to wide public attention.[18]

These discussions regarding a protein crisis overlapped with the founding of CIAT and the development of the research programs described above. In 1966, prior to the creation of CIAT, the maize breeder Lewis M. Roberts and agricultural economist Lowell Hardin outlined a vision for an institute for agricultural research and training in the Latin American tropics.[19] According to their perspective, the institute should focus on improving a few tropical crops with high nutritional value and clear pathways to increased production. Thus, research should target such crops as soy and other beans, as well as livestock, to raise protein availability among impoverished rural populations.

In a 1970 study, Roberts declared the protein consumption deficit as one of the most crucial global nutritional problems of the time.[20] He

---

[16] Richard D. Semba, "The Rise and Fall of Protein Malnutrition in Global Health," *Annals of Nutrition and Metabolism* 69, no. 2 (2016): 79–88.

[17] United Nations, *International Action to Avert the Impending Protein Crisis: Feeding the Expanding World Population* (New York: United Nations, 1968); FAO and OMS, *Informes sobre nutrición*, Informe No. 42 (Rome: FAO-OMS, 1966).

[18] "Starving Children of Biafra," *Life* (July 12, 1968).

[19] Lewis M. Roberts and Lowell S. Hardin, "A Proposal for Creating an International Institute for Agricultural Research and Training to Serve the Lowland Tropical Regions of the Americas," October 1966, https://hdl.handle.net/10568/72329.

[20] Lewis M. Roberts, "The Food Legumes," November 1970, https://hdl.handle.net/10947/1528.

warned that although world food production had generally stabilized in recent years, per capita consumption of protein in the poorest countries continued to decline. According to Roberts, unlike meat with its high production costs, legumes – particularly beans – offered the possibility of tackling this imbalance. Legumes contributed to high protein intake in the diet, were available at low prices, displayed high adaptability to different agro-ecological environments, and were consumed more among peasants. However, governments had prioritized research on agricultural cash crops over research on local consumption staples such as beans. Roberts concluded that this led to declining yields per hectare and limited information on varietal improvement, plant diseases, and pest control.[21]

Early in the 1970s, CIAT surveyed the production and consumption of beans in Latin America and the contribution of the legume to protein intake, aiming to create a bean research program.[22] The results, released in the program's previous drafts and written reports, identified Latin America as the place of genetic origin for *Phaseolus vulgaris* and as the world's largest producer. Furthermore, the studies identified bean consumption as the primary protein source in Latin American rural diets, surpassing animal protein consumption. Notwithstanding these advantages, the region as a whole showed oscillating production with decreasing yields per hectare.

The Bean Program, officially launched in 1974, aimed at the genetic improvement of beans to increase production. It set four goals. To begin with, it created a bean germplasm collection. CIAT's designation as the global gene bank for bean germplasm led to a rapid increase in accessions: from 21,000 in 1978 to 35,000 in 1984.[23] Second, the program established a cooperative network with national research institutes. CIAT's bean program stemmed from a set of projects developed by seventy-four researchers specializing in varietal improvement across Latin America by

[21] Ibid., 132.

[22] Ricardo Bressani et al., "Proposal for the Establishment of a Cooperative Programme for Field Bean Research in Latin America and the Caribbean Zone," November 1973, https://hdl.handle.net/10947/968; Grant M. Scobie, Mario A. Infante, and Uriel Gutiérrez Palacios, "Production and Consumption of Dry Beans and Their Role in Protein Nutrition: A Review," April 1974, https://hdl.handle.net/10568/69739; Uriel Gutiérrez Palacios, Mario Infante, and Antonio Pinchinat, *Situación del cultivo de frijol en América Latina* (Cali, Colombia: CIAT, 1975), https://hdl.handle.net/10568/71869.

[23] CIAT, *Programa de frijol: Informe de 1978* (Cali, Colombia: CIAT, 1979), C-3, https://cgspace.cgiar.org/handle/10568/69042; CIAT, *Informe anual 1984: Programa de frijol* (Cali, Colombia: CIAT, [1984]), p. 9, https://cgspace.cgiar.org/handle/10568/69042. The establishment of global germplasm collections within CGIAR is discussed in Marianna Fenzi, Chapter 11, this volume.

1979.[24] Although these national projects lacked scientific impact and proper funding, CIAT assembled national researchers through training activities such as courses and internships. Varietal improvement became a predominant topic in CIAT's training program: nearly 44 percent of the bean program trainees had become specialized in the subject by 1983.[25]

Third, the program created regional projects for varietal improvement in Latin America and Africa. One of these projects was the Cooperative Regional Project on Beans for Central America, Mexico, and the Caribbean (PROFRIJOL), discussed later in this chapter, which was launched in 1978 and aimed at improving beans to increase production in Mexico, Central America, and the Caribbean. Likewise, CIAT signed a cooperation agreement with the Brazilian government and developed another bean project in the Andean countries. During the 1980s, CIAT established three additional regional improvement projects in Africa.[26]

Finally, CIAT's bean program developed new varieties through international testing and experimental nurseries.[27] CIAT assembled a varietal research team in Colombia comprised of three breeders, each assigned to a geographical region: the first managed Central America and regions of Brazil with significant cultivation of black and red beans; the second oversaw the rest of Brazil, Mexico, Chile, Argentina, and the Middle East; the last supervised the Andean countries and Africa. In addition to their having a geographical specialization, the three breeders developed their varieties paying attention to specific ecological traits, such as diseases, pests, or types of soil. Every year, the breeders supplied CIAT with a selection of promising lines, which were then assessed in three trial cycles. The first cycle, "Bean Team Nursery," consisted of the evaluation of these promising lines in CIAT's experimental stations in Colombia. The second cycle, "Preliminary Performance Testing," assessed the lines' productive performance, as well as additional traits concerning consumers' preferences. The last, known as "Bean Performance and Adaptation

---

[24] N. L. Johnson et al., "The Impact of CIAT's Genetic Improvement Research on Beans," in R. E. Evenson and D. Gollin, eds., *Crop Variety Improvement and Its Effect on Productivity: The Impact of International Agricultural Research* (Cambridge: CABI Publishing, 2003), pp. 257–274, at 260.

[25] This percentage was well above the next-highest percentages of trainees, including "Production" (18%), "Agronomy" (15%), "Entomology" (11%), and "Phytopathology" (11%). CIAT, *Informe anual 1983: Programa de frijol* (Cali, Colombia: CIAT), p. 225, https://cgspace.cgiar.org/handle/10568/69042.

[26] Oswaldo Voysest, *Intercambio de germoplasma dentro de la red de frijol* (Cali, Colombia: CIAT, 1983), https://hdl.handle.net/10568/71967.

[27] Ibid.

International Nursery," evaluated lines selected for their performance in the previous trials. CIAT carried out this cycle in its stations, or abroad at other governments' request. The distribution of testing sites abroad aimed at the international exchange of potentially productive varieties and the evaluation of this material under local ecological conditions. This methodology allowed the implementation of roughly 1,400 experiments for bean improvement across 82 countries.[28]

The bean program widened its scope between 1970 and 1990, expanding its funding from $350,000 at the outset to $14 million by the final year. Moreover, the program's staff increased from two top scientists at the beginning to twenty-six by the end of the period, including seven varietal improvement experts.[29] Furthermore, the impact of varietal improvement, measured by the share of land cultivated with improved varieties, grew over three decades: by 1988 nearly half of the bean fields in Latin America used improved varieties from CIAT's genetic material. But distribution of CIAT material was uneven between regions. In Latin America, such countries as Costa Rica, Argentina, and Bolivia accounted for up to 70 percent of the land hosting CIAT's improved bean varieties, while in Colombia, Peru, and Ecuador their share ranged from 10 to 20 percent. In contrast, Africa accounted for only 15 percent of the land cultivated with varieties released and made available to farmers under the program.[30] During the 1990s, the total number of released varieties in both continents was almost 350.

CIAT's bean programs add additional dimensions to interpretations of the Green Revolution. The foundational accounts of a Green Revolution in Asia presented an economic narrative of agricultural development that regarded technology as a tool for increasing agricultural productivity.[31] According to this narrative, modern technology aimed at the improvement of yields per hectare for peasants, reducing productivity differences between developed and underdeveloped countries. Thus, high-yielding varieties became the suitable solution to end global hunger while enhancing economic growth in the Third World.

Researchers who sought to apply the vision of agricultural development in Central America in the 1970s and 1980s added two new tenets to this narrative. To begin with, observers established an association between

---

[28] Oswaldo Voysest, *Viveros internacionales de rendimiento de frijol: Manual descriptivo: Frijol arbustivo, frijol voluble* (Cali, Colombia: CIAT, 1983), p. 12, https://hdl.handle.net/105 68/69567.

[29] Johnson et al., "The Impact of CIAT's Genetic Improvement Research," p. 259.

[30] Ibid., p. 268.

[31] Yujiro Hayami and Vernon Ruttan, *Agricultural Development: An International Perspective* (Baltimore: Johns Hopkins University Press, 1971).

plant breeding and malnutrition. CIAT and its bean research program unfolded amid the "protein crisis" in the Third World, described in greater detail below. Thus, Central America became a laboratory for boosting protein consumption among the region's impoverished rural populations through bean genetic improvement. In this vein, CIAT's research was innovative: the Green Revolution in beans sought the increase in yields per hectare, as had been the case with crops like wheat and rice, but also the enhancement of the crop's nutritional value. Accordingly, the research interests of agronomists at CIAT and nutrition experts converged under an interdisciplinary association that was unusual at the time.

Finally, the narrative connected malnutrition to physical performance among rural workers. According to some experts, the nutritional deficit among Central America's rural population was not only a matter of public health but also a hindrance to the agricultural workforce. During the 1970s and 1980s, nutrition and agricultural development became the focus of international conferences in which experts analyzed the impact of poor diets on workers' physical performance. A 1974 symposium held in Guatemala produced a study compiling contributions from American universities' scholars and members of world organizations such as FAO, the Nutrition Institute of Central America and Panama (INCAP), CIMMYT, and CIAT.[32] The event was attended by Green Revolution experts and bean-breeding researchers such as Lester R. Brown, Robert F. Chandler, Antonio M. Pinchinat, and Lewis M. Roberts. Moreover, nutrition specialists such as Marina Flores, Leonardo Mata and Fernando E. Viteri participated alongside economists and agronomists. Further, a group of scientists at INCAP and American researchers carried out a field study to gauge the effects of the nutritional condition and caloric intake on rural workers' daily productivity.[33] As this body of research suggests, a narrative based on the tenets of agricultural modernization and public health influenced bean breeding in Central America, a process that sought the improvement of plants and the enhancement of human bodies as well.

---

[32] N. Scrimshaw and M. Behar, eds., *Nutrition and Agricultural Development: Significance and Potential for the Tropics* (New York: Plenum Press, 1976).
[33] Fernando E. Viteri, "Definition of the Nutrition Problem in the Labor Force," in Scrimshaw and Behar, eds., *Nutrition and Agricultural Development*, pp. 87–98; Maarten D. C. Immink et al., "Energy Supplementation and Productivity of Guatemalan Sugar-Cane Cutters: A Longitudinal Approach," *Archivos Latinoamercanos de Nutricion* 36, no. 2 (1986): 247–259.

CIAT's Bean Research and Nutritional Crisis in Central America

After CIAT's inception, Central America quickly emerged as a strategic site for its research programs. Between 1969 and 1982, 16 percent of trainees in any CIAT program came from Central America, a number surpassed only by Colombia, which supplied 21.3 percent.[34] Central America's presence was dominant even compared with countries that had much more extensive agricultural lands, such as Brazil, Mexico, Peru, and Ecuador. Likewise, scientists from the region participated prominently in almost all research programs: they represented 23 percent of participants in the bean program, above Colombia and Brazil. They also accounted for 15 percent of recruits to the pasture program, a number exceeded only by Colombia. Meanwhile, 19 percent of participants in the rice program came from Central America, which was a lower share than came from Brazil, but higher than Colombia. Finally, the region supplied 6 percent of participants in the cassava program.[35]

Central American participants trained to become specialized in the different research programs offered by CIAT. Almost half of the bean trainees came from Honduras and Guatemala, while Costa Rica and Panama led the cassava training program, providing 60 percent of recruits. In rice training, Costa Rica, Panama, Honduras, and Guatemala each supplied between 20 and 25 percent of trainees. Among these, the bean training program was arguably the most influential in circulating knowledge and agricultural research training across Central America. It accounted for more than 30 percent of the total number of Central American professionals trained through CIAT, far above any other program. Beyond Central America, the bean training program encompassed roughly one-fifth of the total professionals trained by CIAT between 1969 and 1983, exceeding the cassava and pasture programs' share of trainees and assembling researchers from thirty-five countries across three continents.[36]

Central America's prominence in CIAT training programs was due to specific conditions on the ground. During the 1960s and 1970s, the nutritional condition of the rural population in Central America was critical. A set of surveys conducted in the region showed a deficit in the population's intake of protein and calories measured according to recommendations by world health organizations. One study reported a decline in the overall consumption of protein and calories between 1965 and

---

[34] CIAT, *Training and Conferences Report*, pp. 8–14.   [35] Ibid.   [36] Ibid., pp. 6, 8–14.

1982.[37] These declines were most severe in El Salvador and Guatemala, which, not coincidentally, experienced significant political turmoil and conflict during this period.

These studies revealed the demographics of unequal nutritional access. In Central America, malnutrition affected mostly the impoverished rural populations, particularly children. A study performed by INCAP found an increase in the prevalence of children with some level of malnutrition between 1965 and 1975, excluding Costa Rica.[38] During the 1970s, rates of malnutrition in children surpassed 40 percent in Guatemala, El Salvador, and Honduras, while this figure came close to 20 percent in Nicaragua and Panama. Further, between 37 and 50 percent of total deaths occurred among children under the age of six in Guatemala, Honduras, El Salvador, and Nicaragua during the same period.

Much of the Central American population's daily diet included high consumption of maize and rice, with limited meat consumption.[39] In Guatemala and El Salvador, maize supplied 60 percent of calories to the diet, while in Costa Rica and Panama, rice provided between 39 and 47 percent of calories. Urban and rural areas within the region differed, as wheat and rice predominated most among urban populations but was considerably less dominant among rural populations.

Beans were crucial to Central American diets. Excluding Panama, nutritional data showed a common trend across the region: beans added more calories to the rural population's diet than beef. In Guatemala, El Salvador, Nicaragua, and Costa Rica, beans provided three times more caloric sustenance than beef. In Honduras, beans doubled beef's consumption rate. Panama was the only country where beef supplied more calories than beans. Beans also surpassed beef as a protein source in the region, excluding Panama. The comparative predominance of beans for both calories and protein was most dramatic among impoverished rural populations.[40]

Despite the dietary importance of beans for the Central American population, regional bean production fluctuated, leading scholars to describe an acute supply crisis.[41] What production growth did occur came as a result of

---

[37] Victor Valverde, Hernán Delgado, and Arnulfo Noguera, "Nutrition in Central America and Panama: Comparative Data and Interpretations," *Food and Nutrition Bulletin* 9, no. 3 (1987): 1–12, at 5.

[38] Charles Teller et al., *Desnutrición, población, desarrollo social y económico: Hacia un marco de referencia* (Guatemala: INCAP, 1980), p. 37.

[39] INCAP, *Nutritional Evaluation of the Population of Central America and Panama: Regional Summary* (Guatemala: INCAP, 1971), pp. 9–22.

[40] Ibid., pp. 12–13.

[41] Antonio M. Pinchinat, "El PCCMF y el fomento del cultivo de frijol en Centroamérica," in PCCMCA, *Frijol: XIV Reunión Anual* (Tegucigalpa, Honduras: IICA, 1968), pp. 63–70, at 68.

the expansion of cultivated land rather than an increase in productivity. Low bean production yields prevailed, even though Central America boasted incredibly rich genetic diversity in bean varieties. In fact, the region's diversity had contributed substantially to the bean gene banks in North America and Europe. Early in the 1980s, CIAT built a collection with roughly 30,000 bean accessions donated by 47 countries. Although North American and European countries were the main donors – providing 33 and 22 percent of accessions, respectively – nearly 30 percent of all accessions of *Phaseolus vulgaris* traced their lineage to Central America[42] (Figure 8.2).

This imbalance – a region with high crop diversity but poor production amidst an escalating protein crisis – prompted CIAT to carry out bean research in Central America. Such programs were not new. In 1954, the Rockefeller Foundation developed the Central American Cooperative Project: Maize Improvement, which was renamed the Central American Cooperative Program for the Cultivation and Improvement of Food Cultivars (PCCMCA) in 1964.[43] The program stemmed from an agreement between Central American governments and the foundation, aiming to improve agricultural practices in maize farming through the exchange of knowledge and technology on farming systems, pests, and diseases, as well as on fertilization and crop varieties. Although the program focused on maize, it added beans (1962) and rice (1965) as strategic crops to the research agenda, as well as other products during the following years.

Institutions such as El Zamorano Pan-American Agricultural School in Honduras and the Inter-American Institute for Cooperation on Agriculture (IICA) in Costa Rica developed bean research programs during this time. IICA – coordinator of PCCMCA since 1965 – administered an alimentary crops unit from its facilities in Turrialba, Costa Rica. The unit coordinated the dissemination of crop varieties among farmers and conducted technical surveys on bean farming. Furthermore, the Bean/Cowpea Collaborative Research Support Program under the auspices of USAID facilitated the creation of a network of bean research projects that included Costa Rica, Honduras, and Guatemala, in collaboration with Michigan State University, the University of Puerto Rico, and Cornell University.[44]

[42] CIAT, *Programa de frijol: Informe anual 1982* (Cali, Colombia: CIAT, 1983), p. 21, https://hdl.handle.net/10568/69042.
[43] PCCMCA, *Frijol: XIII Reunión Anual* (San José, Costa Rica: IICA, 1967); PCCMCA, *Frijol: XIV Reunión Anual* (Tegucigalpa, Honduras: IICA, 1968); PCCMCA, *Frijol: XVI Reunión Anual* (Antigua, Guatemala: IICA, 1970); PCCMCA, *Frijol: XVII Reunión Anual* (Panamá: IICA, 1971).
[44] Bean/Cowpea Collaborative Research Support Program (CRSP), "The Linkage Experience of the Bean/Cowpea CRSP," June 23, 1986, Michigan State University, East Lansing, MI, https://pdf.usaid.gov/pdf_docs/PNABK895.pdf.

Figure 8.2 The bean collections established earlier in CIAT's history
continue. Today, maintaining CIAT's collections of bean germplasm
involves the multiplications of seeds in screenhouses in Colombia's
Central Cordillera, 2017. Photo by Neil Palmer/CIAT. By permission
of Alliance Bioversity–CIAT.

In 1978, CIAT appointed a permanent resident scientist to Central
America to improve coordination with national bean programs.[45] This
step set the foundation of PROFRIJOL. During its first two years, the
project received funding from the United Nations Development
Programme (UNDP); afterwards, in the 1980s, the Swiss Agency for
Development and Cooperation (SDC) provided funds.[46]

PROFRIJOL became an international plant-breeding network integrating programs in Mexico, Central America, and the Caribbean
(Figure 8.3). The project established a regional nursery and testing
system for reviewing new genetic material across different countries,
replicating CIAT's decentralized varietal improvement strategy. One of

[45] CIAT, *Programa de frijol: Informe de 1978*, C-73.
[46] CIAT, *Informe anual del programa de frijol 1980*, p. 83; Rafael Rodríguez, "Evolución
e integración de la investigación del frijol en América Central, México y El Caribe:
PROFRIJOL," in S. P. Singh and O. Voysest, eds., *Taller de mejoramiento de frijol para
el siglo XXI: Bases para una estrategia para América Latina* (Cali, Colombia: CIAT, 1997),
pp. 531–546.

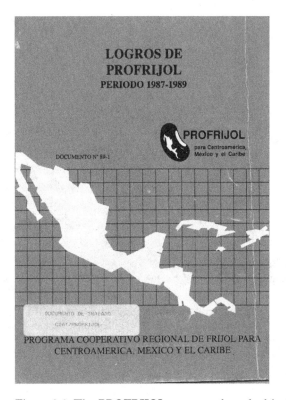

Figure 8.3 The PROFRIJOL program, launched in 1978, sought to coordinate bean research, breeding, and testing across Mexico, Central America, and the Caribbean. *Logros de PROFRIJOL, Periodo 1987–1989* (San Jose, Costa Rica). By permission of Alliance Bioversity–CIAT.

its tasks was avoiding the duplication of projects between countries to save financial and scientific resources for varietal research in the region. In this way, CIAT promoted research and disseminated scientific publications among officials in each country. PROFRIJOL balanced broad participation in the decision-making process through a general coordinator, a directing board, and an assembly consisting of representatives from the associated countries.[47]

PROFRIJOL produced mixed results across the region. Despite the political crises and economic hardships of the 1980s, the project

---

[47] Rodríguez, "Evolución e integración," pp. 532–533.

succeeded in linking national bean programs in a network, mobilizing twenty-three plant-breeding experts. However, during the 1990s, economic liberalization affected national institutions, weakening local programs through staff reductions. By 1999, the network employed only four expert researchers.[48] Regarding varietal improvement, the program developed eighty-one distinct genetic varieties between 1978 and 1997, fifty-six of which were certified and made available to farmers. More than 80 percent of these new varieties originated in genetic material from CIAT or research programs coordinated by the center.[49] However, new varieties' acceptance among farmers was limited: in 1996 roughly 44 percent of bean-cultivated land in Central America grew varieties improved by the program. While countries like Costa Rica accounted for 85 percent of land cultivated with new varieties, noncertified varieties, or *criollas*, remained predominant among peasants in countries such as El Salvador and Nicaragua.[50] Despite the limitations in determining the actual scope of the project in the region, the mild and fluctuating increase in yields per hectare between 1970 and 1990 suggests a limited success of varietal innovation.[51] These limits were owing considerably to political violence. PROFRIJOL created a significant network of people, seeds, and agronomic knowledge. Its campus specialized in the improvement of *Phaseolus vulgaris*, sustained by dozens of scientists from different countries, hundreds of varieties of beans, and thousands of dollars of accumulated investment since the late 1970s. All of these resources were applied to improving the lives of thousands of poor farmers in Central America. This was surely no easy task for CIAT scientists, or for the partner national governments. More complicated still, during its development PROFRIJOL collided with the social realities of Central America, in which the territorial and political scale transcended the bean fields. Specifically, PROFRIJOL contended with the dynamics of the international protein market, new agricultural technology, and, above all, civil war.

---

[48] Johnson et al., "The Impact of CIAT's Genetic Improvement Research," p. 260.
[49] Abelardo Viana Ruano, *Flujo de germoplasma e impacto del PROFRIJOL en Centroamérica: Período 1987–1996* (Guatemala: PROFRIJOL, 1998), pp. 16–19.
[50] Ibid., pp. 21–28.
[51] PROFRIJOL, *Plan quinquenal 1993–1997* (Guatemala: PROFRIJOL, 1992), p. 7; CORECA-IICA, *El mercado mundial del frijol y sus vinculaciones con el mercado centroamericano* (San José, Costa Rica: IICA, 1999), p. 45, https://repositorio.iica.int/handle/11324/9158.

Interpreting Bean Research, the "Protein Crisis," and Civil War in Central America

In a 1974 article in the *Lancet*, Donald McLaren of the Nutrition Research Laboratory at the American University of Beirut disputed the claim made by international organizations (such as FAO, WHO, and UNICEF) a decade prior about the existence of a "protein crisis," instead dubbing the episode the "protein fiasco."[52] According to McLaren, the root causes of childhood malnutrition among Third World impoverished populations were more complex than the protein insufficiency explanation. On the one hand, McLaren argued that a lack of available data hindered the establishment of linkages between protein insufficiency and other health problems associated with "proteic-caloric malnutrition." On the other hand, he labeled concerns about protein as reductionist: poverty and lack of access to food caused malnutrition, rather than food quality itself. McLaren's article showed the cracks in the "protein crisis" narrative. Nor were data limitations and reductionism the only concerns. The food crisis of 1972 and the oil shock of 1973 had demoted the protein crisis within many organizations' agendas, including the dismissal of the once-influential United Nations Protein Advisory Group in 1977.

From the perspective of Central America, concerns regarding protein deficiencies encountered two paradoxes. The first related to decreasing meat consumption amid increasing production. Cattle raising in the region had been growing since the 1950s. Central America became a net beef exporter to US markets, increasing exports from 362,000 kilograms in 1957 to nearly 80 million in 1980.[53] Likewise, the industry expanded from roughly $8 million in exports to the United States in 1960 to $200 million in 1980.[54] These increases were due to the convergence of political, ecological, and market factors. The North American fast-food boom of the 1950s and 1960s substantially increased the demand for beef, while the foot-and-mouth disease quarantine imposed on South America's cattle industry allowed a broadening in export quotas from Central America to the United States.[55] Moreover, national governments, under the auspices of the Alliance for Progress in the 1960s, pursued public infrastructure projects, such as roads and bridges, thereby

[52] Donald S. McLaren, "The Great Protein Fiasco," *Lancet* 304, no. 7872 (1974): 93–96.
[53] Robert G. Williams, *Export Agriculture and the Crisis in Central America* (Chapel Hill: University of North Carolina Press, 1985), p. 204.
[54] Ibid., p. 206.
[55] Alfredo Guerra-Borges, "El desarrollo económico," in Héctor Pérez Brignoli, ed., *Historia General de América Central: De la posguerra a la crisis (1945–1979)* (Madrid: FLACSO, 1993), pp. 13–84, at 32–34.

improving transport to port facilities. These developments led to the growth of nearly thirty meat-packing facilities in the region with refrigerating systems that met the quality standards required by US markets. Thus, in two decades, the region's beef agro-industry experienced significant industrial expansion.

This thriving industry grew despite decreasing meat consumption in Central America. According to one study, beef consumption per capita declined in Central America from 11.9 to 10.2 kilograms during the 1970s and 1980s.[56] In Guatemala, consumption dropped from 5.3 to 3.8 kilograms, while in Nicaragua it drastically decreased from 15.6 to 8.6 kilograms. Demonstrating the effect of civil war and political instability in these countries, comparatively peaceful and stable Costa Rica actually increased beef consumption slightly from 20.4 to 22.9 kilograms per capita. Other factors contributed to decreasing beef consumption in parts of Central America, including regional demographic growth, the impact of the 1980s economic debt crisis on consumer purchasing power, and the rise in the consumption of poultry and other types of meat. Still, for some observers, a contradiction came to define the region: the allocation of ecological and capital resources towards export-led cattle raising contrasted with the nutritional crisis among the region's impoverished rural populations who lacked animal-based protein and depended on beans to fulfill this dietary need.

The second paradox posed by protein deficiency in the region was related to land tenancy. Protein programs focused on the most impoverished rural families, whose agricultural production often took place on sloping lands with highly depleted soils. Since the 1960s, the territorial expansion of monoculture crops, such as sugarcane, cotton, and coffee, and cattle raising contributed to the growing marginalization of sectors of the rural population. Many social scientists understood these populations' economic and social hardships as a root cause for civil war and insurgency in Central America in the 1970s and 1980s. Scholars interpreted the political crisis as resulting from a high concentration of lands in large estates owned by a few landowners.[57] Meanwhile, peasant

---

[56] David Kaimowitz, *Livestock and Deforestation in Central America in the 1980s and 1990s: A Policy Perspective* (Indonesia: Center for International Forestry Research, 1996), pp. 30–31.

[57] Antonio García, "El nuevo problema agrario de América Central," *Anuario de Estudios Centroamericanos* 5, no. 1 (1979): 111–118. Several critical texts in wide circulation during this time more broadly interpreted these inequalities in land tenure as the basis for social conflict historically across Latin America. For example, Alain de Janvry, *The Agrarian Question and Reformism in Latin America* (Baltimore: Johns Hopkins University Press, 1981); Eduardo Galeano, *Open Veins of Latin America: Five Centuries of the Pillage of a Continent* (New York: Monthly Review Press, 1973).

families survived by cultivating less fertile lands for self-consumption, and squatting on state-owned lands or abandoned estates. In 1985, the anthropologist Billie DeWalt argued that the beef-exporting boom fostered the expansion of grasslands, pushing peasants towards marginal lands.[58] The economist Robert Williams asserted that initially lands with the most fertile soils were monopolized by cattle ranchers and cotton farmers, displacing maize farmers towards sloping and frontier lands; cattle raising later colonized sloping lands as well, forcing peasants to move beyond the margins.[59]

Other scholars of the 1980s and early 1990s concurred that the agricultural export boom expelled peasant communities from the best-suited lands, advancing marginalization and poverty in Central America.[60] Agricultural modernization deepened social inequalities in the region.[61] Meanwhile, nutrition researchers studied the relationship between land tenure structures, poverty, and malnutrition. Some studies found the prevalence of moderate child malnutrition was higher among small landowners (between 1 and 2 hectares) than among middle-to-high landowners.[62] As a result, many organizations strove to increase protein intake among Central America's impoverished rural populations. Ironically, those groups had been displaced from their lands by an industry involved in exporting animal-based protein to North American markets.

How did an emphasis on bean breeding affect these conditions? The fundamental aim of CIAT's bean research program was increasing yields per hectare in Central America. However, the particularities of bean crops hindered the possibility of developing a high-yielding variety. Consider rice as a comparison. Rice breeding brought two remarkable achievements. First, it created an "ideal type" of plant embodied in IR8, the Green Revolution's "miracle rice."[63] This semi-dwarf variety

[58] Billie R. DeWalt, "The Agrarian Bases of Conflict in Central America," in Kenneth Coleman and George C. Herring, eds., *The Central American Crisis: Sources of Conflict and the Failure of US Policy* (Wilmington, DE: Scholarly Resources, 1985), pp. 43–54.

[59] Williams, *Export Agriculture*, pp. 155–165. In Chapter 3, this volume, Timothy W. Lorek describes this process of deterritorialization as it related to CIAT in Colombia.

[60] Victor Bulmer-Thomas, *The Political Economy of Central America since 1920* (Cambridge: Cambridge University Press, 1987), p. 207; Alain Rouquié, *Guerras y paz en América Central* (México: Fondo de Cultura Económica, 1994), pp. 98–106; Guerra-Borges, "El desarrollo económico," pp. 19–36.

[61] Edelberto Torres-Rivas, *Revoluciones sin cambios revolucionarios: Ensayos sobre la crisis en Centroamérica* (Guatemala: F&G Editores, 2013), pp. 110–132.

[62] Victor Valverde et al., "Relationship between Family Land Availability and Nutritional Status," *Ecology of Food and Nutrition* 6, no. 1 (1977): 1–7.

[63] Peter Jennings, "Plant Type as a Rice Breeding Objective," *Crop Science* 4, no. 1 (1964): 13–15.

combined the best traits of *japonica* and *indica* varieties, including high yields per hectare, pest and disease resistance, increased response to chemical fertilization, and high consumer acceptance. Arguably, the capacity to adapt to different ecological and social environments was the most outstanding characteristic of varieties genetically associated with IR8.[64] Second, the adoption of high-yielding rice varieties succeeded owing to the development of a farming system based on six conditions, typically affordable only to wealthy farmers: the exploitation of flat fertile lands, the use of farming and harvesting machinery, the application of chemical fertilizers, soil exploitation based on a single crop (a monoculture system), a dependency on pest and disease control, and the availability of irrigation systems.

In contrast to rice, bean improvement posed difficulties. Most obviously, given the variety of shapes, colors, sizes, and flavors of beans, scientists struggled to develop the "ideal type" of plant. CIAT bean researchers Aart van Schoonhoven and Oswaldo Voysest found that populations in Latin America routinely ingested almost fifty different types of bean, which they identified by their colors, sizes, and local names.[65] Variable approaches to cultivation further complicated this encounter with diversity. Depending on local economies and ecologies, peasants often cultivated beans alongside maize, coffee, or other crops. According to van Schoonhoven and Voysest, Latin American peasants produced between 60 and 80 percent of beans in companion plantings with maize. In addition, farmers in different locations practiced culturally specific planting methods. For example, Nicaraguan and Costa Rican peasants relied on nonconventional farming techniques, such as the "frijol tapado," which consisted in spreading the beans over organic waste, such as leaves and branches, left after clearing the forest.[66]

The common conditions of bean cultivation in Central America also hampered the development of a bean monoculture in the region. Peasants generally cultivated legumes in sloping lands, hindering farming mechanization and irrigation. They usually grew beans in acidic and depleted soils with nitrogen deficiencies.[67] Finally, the mixed planting of beans and other crops in small plots inhibited the development of a monoculture system. Meanwhile, chemical fertilization and pest control were limited among peasants. Beans frequently fell prey to numerous diseases and

[64] Robert F. Chandler, *Rice in the Tropics: A Guide to the Development of National Programs* (Colorado: Westview Press, 1979).
[65] Aart van Schoonhoven and Oswaldo Voysest, "El frijol común en América Latina, y sus limitaciones," in Marcial Pastor-Corrales and Howard F. Schwartz, eds., *Problemas de producción del frijol en los trópicos* (Cali, Colombia: CIAT, 1994), pp. 39–66, at 42–44.
[66] Ibid., pp. 48–49.   [67] Ibid., pp. 56–57.

pests. According to experts in the 1990s, between 200 and 450 insect plagues and more than 200 types of malady damaged harvests in the region.[68] Finally, bean production in Central America was most often located in the Dry Pacific Slopes, a subregion frequently affected by droughts. Given the above factors, the development of a Green Revolution in bean production equivalent to that seen in rice seemed an unlikely enterprise.

Political conditions added to these complications. The bean improvement program in Central America overlapped with a period of civil war between 1978 and 1990 that affected most of the region. Although the insurgent movement in Nicaragua started during the 1960s, opposition to Anastacio Somoza's dictatorial regime crystalized in 1974, finally leading to the victory of the Sandinista Revolution in 1979. In El Salvador, guerilla groups and others resisting government repression became one united front, the Farabundo Martí National Liberation Front, in 1980, initiating a decade of civil conflict. In Guatemala, leftist rebel groups seeking land redistribution and an end to repressive government regimes emerged during the 1960s and successfully entrenched in rural areas in the late 1970s. Confrontations between insurgents and the Guatemalan army developed during the 1980s, peaking between 1982 and 1985. Finally, although Honduras and Costa Rica remained free from internal clashes, both countries became the focus of migration for thousands of displaced people. Also, the two countries defended the United States' geopolitical interests in the region indirectly by supporting counter-insurgent strategies[69] (Figures 8.4 and 8.5).

The civil wars affected agriculture in Central America via different avenues. By and large, export farmers avoided the war crisis better than local market producers. The land tenure structure for cash crops (e.g., coffee, sugarcane, bananas) and cattle raising in the region remained unchanged, despite land reform attempts during the 1970s and 1980s.[70] Where it occurred, land redistribution took place in marginal and less fertile lands already occupied by most of the peasant population. Land reform efforts did not necessarily upset the activities of wealthier export farmers. By comparison, peasant agriculture endured far more significant damage on the fields because of conflict than export agriculturists. Peasants grew staple crops, such as maize and beans, in sloping lands and agricultural frontiers, where guerrilla members found shelter

---

[68] Ibid., p. 48.
[69] Stephen G. Rabe, *The Killing Zone: The United States Wages Cold War in Latin America* (Oxford University Press, 2012). One of the best analyses from Central America is Torres-Rivas, *Revoluciones*.
[70] Guerra-Borges, "El desarrollo económico," pp. 57–67.

Figure 8.4 The scorched-earth ferocity of Central America's civil wars affected rural peoples and food production. In this image from 1983, a young woman with the insurgency poses with child and assault rifle in front of maize in Guazapa, El Salvador, a region targeted by the Salvadoran army. Gio Palazzo Collection, Museo de la palabra y la imagen (San Salvador, El Salvador). By permission of Museo de la palabra y la imagen.

from the repressive army forces. Farm fields became battlegrounds, displacing hundreds of thousands of peasants. According to the United Nations Refugee Agency, conflict expelled roughly 20 percent of El Salvador's population, while other estimates indicate that 14 percent of the population in Guatemala and Nicaragua fled from their countries.[71] Some studies estimate that a total of around 2 million

---

[71] Cástor Miguel Díaz, *Los conflictos armados de Centroamérica* (Madrid: Instituto de Estudios Internacionales y Europeos-Universidad Carlos III, 2010), p. 62; Juan Rafael Vargas et al., "El impacto económico y social de las migraciones en Centroamérica (1980–1989)," *Anuario de Estudios Centroamericanos* 21, no. 1/2 (1995): 39–81, at 41.

Figure 8.5   Civilians and army soldiers in front of a building in Perquín,
El Salvador, 1983. Photograph by Richard Cross. Courtesy of the Tom
& Ethel Bradley Center at California State University, Northridge.

people in Central America left their countries during the 1980s, fleeing
violence and persecution.[72]

Across the region, state and military efforts to fight insurgents led to the
reallocation of public budgets to military spending. Guatemala's military
expenditure grew from $67 million in 1975 to $180 million in 1983, while
El Salvador's increased at a similar rate, from $37 to $99 million, during
the same period. Military expenses in both countries reached nearly
15 percent of the public budget in 1983.[73] The economic crisis triggered
by the 1979 oil shock further aggravated the effects of such resource
allocation. Mirroring the broader Latin American economic environment, Central American states underwent financial setbacks due to rising
external debt, fiscal deficit, inflation, and capital outflows.[74] Cuts in
public spending such as farmer subsidies and loans affected peasant
agriculture. For instance, in 1983, coffee and cotton farmers held

---

[72] This calculation is complicated by the lack of reliable data. Abelardo Morales, *La diáspora
de la posguerra: Regionalismo de los migrantes y dinámicas territoriales en América Central* (San
José, Costa Rica: FLACSO, 2007), pp. 117–119.

[73] Alfredo Guerra-Borges, "Reflexiones sobre la economía y la guerra en Centroamérica,"
Anuario de Estudios Centroamericanos 12, no. 2 (1986): 75–88, at 80.

[74] Bulmer-Thomas, *The Political Economy of Central America*, pp. 230–266.

60 percent of farming loans, while rice producers obtained 10 percent. On the other hand, bean producers had access to only 1 percent, despite the dietary importance of the legume.[75] The socioeconomic impact of the civil war on the most vulnerable populations fostered an increase in the United States' food assistance. Between 1979 and 1987, the United States spent more than $700 million on food relief through such programs as PL-480.[76] Most of the aid, almost 70 percent, consisted of wheat, vegetable oils, maize, rice, milk, and beans. This food relief had significant effects beyond the simple provision of calories. For example, the scope of these programs disrupted regional food systems, changing Central Americans' consumption patterns and diets. As the region imported cheap food, such as maize, via donations or low-cost purchases, relief programs discouraged local production and further marginalized peasant farming.

Conclusion

Some US politicians and scholars thought it plausible that Central America would become a "Vietnam of the Americas" during the 1980s. This analogy suggests that the military conflict in Central America, particularly in El Salvador, almost resembled Vietnam before 1975.[77] Indeed, after the 1960s, Central America became one of the many Third World hot spots where the United States led an anti-communist campaign. Most of its population faced poverty and marginalization within agrarian-based economies while enduring political instability from both dictatorial regimes and insurgent guerrilla movements. The success of the Cuban Revolution in 1959 and the island's close relations with the Soviet Union anticipated the radicalization of rebel movements in the region, prompting the United States' military and political intervention in defending its geopolitical interests.

Varietal improvements for increasing food production in Central America during the 1970s and 1980s differed from the Green Revolution in Asia. In contrast to the development of high-yielding varieties of wheat and rice, CIAT's bean research program in Central America went against all the odds. The bean-breeding program unfolded

---

[75] CEPAL, "Centroamérica: Crisis agrícola y perspectivas de un nuevo dinamismo," February 12, 1985, 24, http://hdl.handle.net/11362/26594.
[76] Rachel Garst and Tom Barry, *Feeding the Crisis. US Food Aid and Farm Policy in Central America* (Lincoln: University of Nebraska Press, 1990), p. 61.
[77] George C. Herring, "Vietnam, El Salvador, and the Uses of History," in Coleman and Herring, eds., *Central American Crisis*, pp. 97–110; Susanne Jonas and David Tobis, *Guatemala* (New York: North American Congress on Latin America, 1974), p. 187.

amid civil wars and poverty crises, which were linked to nutritional crises among peasant populations. Yet unlike wheat and rice, the bean was less suited to monoculture and industrial cultivation, maintaining its association with small peasants who farmed mixed plots on sloping lands. These conditions hindered the development of high-yielding varieties and the design of modern technical-package-based farming systems.

Central America was a laboratory for Cold War geopolitics between the 1960s and the 1980s. During this period, this region was the scene of wars between dictators, armies, and guerrillas, all of which were affected by the interests of the United States and the Soviet Union. As a result, Central America became a zone of exchange in military technology, espionage, arms deals, and food assistance from both the capitalist and communist worlds. At the same time, the region was also a laboratory for CIAT efforts to increase the consumption of protein by the poorest classes via the genetic improvement of beans. Although this was a just war against the malnutrition that caused the suffering of thousands of children, it was also paradoxical, in that the program for the improvement of beans occurred in Central America even as thousands of tons of animal protein were exported to the United States or consumed by the wealthiest populations of the region. As the United States and Central American governments combined to invest billions of dollars in the region's civil wars, this episode of the Green Revolution in Central America was characterized not only by the politics of the Cold War but also by the persistent and underlying social inequalities of the region.

Part III

Science in the System

9 Fifty Years of Change in Maize Research at CIMMYT

Derek Byerlee and Greg Edmeades

At the end of the twentieth century, maize became the world's most important crop in terms of tons produced and calories supplied. Originating in the tropics and subtropics of Mexico, maize was spread by Indigenous populations throughout much of the Americas, and then to the Old World from the sixteenth century as part of the Columbian exchange.[1] However, its rise to world dominance began only after World War II as global production leaped to over ten times its pre-1938 level. Much of this growth was due to the use of grain for animal feed and, more recently, for biofuel. However, maize remains the staple food crop in its Latin American center of origin, and in the first half of the twentieth century it became a staple in much of Africa.[2]

In the United States, the almost universal adoption of hybrid maize and associated growth in yields from the 1930s reinforced its position as the world's largest producer of maize. Maize became the focal crop in early US foreign assistance programs, private-sector investment, and international exchange of breeding materials (germplasm).[3] Maize research was also internationalized, with consequences for agricultural research more broadly. In 1950, Ricardo Acosta, a Mexican government official, proposed an international institute for maize research. His proposal was the catalyst for the development of the international center model for agricultural research, which resulted in the creation of the International Rice Research Institute (IRRI) in 1960 in the Philippines, the International Maize and Wheat Improvement Center (CIMMYT) in

Acknowledgments: We thank Dr. B. Prasanna, Director, Global Maize Program, CIMMYT for assistance with information on recent breeding developments at CIMMYT.

[1] Alfred W. Crosby, *The Columbian Exchange: Biological and Cultural Consequences of 1492* (Westport, CT: Greenwood, 1972).

[2] James C. McCann, *Maize and Grace: Africa's Encounter with a New World Crop 1500–2000* (Cambridge, MA: Harvard University Press, 2005).

[3] L. B. Kass, C. Bonneuil, and E. H. Coe, "Cornfests, Cornfabs and Cooperation: The Origins and Beginnings of the Maize Genetics Cooperation Newsletter," *Genetics* 169, no. 4 (2005): 1787–1797; Derek Byerlee, "Globalization of Hybrid Maize, 1921–1970," *Journal of Global History* 15, no. 1 (2020): 101–122.

1966 in Mexico, and the Consultative Group on International Agricultural Research (CGIAR) in 1971.[4]

Although maize appeared to be at the forefront, we argue in this chapter that the development of an international maize program at CIMMYT took place in the shadow of experiences with rice and wheat that were already attracting global attention as part of the Green Revolution. The key design element of international research on rice and wheat was a centralized breeding program linked to a network of public-sector research systems at the national level where new varieties were adapted and tested.[5] A fundamental characteristic of the model was its "open-source" approach, in which countries were free to directly release varieties from the testing program or use these as inputs into their own breeding programs. Nonetheless, the first international centers aspired to realize quick payoffs by developing widely adapted varieties that could be immediately used in multiple countries to help meet the food needs of rapidly growing populations.[6]

In applying this model to maize, researchers confronted three characteristics that distinguished this crop from rice and wheat. First, given Malthusian famine scares, attention in the 1960s was firmly focused on Asia, where maize was not a staple food except for marginalized populations in hill areas. It was therefore not a "political crop."[7] Maize was a staple in eastern and southern Africa, but that region only became a major CGIAR priority much later. With the exception of Latin American countries and a handful of white settler economies in Africa, maize research remained a low priority in low- and middle-income country contexts.

Second, nearly all maize in low- and middle-income countries outside of China, Argentina, and South Africa was grown in tropical and subtropical ecologies.[8] CIMMYT naturally focused on these ecologies, but, unlike rice and wheat, which were often grown in relatively uniform

---

[4] Derek Byerlee and John K. Lynam, "The Development of the International Center Model for Agricultural Research: A Prehistory of the CGIAR," *World Development* 135 (2020): 105080.

[5] Ibid.

[6] Robert S. Anderson, Edwin Levy, and Barrie M. Morrison, *Rice Science and Development Politics: Research Strategies and IRRI's Technologies Confront Asian Diversity, 1950–1980* (Oxford: Clarendon Press, 1991); Marci R. Baranski, "Wide Adaptation of Green Revolution Wheat: International Roots and the Indian Context of a New Plant Breeding Ideal, 1960–1970," *Studies in History and Philosophy of Science* 50 (2015): 41–50.

[7] Several largely unsuccessful attempts to transfer hybrid maize to Asia reflected scientific interest in the hybrid technology rather than a high priority within maize research. See Byerlee, "Globalization of Hybrid Maize."

[8] In the 1960s, nearly half of the global area planted to maize was in the tropics and subtropics. With the exception of Brazil and some commercial farming areas of eastern and southern Africa, nearly all of this area was planted by small-scale farmers.

irrigated areas, nearly all tropical and subtropical maize was grown under rainfed conditions that were highly diverse with respect to altitude, soils, and rainfall.[9] Further, and in contrast to wheat, farmers' preferences for maize grain type and color also varied widely, partly reflecting its varied uses in foods – from tortillas and porridges to fresh corn on the cob and snack foods – and for animal feed. This diversity challenged the established centralized breeding model employed for rice and wheat. It required considerable innovation and learning to develop an appropriate model for international maize research.

Finally, CIMMYT had to deal with significant private-sector involvement in maize research and seed production, a circumstance that did not apply to rice and wheat. In maize, the male (tassels) and female flowers (ears) are separated, making it relatively easy and cheap to produce hybrids by inbreeding parental lines for several generations and then crossing the inbreds to express heterosis (also known as hybrid vigor). These hybrids provide a significant yield advantage under a range of growing conditions; however, farmers need to buy seed annually to maintain this advantage. These characteristics of maize incentivized private firms to invest in the production and promotion of hybrid maize seed and for larger seed companies to invest in their own breeding programs.[10] By 1970, maize farmers in high-income countries had almost completely switched to hybrid seed developed and sold by private firms, and some of these firms had evolved into large multinational corporations.[11]

Some earlier maize-breeding programs had explored the option of improved open-pollinated varieties that allowed farmers to save seed.[12] CIMMYT could pursue this option, too, and focus on open-pollinated varieties at the cost of potentially lower yields, or it could develop hybrids and partner with the public and private sector to deliver its seed. Working with the private sector naturally introduced tensions for an international center set up to produce "international public goods" – that is, products that could be freely exchanged and used across countries (see the discussion of these issues in David J. Jefferson, Chapter 12, this volume).

---

[9] For a fuller comparison of these crops, see D. Byerlee and G. O. Edmeades, *Fifty Years of Maize Research in the CGIAR: Diversity, Change, and Ultimate Success* (Mexico City: CIMMYT, 2021), https://hdl.handle.net/10883/21633.

[10] M. L. Morris, ed., *Maize Seed Industries in Developing Countries* (Boulder, CO: Lynne Rienner, 1998).

[11] Byerlee, "Globalization of Hybrid Maize"; Jack R. Kloppenburg, *First the Seed: The Political Economy of Plant Biotechnology, 1492–2000*, 2nd edn. (Madison: University of Wisconsin Press, 2004).

[12] For a review of open-pollinated varieties and hybrids in Mexican maize-breeding programs prior to CIMMYT, see Karen E. Matchett, "Untold Innovation: Scientific Practice and Corn Improvement in Mexico, 1935–1965," Ph.D. dissertation, University of Minnesota (2002).

With these facets of maize history and biology in mind, this chapter aims to describe and analyze the design and evolution of international maize research at CIMMYT during its first fifty years. We identify three distinct periods in this research between 1966 and 2020, recognizing that the transition between periods is often blurred. Our focus is on breeding research, although CIMMYT invested significant resources in maize agronomic and social science research that mostly complemented its breeding efforts. We do not consider West Africa, where a strong maize research program of the International Institute of Tropical Agriculture (IITA), another CGIAR center, focused its work with varying degrees of collaboration and sometimes competition with CIMMYT.[13]

Building a Global Program with Scientists in the Lead, 1966–85

CIMMYT was formally established in 1966 in the context of widespread concern over global population growth and impending food and resource shortages. Its founders enthusiastically embraced the food-population challenge and defined its mission as increasing the "quantity of food produced." However, CIMMYT, along with other development actors at the time, poorly articulated the pathway from increasing the "pile of food" to reducing hunger.[14] This narrow focus on production would dominate CIMMYT's narrative for the next fifteen years.

Given the high priority assigned to increasing food supply by international organizations, CIMMYT in this period enjoyed strong initial financial support from the Rockefeller and Ford Foundations, joined by the United States Agency for International Development (USAID) and several other multinational and bilateral donors after CGIAR was created in 1971. Stable and largely unrestricted financial support provided CIMMYT scientists substantial freedom to set priorities, as well as to pursue research objectives with potentially high but uncertain long-term payoffs. The eminent scientists on CGIAR's Technical Advisory Committee (TAC) exercised considerable influence over donors in allocating funds and consistently endorsed a high priority for maize research.[15]

---

[13] For a fuller treatment of maize research in CGIAR, see Byerlee and Edmeades, *Fifty Years of Maize Research in the CGIAR*.

[14] Bruce H. Jennings, *Foundations of International Agricultural Research: Science and Politics in Mexican Agriculture* (Boulder, CO: Westview Press, 1988).

[15] Technical Advisory Committee, *CGIAR Priorities and Future Strategies* (Rome: CGIAR, 1987), https://hdl.handle.net/10947/324.

International maize research was not new in the 1960s. Indeed, CIMMYT's maize program was built from the legacy of eight country and regional programs of the Rockefeller and Ford Foundations that operated relatively independently of each other across Latin America, Asia, and Africa.[16] Transforming these legacy research programs and networks into an integrated, coordinated international program was challenging. As noted in the introduction, the centralized breeding model employed for rice and wheat had to be adapted to the diversity of maize types and growing conditions, as well as to accommodate the sensitivity of maize varieties to changes in day length as they moved across latitudinal zones. The centralized model was further challenged by the narrow adaptation of maize to local conditions, which stood in contrast to the relatively wide adaptation of CIMMYT's wheat varieties.[17] Many maize landraces had been developed through millennia of farmer selection in geographically isolated areas where they performed well, but were susceptible to diseases, pests, and other problems when sown in other locations. In the widely publicized Plan Puebla project, established by CIMMYT and the Mexican Colegio de Postgraduados in 1967 to improve farmers' maize yields in the Mexican highlands, scientists were unable to identify a single improved open-pollinated variety or hybrid that was superior to the varieties developed by farmers in their specific locations, despite more than two decades of prior investment in maize research in Mexico.[18]

When Ernest W. Sprague, the leader of the Inter-Asian Corn Program (one of the legacy programs of the Rockefeller Foundation) was transferred to become director of CIMMYT's Maize Program in 1970, he began to design a well-coordinated global maize program (Figure 9.1). Under his leadership, CIMMYT hosted two international maize conferences, one to assess national demands for its products and a second to review the work of all maize staff from across its legacy programs.[19] These efforts led to the first systematic approach to international maize breeding

---

[16] The legacy programs included the Rockefeller Foundation programs in Mexico, Kenya, and Nigeria, its regional networks in Central America, the Andes, and Asia, and the Ford Foundation's maize programs in Egypt and Pakistan (from 1967).

[17] The maize biologist Paul Mangelsdorf, an advisor to the Rockefeller Foundation's agricultural program, had argued "emphatically" against an international institute for maize because of the local specificity of maize varieties. Warren Weaver, diary, October 11, 1950, Rockefeller Foundation Archives, Rockefeller Archive Center, RG12, S-Z (FA394).

[18] Donald L. Winkelmann, *The Adoption of New Maize Technology in Plan Puebla, Mexico* (Mexico City: CIMMYT, 1976).

[19] CIMMYT, *Proceedings of the First Maize Workshop* (El Batan: CIMMYT, 1971); CIMMYT, *World Wide Maize Improvement and the Role of CIMMYT: Symposium Proceedings* (El Batan: CIMMYT, 1974).

Figure 9.1 Ernest Sprague lecturing to visitors in Poza Rica, Veracruz, 1979. CIMMYT Repository. © CIMMYT.

and testing. The geographic location of CIMMYT headquarters in the highlands of central Mexico meant that its staff could conduct maize breeding across varied tropical and subtropical growing environments within a 250-kilometer range of the institute. Another core asset inherited by CIMMYT was the extensive collections of Latin American maize landraces assembled under the auspices of the US National Academy of Sciences during the 1950s and 1960s.[20] Twenty-eight "populations" were developed from these landraces to represent diversity in grain color, texture, ecological adaptation, and maturity. Each population was then evaluated at dozens of sites across the world to identify its suitability for that location. A small subset (six) of these sites' 250

---

[20] Helen Anne Curry, "From Working Collections to the World Germplasm Project: Agricultural Modernization and Genetic Conservation at the Rockefeller Foundation," *History and Philosophy of the Life Sciences* 39, no. 2 (2017): 5; Diana Alejandra Méndez Rojas, "Los libros del maíz: Revolución Verde y diversidad biológica en América Latina, 1951–1970," *Letras Históricas* 24 (spring–summer 2021): 149–182.

"families" of each population were evaluated to identify the best families for further improvement of that population.[21] Through the testing network, national scientists gained access to an array of new, tropically adapted breeding materials. International testing also helped to broaden the adaptation of these populations. However, overall progress was slowed by some mismatches between populations and testing environments, the two-year cycle needed to receive results from both hemispheres, and the reality that many national programs had limited capacity to conduct precise field trials. Although CIMMYT's international testing program provided a well-structured way to expose CIMMYT's germplasm to national scientists and vice versa, it was an inefficient route to genetic improvement.[22]

The relative freedom given to CIMMYT in this early period allowed its maize scientists to go against the grain with respect to the prevailing orthodoxy in maize breeding that emphasized hybrids. Instead CIMMYT focused all its breeding and testing work in the early years on open-pollinated varieties. The decades prior to CIMMYT's founding had seen many attempts to extend hybrid technology to the tropics and frequent failures.[23] Many researchers believed hybrids to be unsuitable for small-scale farmers producing maize in risky rainfed areas, given the need for farmers to buy relatively expensive seed annually and the national resources and skills required to develop an effective hybrid seed industry.[24]

CIMMYT's focus on open-pollinated varieties was led by Sprague. In 1958 Sprague had been posted by the Rockefeller Foundation to India, where he initially worked exclusively on hybrids. However, by 1964 he was actively promoting open-pollinated varieties. It seems that his frustration with the slow pace and inconsistent quality of hybrid seed production in India, mostly in the public sector, together with his visits to Thailand to establish the Inter-Asian Corn Program, were important in this transition. Thailand had become a leading maize producer and exporter in the 1950s, based on the widespread adoption of an open-pollinated variety imported from Guatemala.[25] When Sprague moved to Mexico in 1970 to head CIMMYT's maize program, he vigorously championed the role of open-pollinated varieties over hybrids, asserting

---

[21] S. Pandey and C. O. Gardner, "Recurrent Selection for Population, Variety, and Hybrid Improvement in Tropical Maize," *Advances in Agronomy* 48 (1992): 1–87.
[22] Ibid.   [23] Byerlee, "Globalization of Hybrid Maize."
[24] Matchett, "Untold Innovation"; P. W Heisey, M. L. Morris, D. Byerlee, and M. A. Lopez-Pereira, "Economics of Hybrid Maize Adoption," in Morris, ed., *Maize Seed Industries in Developing Countries*, pp. 143–158.
[25] Byerlee, "Globalization of Hybrid Maize."

that "none of the developing countries with small farm holdings should be working with hybrid development . . . fortunately, a number of countries with more advanced programs abandoned their work on hybrids."[26] Although his views prevailed in CIMMYT, they were questioned by others. The distinguished maize geneticist George F. Sprague (no relation to E. Sprague) of the US Department of Agriculture and Iowa State University disagreed, citing Kenya as an example of smallholder adoption of hybrid seed.[27]

During its first two decades, CIMMYT focused almost exclusively on working with the public sector to develop and promote varieties. This strategy was in line with the prevailing view among foreign assistance agencies and governments of the leading role of the "development state."[28] Given CIMMYT's close relations with national programs, especially through its extensive training of their scientists, most countries that did not already have a well-developed hybrid program followed CIMMYT's policy of developing open-pollinated varieties. The share of these varieties among all public-sector releases in the tropics and subtropics increased steadily, peaking at two-thirds of the total in the 1980s.[29]

Stable and flexible funding also allowed CIMMYT scientists to pursue several risky, long-term research ventures that would have lasting influence on breeding strategies for tropical maize. The first was an effort to reduce plant height. Especially when fertilized, many landraces grew very tall, to over 3 meters, and their grain yield was modest because of their low harvest index (the ratio of grain to total dry matter) and susceptibility to lodging (the tendency to fall over before harvest). Breeders' initial efforts to duplicate the Green Revolution approach by introducing a dwarfing gene to tropical maize populations were not successful because the process resulted in variable height reduction and introduced other undesirable traits. As an alternative, CIMMYT breeders started selection for shorter plants with a higher harvest index. After fifteen seasons, they had spectacularly reduced plant height by 1 meter and increased yield potential by 60 percent at the higher planting densities made possible by shorter plants.[30] This process that concentrated many genes with small negative

---

[26] Ernest W. Sprague, "What Limits World Maize Production," in CIMMYT, *World Wide Maize Improvement*, pp. 2–1 to 2–22, at 2–7.

[27] CIMMYT, *World Wide Maize Improvement*, p. 14–4.

[28] For example, M. C. Saeteurn, *Cultivating Their Own: Agriculture in Western Kenya during the "Development" Era* (Rochester, NY: University of Rochester Press, 2020).

[29] M. A. López-Pereira, and M. L. Morris, *Impacts of International Maize Breeding Research in the Developing World, 1966–1990* (Mexico City: CIMMYT, 1998).

[30] E. C. Johnson, K. S. Fischer, G. O. Edmeades, and A. F. E. Palmer, "Recurrent Selection for Reduced Plant Height in Lowland Tropical Maize," *Crop Science* 26, no. 2 (1986): 253–260.

effects on height within a breeding population provided basic directions for tropical maize breeding over the following decades.

CIMMYT's maize physiologists were also among the first in CGIAR to challenge the prevailing belief that varieties bred for high-yield potential in favorable environments using high levels of inputs would also perform well in less favorable growing environments where use of external inputs was risky.[31] In the 1970s, CIMMYT began a pilot program of selecting under controlled drought conditions within the most important maize type of the lowland tropics, Tuxpeño, seeking at the same time to generate varieties that could yield well in favorable seasons. Initial promising results encouraged an increased focus on breeding for drought tolerance in CIMMYT's maize programs.[32] Using similar methods, CIMMYT researchers began screening for tolerance to low soil fertility (nitrogen) in 1987, seeking to produce better-performing varieties for areas where synthetic fertilizers were not available or their use was unprofitable.[33] These exploratory efforts laid the basis for a later mainstreaming of these methods after 2000 when CIMMYT shifted focus to Africa.

Another risky, long-term program initiated in this period was breeding maize with high levels of the amino acid lysine to enhance protein quality. As Wilson Picado-Umaña and Lucas M. Mueller discuss in Chapters 8 and 5 respectively, this volume, an emerging consensus within the United Nations Food and Agriculture Organization (FAO) and World Health Organization (WHO) in the 1950s identified protein malnutrition as the leading nutritional problem in much of the developing world. The 1960s became the "protein decade" as FAO declared that "the greatest [nutritional] problem ... results from inadequate protein in the diets of a large proportion of the population."[34] It was in this context that in 1963 scientists at Purdue University discovered the *opaque-2* gene in maize, which increased lysine content by 69 percent over normal maize.[35] This

---

[31] This belief was strongly promoted by Norman Borlaug as head of CIMMYT's wheat program. The debate is evident in CGIAR Technical Advisory Committee, "Report on the TAC Quinquennial Review Mission to CIMMYT, 1976," September 1976, https://hdl.handle.net/10947/1385.

[32] G. O. Edmeades, W. Trevisan, B. N. Prasanna, and H. Campos, "Tropical Maize," in H. Campos and P. Caligari, eds., *Genetic Improvement of Tropical Crops* (Switzerland: Springer, 2017), pp. 57–109.

[33] M. Bänziger, G. O. Edmeades, and H. R. Lafitte, "Selection for Drought Tolerance Increases Maize Yields across a Range of Nitrogen Levels," *Crop Science* 39, no. 4 (1999): 1035–1040.

[34] FAO, *The State of Food and Agriculture* (Rome: FAO, 1964), p. 98. See also Kenneth Carpenter, *Protein and Energy: A Study of Changing Ideas in Nutrition* (Cambridge, UK: Cambridge University Press, 1994).

[35] E. T. Mertz, L. S. Bates, and O. E. Nelson, "Mutant Gene That Changes Protein Composition and Increases Lysine Content of Maize Endosperm," *Science* 145, no. 3629 (1964): 279–280.

discovery gave rise to visions of a single gene being incorporated into all new maize varieties to boost protein intake worldwide. The opening speaker at a 1966 conference enthused that "within the next five years millions of undernourished people ... would find their diets improved markedly due to the availability of high lysine corn."[36] Norman Borlaug, a wheat breeder for the Rockefeller Foundation and then CIMMYT, also quickly endorsed the potential of high-lysine maize and became an enthusiastic advocate in the following decades.[37]

The new high-lysine varieties manifested undesirable traits associated with the *opaque-2* gene, such as dull grain type, soft endosperm, low yields, and higher pest losses in production and storage. The recessive nature of the gene meant that open-pollinated varieties quickly lost their quality advantage. However, after a meeting with Borlaug in 1971, the United Nations Development Programme (UNDP) invested heavily in research on high-lysine maize at CIMMYT over the next two decades to produce acceptable varieties. (The investment totaled $64 million in 2020 US dollars for 1971–84 alone.)[38] Buoyed by the additional resources from UNDP, CIMMYT enthusiastically promoted the potential of what it called "quality protein maize," projecting that "mankind will have available a super grain which contains everything for complete human nutrition."[39] CIMMYT explicitly aimed to produce quality protein varieties with grain visually indistinguishable from that of normal maize.[40] Meanwhile, the majority view in the nutritional community by 1975 had revised its minimum protein requirements downward and moved decisively towards energy intake as the major problem of hunger. The influential nutritionist John C. Waterlow firmly stated in 1975 that "the concept of a worldwide protein gap is no longer tenable" and that the "protein gap is a myth."[41] UNDP and CIMMYT were aware of these changes in nutritional priorities, but, as described by the CIMMYT social scientist Robert Tripp, "the train was already rolling down the track," and

[36] E. T. Mertz and O. E. Nelson, eds., *Proceedings of the High Lysine Conference, June 21–22, Purdue University* (Washington, DC: Corn Industries Research Foundation, 1966).

[37] N. E. Borlaug, "Weak Spots in the Rockefeller Foundation's Agricultural Programs Considering the Great Need for Expansion of Plant Protein Production to Human Needs," memo to E. Wellhausen, 1966, John Wooston Library, CIMMYT, Mexico City.

[38] P. G. Hoffman, "Development Co-operation: A Fact of Modern Life," *Virginia Quarterly Review* 47, no. 3 (1971): 321–335, at 330.

[39] T. Wolf, "Quality Protein Maize," *CIMMYT Today*, no. 1 (1975).

[40] G. N. Atlin et al., "Quality Protein Maize: Progress and Prospects," *Plant Breeding Reviews* 34 (2011): 83–131.

[41] J. C. Waterlow and P. R. Payne, "The Protein Gap," *Nature* 258, no. 5531 (1975): 117.

Figure 9.2 Postweaning children and their families, such as this Ghanaian father and his children, were the stated target consumers for Quality Protein Maize, 1995. QPM Program in South Africa, CIMMYT Repository. © CIMMYT.

CIMMYT's breeding for protein quality continued at full speed[42] (Figure 9.2).

In the 1980s, UNDP claimed that the development of quality protein maize with normal grain type was a "spectacular achievement," and that the main problem was "how farmers can be persuaded to use the new varieties."[43] In fact, after a decade of intensive breeding, adoption of the new varieties remained low because of reduced yields and susceptibility to insects, kernel rot, and loss of quality in open-pollinated varieties. By this time, experts also recognized several practical problems that further impeded uptake. A high-lysine grain that was visually indistinguishable from normal maize would not have a price premium in the market and therefore carry no incentive for farmers to adopt it. Farmers also lacked interest in growing the varieties for their own subsistence, since little effort was made to complement varietal introduction with nutrition education programs or even to conduct field trials with farmers to evaluate the nutritional benefits.[44] In short, there was no demand for the product

---

[42] Robert Tripp, email communication to Derek Byerlee, October 22, 2020.
[43] UNDP, "Evaluation of Global Programs," Report of the Administrator to the Governing Council, DP/456, March 20, 1984, 14–15, http://web.undp.org/execbrd/archives/ses sions/gc/27th-1980/DP-456.pdf.
[44] Robert Tripp, "Does Nutrition Have a Place in Agricultural Research?" *Food Policy* 15, no. 6 (1990): 467–474.

and, even if one were created, there was no way to distinguish high-lysine maize from normal maize in the market.

Faced with growing funding stress, CIMMYT closed the quality protein maize program in the 1990s. However, this research was kept alive by Borlaug after he retired from CIMMYT and became the chief technical advisor to the nongovernmental organization (NGO) Sasakawa Global 2000. With leadership from Borlaug and former US President Jimmy Carter, and philanthropic support from Ryōichi Sasakawa of the Nippon Foundation in Japan, Sasakawa Global 2000's mission was to bring the Green Revolution to Africa. In 2000, the award of the World Food Prize to CIMMYT's S. K. Vasal and Evangelina Villegas for their development of quality protein varieties with "normal" grain type helped to revive donor support for CIMMYT's quality protein maize program in Africa, this time mostly to develop hybrids. Although this later phase of research included much-needed investment in nutritional field trials, the problem of creating demand persisted. Without concrete results on the ground, support for quality protein maize was again reduced to a trickle.[45]

In summary, the initial period of CIMMYT's international maize research was characterized by efforts to develop a systematic approach to breeding and testing open-pollinated varieties adapted to highly diverse maize-growing environments around the world. Research products were provided freely to all, and one of the major accomplishments was the increased scale and reach in international maize germplasm exchange. It was also a period of stable and flexible funding that encouraged long-term research with uncertain payoffs, which in turn led to breakthroughs in breeding for stress tolerance that would have lasting value. In contrast, despite generous funding and sound scientific breeding, the large investment in quality protein maize did not pay off because the responses of farmers, consumers, and the market were not adequately considered.

A Sharpened Focus and Pivot to the Private Sector, 1985–2000

From the mid 1980s, factors external to CIMMYT began to play a larger role in shaping its maize research agenda. With the end of the Cold War, foreign assistance to agriculture sharply declined, and funding for international crop research tightened.[46] In CIMMYT, funding and staffing

---

[45] Byerlee and Edmeades, *Fifty Years of Maize Research in the CGIAR.*

[46] P. Pingali and T. Kelley, "The Role of International Agricultural Research in Contributing to Global Food Security and Poverty Alleviation: The Case of the CGIAR," in R. Evenson and P. Pingali, eds., *Handbook of Agricultural Economics*, vol. III (Amsterdam: Elsevier, 2007), pp. 2381–2418.

peaked around 1990, and maize-specific budgets and staff were cut by almost half by the end of the decade. In this new funding environment CIMMYT had to focus its limited resources more carefully. Responding to pressure from the development assistance community and reflecting a more nuanced understanding of the causes of hunger, CIMMYT also changed its mission from increasing food production to reducing poverty, prioritized research in Africa, and introduced the role of gender and sustainable management of natural resources. These are still major elements of CIMMYT's research today.

Experience and feedback from national systems indicated that CIMMYT's international testing sites were not well targeted, especially in the very diverse African environments.[47] Testing and breeding priorities were sharpened in the 1980s through the concept of mega-environments – areas of more than 1 million maize hectares often distributed over several countries and perhaps continents, where crop performance, climate, disease and pest incidence, and grain preferences were similar. This was a significant advance over previous extensive efforts by FAO and others to define world agro-ecological zones, because CIMMYT included crop-specific criteria to define environments. Although agro-ecological zones had been used to organize research, crop-specific mega-environments specifically aimed to make international breeding programs and germplasm exchange more effective. By the late 1990s, CIMMYT's maize mega-environments were further refined through the emerging science of geographical information systems, which facilitated the overlay of several types of spatial data.[48]

These changes were accompanied by increasing decentralization of the CIMMYT breeding program to regions that better represented diverse growing conditions. This shift also placed breeders closer to their "clients" where they could better assess demand for new varieties. Breeders had learned that although stable performance over a range of conditions remained key goals, one centralized program could not serve all regions.[49] The Inter-Asian Corn Program, started in 1963, had maintained its own breeding program in Thailand, closely linked with the Thai national program led by Sujin Sriwatanapongse. It focused on downy mildew resistance – largely an Asian problem – and produced the Suwan varieties that became one of the most widely grown varieties in the tropics. An even older Central American maize program, started by the Rockefeller

[47] CGIAR, *1988–1989 Annual Report* (Washington, DC: CGIAR Secretariat, 1989).
[48] A. D. Hartkamp, *Maize Production Environments Revisited: A GIS-based Approach* (Mexico City: CIMMYT, 2001).
[49] Haldore Hanson, "The Role of Maize in World Food Needs to 1980," in CIMMYT, *World Wide Maize Improvement*, pp. 1–1 to 1–19.

Foundation and initially headquartered in Mexico, was shifted to Guatemala in the mid 1980s. In 1985, CIMMYT also built its own breeding station for eastern and southern Africa near Harare, Zimbabwe. As in Asia, region-specific diseases were decisive in developing regional breeding programs for Central America and Africa, although a regional program in the Andes focused on products that would have the floury-grain type typical of that region.

These moves to greater decentralization were still not sufficient to address the considerable microvariation in many rainfed maize environments and differences in farmers' grain preferences. To accommodate local variations, breeders began to engage farmers in testing varieties under their own field conditions and in selecting varieties to fit their specific farm management and consumer preferences. From the late 1970s, CIMMYT social scientists had employed research methods involving farmer participation in the design and testing of maize practices and systems, and the results often provided important feedback to maize breeders. For example, participatory research in southern Africa emphasized the need for early maturing varieties to accommodate farmers' seasonal food needs and delayed planting due to labor or draft power constraints.[50] In Malawi, a participatory study identified strong local preferences for grain texture and ease of shelling that affected adoption by women farmers and processors.[51] Farmer participatory methods were further mainstreamed in maize breeding through "mother-baby trials," where small subsets of varieties were tested by men and women farmers under their management, post-harvest processing, and use. The farmers' ratings were then used in decisions on varietal release.[52]

During this period there was also a sharp shift away from the "development state" towards market-oriented approaches to development in what became known as the Washington consensus. In this new environment, multinational seed companies began to invest in middle-income countries led by Pioneer Hi-Bred, then the world's largest seed company. By 1985 these companies worked at twenty-nine stations in seven tropical and subtropical countries.[53] Regional and local seed companies also held

---

[50] Angelique Haugerud and Michael P. Collinson, "Plants, Genes, and People: Improving the Relevance of Plant Breeding in Africa," *Experimental Agriculture* 26, no. 3 (1990): 341–362.
[51] M. Smale, "'Maize Is Life': Malawi's Delayed Green Revolution," *World Development* 23, no. 5 (1995): 819–831; McCann, *Maize and Grace*.
[52] M. Bänziger, P. S Setimela, D. Hodson, and B. Vivek, "Breeding for Improved Abiotic Stress Tolerance in Maize Adapted to Southern Africa," *Agricultural Water Management* 80 (2006): 212–224.
[53] C. E. Pray and R. G. Echeverria, "Transferring Hybrid Maize Technology: The Role of the Private Sector," *Food Policy* 13, no. 4 (1988): 366–374.

significant market share, although some were taken over by the expanding multinationals. Private seed companies naturally emphasized hybrid seed, and most of them, especially regional and local companies, used some CIMMYT germplasm in their breeding programs.

Internal forces were also driving CIMMYT towards greater emphasis on hybrids over open-pollinated varieties. By 1986, two decades after CIMMYT's founding, only 11 percent of the tropical and subtropical maize area (excluding large commercial farms in Brazil) was sown to improved open-pollinated varieties, compared with 16 percent sown to hybrids, most of which were developed independently of CIMMYT.[54] Ironically, given that one of the original motivations for CIMMYT's focus on open-pollinated varieties was to allow farmers to save seed, their slow spread was largely due to the difficulty of developing sustainable seed systems. A few seed companies did sell open-pollinated varieties as a sideline to their main business of hybrid seed, as in Zimbabwe, or as an entry point for hybrid sales, as in Thailand. A handful of countries, notably Thailand, successfully produced and disseminated seed of open-pollinated varieties largely through the public sector, but most was supplied through ad hoc arrangements such as development projects and was of variable quality. As early as 1978, Edwin Wellhausen, the original leader of maize research for the Rockefeller Foundation in Mexico and the first director general of CIMMYT, concluded:

During my 32 years of promotion of maize production in the tropics, I have been unable to interest either the public sector or the private sector in the production of large volumes of seed of OPVs. Where it [open-pollinated variety seed] is produced, it is produced by individual farmers or as a stopgap by commercial seed producers, until some kind of hybrid can be developed.[55]

At the same time, there was mounting evidence of the willingness of smallholders to adopt hybrids even under marginal growing conditions.[56] This was especially true in eastern and southern Africa, where much of the extensive hybrid maize area was sown by smallholders with limited or no fertilizer and was subject to frequent drought. Their choice reflected the development of superior hybrids by strong national programs in Zimbabwe and Kenya, the emergence of an efficient private seed industry producing affordable hybrid seed, and effective public extension programs to promote the initial adoption of hybrids. Elsewhere, national programs were also

---

[54] CIMMYT, *Maize Facts and Trends: The Economics of Commercial Maize Seed Production in Developing Countries* (Mexico City: CIMMYT, 1987).
[55] Edwin J. Wellhausen, "Recent Developments in Maize Breeding in the Tropics," in D. B. Walden, ed., *Maize Breeding and Genetics* (Chichester: John Wiley & Sons, 1978), p. 81.
[56] Heisey et al., "Economics of Hybrid Maize Adoption."

converting to hybrids and ending their reliance on public-sector seed production.[57] Thailand, the star in the adoption of open-pollinated varieties, had by the 1990s become a leader in hybrid maize. In 2003, the CIMMYT economist Roberta Gerpacio concluded that "the primary locus of maize breeding research in Asia has shifted from the public to the private sector."[58] She also noted the "strong likelihood that the private sector will be reluctant" to "address the needs of farmers in marginal areas."[59]

The 1984 departure of Sprague, the champion of open-pollinated varieties in CIMMYT, opened the way for the center's breeders to turn back to hybrids after a hiatus of twenty years. Resources were shifted from open-pollinated varieties to hybrids, and the international testing program gradually converted to testing inbred lines and hybrids. These materials were made available to both public and private seed companies; however, CIMMYT clearly saw small- and medium-sized local and regional seed companies as its main partners for delivering hybrid seed to smallholders, especially in more marginal environments.[60] In contrast with the multinational companies, these companies were generally nationally owned, served local markets, and had, at best, minimal research capacity to produce their own inbreds and hybrids.[61] By 1988, the first 100 inbreds were made available, with free access to both the public and private sectors. Ten years later, 58 percent of hybrids released by the private sector in the tropics and subtropics contained some CIMMYT germplasm.[62] This transition was overseen by Ripsudan Paliwal, the long-serving deputy director and later program director of the Maize Program, who was experienced in hybrid seed production in India.

The partnership of an international research program established to produce public goods with private-sector actors was not without

---

[57] The public sector was generally even more ineffective in producing hybrid seed than seed of open-pollinated varieties. See Byerlee, "Globalization of Hybrid Maize."

[58] R. V. Gerpacio, "The Roles of Public Sector versus Private Sector in R&D and Technology Generation: The Case of Maize in Asia," *Agricultural Economics* 29, no. 3 (2003): 319–330, at 328.

[59] Ibid., 320.

[60] CIMMYT, *Seeds of Innovation: CIMMYT's Strategy for Helping to Reduce Poverty and Hunger by 2020* (Mexico City: CIMMYT, 2004). CIMMYT defines small- and medium-sized companies as worth less than $2 million, and between $2 and $5 million, respectively, in terms of annual sales; B. Prasanna, email communication to Greg Edmeades, September 9, 2021.

[61] To facilitate its changing priorities and partnerships, CIMMYT added the director of research at Pioneer Hi-Bred International to its governing board and hired a maize director from the private sector.

[62] M. L. Morris, *Impacts of International Maize Breeding Research in Developing Countries, 1966–98* (Mexico City: CIMMYT, 2002).

controversy in a period when the growing power of large seed companies in research and the ownership of intellectual property was attracting attention.[63] CIMMYT countered critiques by focusing on the development of local seed companies with limited research capacity. Evidence indicated that these companies, with support from CIMMYT and national, public-sector research, could provide hybrid seed at lower prices than the large companies and serve markets that were not attractive to large companies, especially in more marginal areas.[64] Some evidence also suggested that farmers received more than half of the "surplus" generated by use of hybrid seed, with the remainder going to the seed company.[65] This pattern at the international level followed the example in the United States where public development of inbreds for private-sector use continued long after large private companies had developed strong in-house research and development programs.[66]

In recent years, CIMMYT has experimented with other models to incentivize delivery of its products through small- and medium-sized seed companies. In Africa it employs royalty-free licenses to supply hybrids to seed companies that then enjoy exclusive rights for a specific region and duration. This approach recognizes that testing and developing markets for new hybrids entails significant fixed costs, especially for smaller companies.[67] CIMMYT also has established International Maize Improvement Consortia, groups of companies with some research capacity that have first right of access to selected inbreds from CIMMYT and receive services to support hybrid development and seed production in exchange for a modest membership fee.[68]

In this new environment, the seed market has further diversified. For example, the number of seed companies in eastern and southern Africa increased fourfold between 1997 and 2007.[69] Similarly, locally owned seed companies in Mexico increased from 20 companies in 1995 to 114 in

---

[63] Kloppenburg, *First the Seed*, p. 81.   [64] CIMMYT, *Maize Facts and Trends*.

[65] Donald N. Duvick, "The United States," in Morris, ed., *Maize Seed Industries*, pp. 193–211.

[66] The early years of hybrid development in the United States saw lively debate on whether the public sector should continue to develop "open source" inbreds or leave this to the private sector. See Deborah K. Fitzgerald, *The Business of Breeding: Hybrid Corn in Illinois, 1890–1940* (Ithaca, NY: Cornell University Press, 1990).

[67] CIMMYT, "New Pre-commercial Hybrids for Southern Africa," November 29, 2018, www.cimmyt.org/news/new-cimmyt-pre-commercial-hybrids-for-southern-africa.

[68] FAO, "Views, Experiences and Best Practices as an Example of Possible Options for the National Implementation of Article 9 of the International Treaty," July 23, 2019, www.fao.org/3/ca7857en/ca7857en.pdf.

[69] A. S. Langyintuo, W. Mwangi, and A. O. Diallo, "Challenges of the Maize Seed Industry in Eastern and Southern Africa: A Compelling Case for Private–Public Intervention to Promote Growth," *Food Policy* 35, no. 4 (2010): 323–331.

2015, and the share of these companies in maize seed sales rose from 5 percent in 2009 to 31 percent in 2016.[70] In addition, most of these companies serve farmers in rainfed regions where hybrid seed adoption has now reached 40 percent of the area planted to maize, effectively reversing decades of failure to reach these farmers.[71] Even so, it is not clear that seed companies are reaching a significant share of Mexico's poorest farmers in the south of the country.[72]

In retrospect, the early CIMMYT dogma with respect to an exclusive focus on open-pollinated varieties was well meaning but patronizing in terms of small farmers' willingness to adopt hybrid seed and countries' abilities to develop private seed industries. CIMMYT also overestimated the capacity and willingness of the public sector to deliver high-quality seed of open-pollinated varieties. Our assessment is that CIMMYT's single-minded dedication to these varieties in the 1970s delayed the development of hybrids by the public sector and the emergence of small- and medium-sized seed enterprises by about a decade. At the same time, with the development of hybrids and associated private-sector partnerships, CIMMYT has compromised on its original policy of unrestricted access to all its products in the interest of engaging the private sector to quickly increase the number of farmers it reaches.

Scaling up in Africa and Accessing Proprietary Science, 2000–20

From the 1980s, CGIAR increasingly focused on sub-Saharan Africa. Africa was the only region where the prevalence of undernutrition and poverty continued to grow and yields of food staples were low and stagnant. It was widely recognized that Africa had been bypassed by the Green Revolution, and donors, national governments, and CGIAR set out to ignite an "African Green Revolution." Their ambitions echoed the rhetoric of 1970 when the new headquarters of IITA was opened with much fanfare in Nigeria, aiming to bring the Green Revolution to Africa.[73]

---

[70] Prior to market liberalization, public research organizations in Mexico were required to "commercialize" their products through the public-sector seed company PRONASE, stifling the growth of local companies.

[71] M. L. Donnet, I. D. López-Becerril, C. Dominguez, and J. Arista-Cortés, "Análisis de la estructura del sector y la asociación público-privada de semillas de maíz en México," *Agronomía Mesoamericana* 31, no. 2 (2020): 367–383.

[72] A. Turrent Fernandez, A. Espinosa Calderón, J. I. Cortés Flores, and H. Mejía Andrade, "Análisis de la estrategia MasAgro-maíz," *Revista Mexicana de Ciencias Agrícolas* 5, no. 8 (2014): 1531–1547.

[73] Ford Foundation, *Sowing the Green Revolution: The International Institute of Tropical Agriculture, Ibadan, Nigeria* (New York: Ford Foundation, 1970). Haldore Hanson, the

The 2008–12 world food crisis also stimulated a doubling of funding for international agricultural research, ending a funding plateau that had lasted nearly two decades. Unlike the first period of strong financial support, funding was now largely restricted to specific projects, and for maize these mostly focused on Africa. The Bill & Melinda Gates Foundation became a major donor to large projects on stress-tolerant maize starting in 2007, and its support has continued until today with the addition of research on disease- and insect-resistance and efficiency in breeding. The Gates Foundation was well aware of the scientific advances in breeding for stress tolerance at CIMMYT; indeed, three of the Foundation's senior scientific staff in this period had prior experience in CIMMYT's maize program.

Against this background, CIMMYT relocated its first female maize director, Marianne Bänziger, to Nairobi in 2004. By 2010, its maize research effort was firmly centered in sub-Saharan Africa, with over half of its staff located there. The prevalence of drought stress, infertile and often degraded soils, and low use of external inputs in much of Africa demanded that priority be given to breeding for stress tolerance (Figure 9.3). CIMMYT's stress-breeding methods, developed earlier in Mexico, had been judged sufficiently mature to make screening for drought tolerance routine in maize breeding in Africa by 1995. Experiment stations were established at Chiredze, Zimbabwe and Kiboko, Kenya, where research under limited irrigation to simulate drought stress could be conducted on a large scale. This research was accompanied by testing at up to sixty largely rainfed locations across eastern and southern Africa, and a smaller number of sites across West Africa. Between 2016 and 2019 alone, over 230 open-pollinated varieties and hybrids with stress tolerance were released across Africa.[74]

Two further factors influenced the focus and reach of CIMMYT in Africa in the early twenty-first century. First, the development pendulum that had swung to market-based approaches in the 1990s now reversed and explicitly recognized the "visible hand of the state" and the "entrepreneurial state" in facilitating change.[75] In Africa, donors supported the development of local, private seed companies, and most countries

---

Ford Foundation representative in Nigeria and soon-to-become CIMMYT's second director general, was much more thoughtful about the difficulty of translating Asian experiences to Africa. See H. Hanson, "Agricultural Development in Tropical Africa and the Role of the Ford Foundation," December 1970, Ford Foundation Archives, Rockefeller Archive Center, Ford Foundation document 0002799.
[74] Vijesh V. Krishna, Maximina A. Lantican, B. M. Prasanna et al., "Impact of CGIAR Maize Germplasm in Sub-Saharan Africa," *Field Crops Research* 290 (2023): 108756.
[75] World Bank, *World Development Report: Agriculture for Development* (Washington, DC: World Bank, 2007); M. Mazzucato, *The Entrepreneurial State* (London: Demos, 2011).

Figure 9.3 CIMMYT maize breeder Dr. Cosmos Magorokosho with several drought-tolerant maize hybrids developed under managed drought stress and confirmed in on-farm trials, Harare, Zimbabwe, 2011. Photo by Gregory Edmeades.

reintroduced policies to promote technology adoption through subsidies to farmers to purchase seed and fertilizers.[76] Second, donors operating within the context of the new UN Millennium Development Targets began to promote an "impact culture," requiring CIMMYT to establish explicit, time-bound metrics for the adoption and impact of its work. This moved CIMMYT to invest more effort on delivery of its products by working closely with seed companies through training and technical assistance. By 2023 CIMMYT claimed that 165,000 tons of seed of its stress tolerant varieties and hybrids were being produced annually in East and Southern Africa, enough to reach 7.4 million households. Studies of the adoption of stress-tolerant varieties also suggested accelerated uptake of CIMMYT's products, stimulated by input subsidies in some countries.[77] However, in contrast to the first years of CIMMYT's maize program, the focus on short-term impacts and the restricted nature of most funding left little time, resources, and incentives for CIMMYT

[76] T. S. Jayne and S. Rashid, "Input Subsidy Programs in Sub-Saharan Africa: A Synthesis of Recent Evidence," *Agricultural Economics* 44, no. 6 (2013): 547–562.
[77] Krishna et al., "Impact of CGIAR Maize Germplasm."

scientists to pursue longer-term research with more uncertain payoffs. Although too early to assess in 2022, these shifts in maize research funding, which mirror circumstances elsewhere in CGIAR, may undermine the chances of future research breakthroughs.

A second important influence on CIMMYT's maize agenda in the 2000s was biotechnology and its concentration in the private sector. Most of the capacity to apply advances in molecular biological research rested in companies that, protected by stronger intellectual property rights, invested an estimated $1.6 billion in maize research in 2010, compared with CIMMYT's investment of about $28 million in the same year.[78] The quest to gain access to patented technologies stimulated a surge of mergers and acquisitions among seed, chemical, and biotechnology companies. By the 2010s, the top four companies were multibillion-dollar operations accounting for an estimated 82 percent of maize seed sales in the USA (up from 52 percent in 1988). Monsanto alone owned an estimated 85 percent of patents on traits for genetically modified (GM) maize, weighted by area planted in 2010.[79] The growing concentration of intellectual property ownership in the "gene giants" caused an uproar from NGOs, academics, and international organizations.[80] Many argued that genetic resources were the result of millennia of selection and conservation by small-scale farmers who were their real owners.

At CIMMYT, and within CGIAR more generally (see David J. Jefferson, Chapter 12, this volume), scientists and administrators were concerned about their freedom to operate in a world increasingly dominated by patented technologies, some of which they considered relevant to solving intractable problems of poor farmers. CIMMYT did not have the time, funds, or laboratories to "invent around" patents, so it elected to negotiate with private companies to access the most relevant technologies. As CIMMYT concluded in 2002, "the continuing relevance of the international agricultural research centers will depend critically on their ability to forge effective partnerships with the private firms that now control many critical technologies."[81] This view was echoed by

---

[78] P. W. Heisey and K. O. Fuglie, "Private Research and Development for Crop Genetic Improvement," in K. Fuglie et al., eds., *Research Investments and Market Structure in the Food Processing, Agricultural Input, and Biofuel Industries Worldwide*, USDA Economic Research Report 130 (Washington, DC: USDA, 2011), pp. 25–48.

[79] Ibid. In 2018, Monsanto was acquired by Bayer.

[80] Kloppenburg, *First the Seed*; C. Fowler, *Unnatural Selection: Technology, Politics and Plant Evolution* (Yverdon, Switzerland: Gordon and Breach, 1994); UNDP, *Human Development Report 2001: Making New Technologies Work for Human Development* (New York: Oxford University Press, 2001).

[81] M. L. Morris and B. Ekasingh, "Plant Breeding Research in Developing Countries: What Roles for the Public and Private Sectors?" in D. Byerlee and R. G. Echeverría, eds.,

CIMMYT's consultations with national scientists. Maize was the crop most affected by developments in biotechnology and private-sector control, and in 2002 CIMMYT arranged a small meeting with private companies and international agencies to agree on some common principles for public–private partnerships.[82] The CIMMYT policy of 2012 on GM maize summed up the approach:

In line with its continued role to develop, use, and share global public goods, CIMMYT sees its role to focus on serving its primary customer base of small and marginal farmers who may not otherwise have access to such innovations/technologies. To this end, CIMMYT strategically uses intellectual property protection systems, including ascertaining and gaining freedom to operate to ensure and further its capacity to serve farmers and R&D organizations in the developing world.[83]

In addition to grappling with intellectual property rights, CIMMYT had to wrestle with the merits of becoming involved in the development of GM maize, considering the acrimonious debate about the value and possible risks of GM crops. Engaging with this technology would also necessitate appropriate biosafety regulatory environments in order to make GM maize available on a country-by-country basis.

Given widespread attention to the role of the private sector and intellectual property protections in limiting farmer seed-saving, one of CIMMYT's first public–private partnerships was an attempt to develop apomictic tropical maize. Allowing asexual reproduction (apomixis) would enable hybrids to retain their yield advantage from one generation to the next even when farmers saved their seed. The partnership included the (then French) Office for Overseas Scientific and Technological Research (ORSTOM) and three private multinational seed companies. It ran for over a decade without achieving its objective. However, it was an important learning experience for CIMMYT in balancing public interest in free access to technologies versus private interest in proprietary technologies for profit.[84]

From the 2000s, partnerships with the private sector to access technology were often funded by the Bill & Melinda Gates Foundation with

---

Agricultural Research Policy in an Era of Privatization (Wallingford, UK: CABI, 2002), pp. 199–225, at 223.

[82] CIMMYT, "Tlaxcala Statement on Public–Private Sector Alliances in Agricultural Research: Opportunities, Mechanisms, and Limits," November 1999, http://hdl.han dle.net/10883/3827.

[83] CIMMYT, "Position Statement on Genetically Modified Crop Varieties," January 2012, http://hdl.handle.net/10883/4393.

[84] M. Hodges, "The Politics of Emergence: Public–Private Partnerships and the Conflictive Timescapes of Apomixis Technology Development," *BioSocieties* 7, no. 1 (2012): 23–49.

a special focus on Africa.[85] The largest and longest-running project, Water Efficient Maize for Africa, supported breeding and testing facilities for drought tolerance. It operated under an agreement between Monsanto, CIMMYT, and the African Agricultural Technology Foundation (an NGO in Nairobi supported initially by the Rockefeller Foundation to broker access by African farmers to proprietary technologies) as the executing agent. The project, regarded as controversial given the partnership with Monsanto, the icon of the "gene giants," invested over $100 million from the Gates Foundation between 2008 and 2018. Monsanto provided royalty-free access for five countries in sub-Saharan Africa to its commercial drought transgene, which researchers subsequently combined with a Monsanto insect-resistance transgene. The insect resistance work built on an earlier CIMMYT partnership with the Novartis Foundation from the late 1990s that was halted when CIMMYT was unable to gain access to intellectual property rights for its commercial use.[86]

As of 2022 none of these transgenic options had been released outside of South Africa because of delays in implementing national biosafety regulations and, in the case of the drought transgene, lack of evidence of its value added over CIMMYT's conventionally bred drought-tolerant varieties. A twenty-year effort in East Africa to incorporate Bt (*Bacillus thuringiensis*) genes for stem-borer resistance in maize, although very costly and time-consuming, may eventually pay off, given serious losses caused by the invasion of fall armyworm from the Americas in the late 2010s.[87]

After more than two decades of experience, CIMMYT's partnerships with multinational companies to access new technologies remained marginal to its impacts.[88] More important has been an agreement with the University of Hohenheim, Germany for CIMMYT to "tropicalize" the university's proprietary double-haploid technology, a process that makes the development of its tropical hybrids more efficient and faster.[89] The

[85] M. A. Schnurr, *Africa's Gene Revolution: Genetically Modified Crops and the Future of African Agriculture* (Montreal: Mcgill Queens University Press, 2019).

[86] J. Mabeya and O. C. Ezezika, "Unfulfilled Farmer Expectations: The Case of the Insect Resistant Maize for Africa (IRMA) project in Kenya," *Agriculture & Food Security* 1, suppl. 1 (2012): S6.

[87] J. Wesseler, R. D. Smart, J. Thomson, and D. Zilberman, "Foregone Benefits of Important Food Crop Improvements in Sub-Saharan Africa," *PLoS One* 12, no. 7 (2017): e0181353.

[88] For a review of these partnerships, see Byerlee and Edmeades, *Fifty Years of Maize Research in the CGIAR*.

[89] With this technology, a single set of maize chromosomes (the haploid set) is generated and then doubled in the laboratory to produce the normal diploid in which both sets of chromosomes are identical. It thereby reduced the time to produce inbreds by half. See

technology, which does not involve transgenes and therefore does not invoke concerns about GM crops, is patented, and seed companies pay a license fee for its use to the university. CIMMYT now routinely uses the technology in its breeding program, making its products more rapidly available to public research systems and seed companies.

Conclusion

CIMMYT's maize research has undergone profound shifts over fifty years, probably more than any other CGIAR crop program. The type of product, geographical scope, and partnerships of the 2020s are quite different from those seen in the first two decades in which the international maize research program was designed and established. The main product has shifted from open-pollinated varieties for public-sector programs towards mostly inbreds and hybrids for national programs and private-sector use. This was driven by the rapid rise of the private seed sector and the development of public–private partnerships between small- to medium-sized seed enterprises and CIMMYT and/or publicly funded national programs. It reflected mounting evidence of the willingness of smallholders to pay for yield advantages of hybrids even in risky environments. While much of CIMMYT's engagement with the private sector was with local and regional seed companies possessing limited research capacity, the growing dominance of large multinationals in biotechnology pressured CIMMYT to seek further high-level partnerships to access these companies' patented tools and technologies. These partnerships have had a cost, moving CIMMYT away from the "open source" system of its early decades to one more constrained by intellectual property and some limits on access to its products.

Departing from the centralized breeding model that predominated within the early CGIAR, CIMMYT's maize-breeding research steadily became more decentralized as it attempted to serve the wide diversity of growing conditions and grain types found in tropical maize farming. Even with the more decentralized programs, rigorous testing was still required. In recent years, this testing was often performed collaboratively by private seed companies, as well as by CIMMYT's traditional public-sector partners. As it decentralized, the locus of CIMMYT maize research also shifted, moving from Latin America and Asia to eastern and southern Africa. This move reflected high levels of food insecurity in Africa, the

CIMMYT, "Tropicalized Maize Haploid Inducers for Doubled Haploid-Based Breeding," December 28, 2012, www.cimmyt.org/news/tropicalized-maize-haploid-ind ucers-for-doubled-haploid-based-breeding.

preeminent role of maize as a food staple in the region, and the focus of donor funding on Africa.

There were also important continuities throughout this history. As one example, CIMMYT scientists initiated breeding for stress-tolerant maize early, and against prevailing conventions. This work was maintained and expanded, eventually becoming the mainstream of CIMMYT breeding efforts, especially in Africa, where drought and low soil fertility are pervasive. The stress-tolerant hybrids and open-pollinated varieties produced through these efforts were widely accepted by smallholders operating in risky rainfed environments. By comparison, a long-term effort on quality protein maize, despite strong scientific underpinnings, met with only modest results on the ground. This was largely because the "demand side" of the program was missing, in which farmers' interest in growing quality protein maize and consumer interest in eating it would be assessed and encouraged.

The evolution of CIMMYT's maize program at first sight suggests that the freedom of scientists to set their agenda has been steadily narrowed as "donor sovereignty," restricted funding, and a short-term impact culture have taken center stage in the twenty-first century (as Rebekah Thompson and James Smith highlight in their analysis of the International Livestock Research Institute [ILRI], Chapter 7, this volume). Yet the growing emphasis on achieving "outcome milestones" also underlies breakthroughs in the adoption of maize hybrids and open-pollinated varieties and yield takeoff in several African countries, achievements that have made maize the leading crop in generating CGIAR impacts in Africa in the 2010s.[90] Our history suggests that a better question is whether CIMMYT's funding environment supports sufficient longer-term research needed to tackle emerging and recalcitrant problems of the twenty-first century, such as new pests and diseases or building resilience to climate change.

---

[90] See, for example, "Climate-Smart Maize," in CGIAR, "50 Years of Innovation That Changed the World" (n.d.), www.cgiar.org/innovations/climate-smart-maize.

10 Crop Descriptors and the Forging of "System-Wide" Research in CGIAR

Helen Anne Curry and Sabina Leonelli

The circulation of data – "full exchange of information among national, regional and international agricultural research centers" – ranked high among the objectives adopted by representatives to the Consultative Group on International Agricultural Research (CGIAR) at its first meeting in 1971.[1] It was considered essential to CGIAR's most important goals, from identifying the needs of individual countries or regions, to ensuring the coordination of research among different institutions, to allocating funds. In this chapter, we look at how agricultural experts attempted to realize this "full exchange of information" among scientists working at geographically distant sites, in different languages and cultural contexts, and with different organisms and research interests, in the four decades after the founding of CGIAR. Our focus is the historical development of crop descriptors, which CGIAR today defines as providing an "international format and a universally understood language for plant genetic resources data."[2] We examine crop descriptors as a critical component of CGIAR's earliest efforts to create "system-wide" research tools and agendas, emphasizing the scientific and political agendas that shaped centralizing, systematizing work orchestrated as a top-down enterprise.

Developers of descriptors aspire to agree on specific characteristics of crops, such as plant height or fruit shape, and exact terms for describing

Acknowledgments: We gratefully acknowledge the financial support of the Wellcome Trust (grant number 217968/Z/19/Z) for Helen Anne Curry's research and the intellectual support of the "From Collection to Cultivation" research team at the University of Cambridge. We are grateful to Adriana Alercia at Bioversity, Elizabeth Arnaud at CGIAR, and the Plant Life group at Exeter for helpful discussions; and to the Alan Turing Institute (EPSRC grant EP/N510129/1) and the European Research Council (award number 101001145) for funding Sabina Leonelli's research.

[1] Summary of Proceedings, Consultative Group on International Agricultural Research, First Meeting, May 19, 1971, Washington, DC, Annex III, https://hdl.handle.net/10947/260.

[2] CGIAR Genebank Platform, "Crop Descriptors," www.genebanks.org/resources/crop-descriptors/.

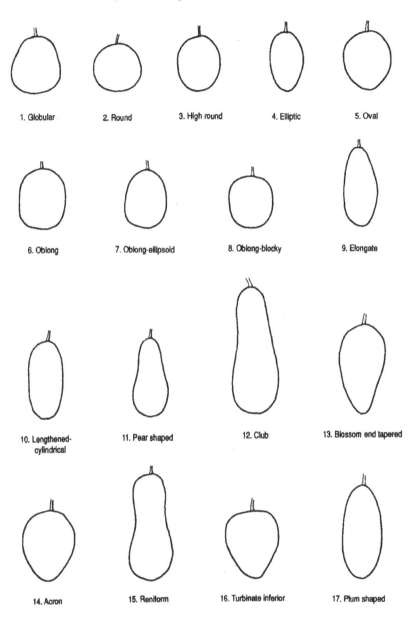

1. Globular   2. Round   3. High round   4. Elliptic   5. Oval

6. Oblong   7. Oblong-ellipsoid   8. Oblong-blocky   9. Elongate

10. Lengthened-cylindrical   11. Pear shaped   12. Club   13. Blossom end tapered

14. Acron   15. Reniform   16. Turbinate inferior   17. Plum shaped

Figure 10.1 This list of possible fruit shapes was intended to guide researchers working with papaya in systematic description of this trait in their collections and field trials. From IBPGR, *Descriptors for Papaya* (Rome: IBPGR, 1988), p. 17. Reprinted by permission of Alliance Bioversity–CIAT.

4.2.10 Fruit shape (fruits from hermaphrodite flowers)

Scored at full development. See Fig. 2

1	Globular
2	Round
3	High round
4	Elliptic
5	Oval
6	Oblong
7	Oblong-ellipsoid
8	Oblong-blocky
9	Elongate
10	Lengthened cylindrical
11	Pear shaped (pyriform)
12	Club
13	Blossom end tapered
14	Acron (heart shaped)
15	Reniform
16	Turbinate inferior
17	Plum shaped
18	Other (specify/describe)

4.2.11 Fruit shape (fruits from female flowers)

Scored at full development. See Fig. 2

1	Globular
2	Round
3	High round
4	Elliptic
5	Oval
6	Oblong
7	Oblong-ellipsoid
8	Oblong-blocky
9	Elongate
10	Lengthened cylindrical
11	Pear shaped (pyriform)
12	Club
13	Blossom end tapered
14	Acron (heart shaped)
15	Reniform
16	Turbinate inferior

4.2.12 Fruit skin colour

Overall colour of the skin of ripe fruits

1	Yellow
2	Deep yellow to orange
3	Red/purple
4	Yellowish green
5	Green
6	Other (specify in the NOTES descriptor, 11)

4.2.13 Fruit flesh colour

Observe on ripe fruits

1	Light yellow
2	Bright yellow
3	Deep yellow to orange
4	Reddish orange
5	Scarlet
6	Other (specify in the NOTES descriptor, 11)

4.2.14 Tree fruit productivity [kg per annum]

3	Low (approximately 20 kg)
5	Intermediate (approximately 50 kg)
7	High (approximately 80 kg)
9	Extremely high (approximately 120 kg)

| SEED

4.3.1 Seed colour

1	Generally tan
2	Generally grey-yellow
3	Generally grey
4	Generally brown black
5	Generally black
6	Variable

4.3.2 Seed germinating in ripe fruit

| 0 | Absent |
| + | Present |

Figure 10.2 Fruit shape, skin color, flesh color, and productivity were just a few of the several dozen traits and other identifying data that papaya researchers were encouraged to track in standardized form. From IBPGR, *Descriptors for Papaya* (Rome: IBPGR, 1988), pp. 16–18. By permission of Alliance Bioversity–CIAT.

these – for example, "height of plant at maturity, measured in centimeters from ground to top of spike, excluding awns" or "plum-shaped" (Figures 10.1 and 10.2). Historically this work has been motivated by the idea that widely agreed descriptors will allow diverse and globally dispersed users to share plant materials and information. It has sought especially to make it easy to manage and communicate the data associated with samples of plant genetic material tested in field trials or stored in research institutions and gene banks. Finding common labels and formats for such data has long been a challenge for agronomists, plant scientists, and curators, not least because the characteristics of interest to these researchers extend beyond those deemed to be relevant in traditional taxonomy. Today they may include everything from genomic data to breeders' assessments to ethnobotanical context to market uses.

Our historical reconstruction of the technically and culturally complex project of descriptor creation shows how, in addition to bridging expert domains, including botany, agronomy, genetics, breeding, and farming, it provided an opportunity for CGIAR to instantiate and consolidate its central position in a larger web of international agricultural research initiatives. Providing descriptors served to advance CGIAR's identity as an essential resource for globalized development. As we show, descriptors acquired increasing strategic importance within CGIAR over time, serving as evidence of the organization's role in enabling global agricultural research and as instruments for shaping related policies and strategic objectives. Descriptors fulfilled a politically significant social function, establishing CGIAR as a necessary passage point in coordinating the exchange of data and expertise about plant genetic resources and constraining alternative approaches.

We argue that the project of producing descriptors both defined and embodied CGIAR institutional identity and objectives as these evolved from the 1970s to the 2000s. On the one hand, descriptors were intended as generalizable tools for agricultural development. Well-defined and widely used descriptors would not only enable CGIAR institutions to work together by pooling data and related materials and methods, but also allow CGIAR to respond to – and to some extent shape – the key institutions and regulations, both national and international, of global agricultural development. On the other hand, for universalized descriptors to be adopted and effective in research, they needed to be locally meaningful. This meant identifying descriptors able to encompass different crops, users, research agendas, and even diverging agricultural strategies. This was (and continues to be) a complex challenge, especially given the enduring tensions between a regulatory and scientific sphere dominated by Euro-American interests and expertise and the heterogeneous demands for and understandings of agricultural development emerging from the Global South. In reconstructing this history, our analysis complements studies that approach the history of CGIAR via local experiences of top-down research agendas (see the contributions to Parts I and II of this volume). We show how "the center" – and not "the Centers" – responded to changing circumstances, including frictions felt at the local level.

The agricultural technologies that precipitated the creation of CGIAR, and which remain central to the work of many of its centers, are the seeds of novel crop varieties and the systems of production that sustain their cultivation. As Marianna Fenzi (Chapter 11, this volume) describes, CGIAR's position as the steward of some of the world's most extensive collections of "plant genetic resources" – that is, seeds and other crop genetic materials held in gene banks and related facilities – has placed it at the heart of international controversies regarding ownership of and control over these resources.[3] As a result, seeds and the power associated with the possession and dissemination of these have been central to historical investigations of CGIAR and the research institutions associated with it.[4] By comparison, the history of data management, sharing, and reuse within CGIAR has mostly escaped close observation, although it will come as no surprise to historians of science and agriculture that information about seeds has been as important as the seeds themselves. The infrastructures needed to shuttle seeds from site to site without losing the identifiers and data attached to these have been crucial to CGIAR gaining and retaining power in international agriculture. They are becoming ever more influential within digitalized, data-intensive, and increasingly automated approaches to biology and breeding, in contexts where ownership of seeds and data remains hotly contested (see David J. Jefferson, Chapter 12, this volume).[5] A historical understanding of the role played by CGIAR in this domain is therefore essential to understanding the present and possible futures of global agricultural development.

---

[3] See Robin Pistorius, *Scientists, Plants and Politics: A History of the Plant Genetic Resources Movement* (Rome: IPGRI, 1997); Marianna Fenzi and Christophe Bonneuil, "From 'Genetic Resources' to 'Ecosystems Services': A Century of Science and Global Policies for Crop Diversity Conservation," *Culture, Agriculture, Food and Environment* 38, no. 2 (2016): 72–83.

[4] E.g., Deborah Fitzgerald, "Exporting American Agriculture: The Rockefeller Foundation in Mexico, 1943–1953," *Social Studies of Science* 16, no. 3 (1986): 457–483; John H. Perkins, *Geopolitics and the Green Revolution: Wheat, Genes, and the Cold War* (Oxford: Oxford University Press, 1997); Nick Cullather, *The Hungry World: America's Cold War Battle against Poverty in Asia* (Cambridge, MA: Harvard University Press, 2010); Helen Anne Curry, "From Working Collections to the World Germplasm Project: Agricultural Modernization and Genetic Conservation at the Rockefeller Foundation," *History and Philosophy of the Life Sciences* 39, no. 2 (2017): 5; Sara Peres, "Seed Banking as Cryopower: A Cryopolitical Account of the Work of the International Board of Plant Genetic Resources, 1973–1984," *Culture, Agriculture, Food and Environment* 41, no. 2 (2019): 76–86.

[5] Sabina Leonelli, *Data-Centric Biology: A Philosophical Study* (Chicago: University of Chicago Press, 2016); Christopher Miles, "The Combine Will Tell the Truth: On Precision Agriculture and Algorithmic Rationality," *Big Data & Society* 6, no. 1 (2019), https://doi.org/10.1177/2053951719849444; Sabina Leonelli and Hugh Williamson, "Towards Responsible Plant Data Linkage," in H. Williamson and S. Leonelli, eds., *Towards Responsible Plant Data Linkage* (Cham: Springer, 2023), pp. 1–24.

Building a Network

Since its inception, the work of crafting descriptors has been tied up with the management and use of crop genetic resources, especially by breeders. Descriptors initially aimed to identify useful seeds in collections and facilitate their exchange among an ever-growing number of researchers. From the late 1960s, several different crop research communities attempted to coordinate the methods and language they used to document information about breeding materials held in collections or used by researchers. Rice scientists saw standardization in documentation as a way to deal with their own diversity as much as crop diversity.[6] Other researchers were motivated to study standardization by the possibility of using new automated data storage and retrieval systems to coordinate international breeding activities.[7] By the early 1970s, dealing with a surfeit of seeds provided additional impetus. With millions of seeds already in seed banks and more anticipated, only clear and consistent modes of description would enable researchers to navigate these collections.[8]

CGIAR's entry into data standardization initiatives came via its early focus on disseminating new crop varieties, a task that both generated collections of crop diversity and made conservation of these imperative. At their first-ever meeting, in the summer of 1971, members of the CGIAR's Technical Advisory Committee (TAC) debated what research activities would best ensure that the "promise already shown by the 'Green Revolution'" could be extended geographically. Towards the end of their deliberations, which mainly focused on what new international research centers would complement the four existing institutes, several participants relayed their concern that the accelerated spread of "modern" crop varieties was causing "the progressive erosion of natural genetic resources." In other words, they believed that genetically heterogeneous farmers' varieties were giving way to more uniform breeders' varieties. As Marianna Fenzi (Chapter 11, this volume) recounts, this concern eventually precipitated a new CGIAR institute, the International Board for Plant Genetic Resources (IBPGR), with the mandate to "promote an international network of genetic resources

---

[6] See Tze-Tu Chang and Eliseo A. Bardenas, *The Morphology and Varietal Characteristics of the Rice Plant*, Technical Bulletin No. 4 (Los Baños, Philippines: IRRI, December 1965); IRRI, *Rice Genetics and Cytogenetics* (Amsterdam: Elsevier, 1964).

[7] C. F. Konzak and S. M. Dietz, "Documentation for the Conservation, Management, and Use of Plant Genetic Resources," *Economic Botany* 23, no. 4 (1969): 299–308, at 306.

[8] J. G. Hawkes, "Workshop on Information Systems for World Genetic Resources" (workshop documents, Birmingham, England, July 4–5, 1972), Archives of the International Center for Maize and Wheat Improvement (CIMMYT), El Batán, Mexico, Folder 3–10 1972 Germplasm World Project, Box 56.

activities to further the collection, conservation, documentation, evaluation and utilization of plant germplasm."[9]

IBPGR was unusual for a CGIAR center in that it was not a physical facility, but instead a group of geographically dispersed experts who convened at regular intervals and were supported by a secretariat at the United Nations Food and Agriculture Organization (FAO) headquarters in Rome. It was also unusual in that, initially, it didn't conduct any research itself, but served chiefly to manage funds and – more aspirationally – to coordinate the actions of many widely dispersed and independently motivated researchers and institutions. These two features of IBPGR explain the largest line-item in its budget during its earliest years: investment in the creation of a "Communication, Information and Documentation System."[10] This "integrated system" would, it was hoped, "support all phases of management of genetic resources data," from collection in a farmer's field, to filing in seed bank storage, to evaluation by a crop scientist in an experimental plot. In addition to being flexible enough to accommodate these different scientists, it would be adaptable to the different computing capacities found at different institutions managing genetic resources.[11]

Since IBPGR didn't have its own in-house research and development capacity (beyond desk studies conducted by the small staff of the secretariat at FAO), it contracted out the work of creating its information infrastructure to a research group, the Taximetrics Laboratory, at the University of Colorado, Boulder. A few years earlier, FAO-led efforts to orchestrate collaboration in crop exploration and conservation had prompted an assessment of the Taximetrics Laboratory's Taxonomic Information Retrieval system (TAXIR), for this purpose.[12] With the influx of money from IBPGR, the Taximetrics Laboratory turned its attention to developing an information system to be used in managing genetic materials held at CGIAR centers and national collaborators of IBPGR. TAXIR, which was a product of US National Science Foundation funding, was adapted into a new system for managing genetic resources data, called EXIR.[13]

[9] International Board for Plant Genetic Resources (IBPGR), *IBPGR Annual Report 1974* (Rome: IBPGR, 1975), p. 1. All annual reports of IBPGR, IPGRI, and Bioversity cited in this chapter are archived at https://alliancebioversityciat.org/publications-data.
[10] See IBPGR's *Annual Reports* for 1974 through 1978.
[11] IBPGR, *Annual Report 1974*, pp. 2–3.
[12] G. N. Hersh and D. J. Rodgers, "Documentation and Information Requirements for Genetic Resources Application," in Otto Herzberg Frankel and John Gregory Hawkes, eds., *Crop Genetic Resources for Today and Tomorrow* (Cambridge: Cambridge University Press, 1975), pp. 407–446, at 408; Hawkes, "Workshop on Information Systems."
[13] TAC Secretariat, "Report of the TAC Mission to the IBPGR Programme at Boulder, Colorado," April 1979, https://hdl.handle.net/10947/1157; TAC Secretariat, "Report of the TAC Quinquennial Review of IBPGR," May 1980, https://hdl.handle.net/10947/1

Having an agreed-upon set of identifying information to describe samples in collections – and consistent terms for communicating it – was considered crucial to the operation of this system. Echoing a view already circulating among crop scientists, IBPGR maintained that collections of plant genetic materials were "only as good as the use that can be made of them, and without information they can hardly be used at all."[14] Its planned system aimed to ensure that essential information accompany all samples in its affiliated *ex situ* collections. This was not simply a matter of creating means for data storage and access. It was also one of dictating the nature of the data stored. As an IBPGR report explained, "Different collections of the same species have been made by different people and for different purposes and so they have been described in dissimilar ways." Facilitating management of, communication about, and access to collections therefore required an "internationally accepted system" for describing their contents.[15]

The stakes for setting descriptors and the challenges to agreeing on these were evident in IBPGR's attempt to enlist researchers in setting "a minimum list of taxonomic, morphological, physiological, resistance, and quality characteristics" for wheat and its relatives, and an "inventory of descriptors" to capture these, in the mid-1970s.[16] In Boulder, where researchers were central in forging the very concept of descriptor, a group of maize and wheat experts gathered in 1975 to develop initial lists of types of descriptors.[17] These proposals informed discussions at an international symposium on wheat later that year in Leningrad (St Petersburg), where attendees agreed on the minimum information that should be attached to every item accessioned into a collection.[18] This list went through further refinement in 1977 when Japanese, West German, Soviet, and US wheat scientists, in consultation with the team of Boulder-based data scientists, drew on the Leningrad recommendations, data collated from world wheat collections, a glossary of wheat characteristics, and other data to propose "a list of minimum descriptors." The purpose of this list was to facilitate an international evaluation of wheat varieties.[19] If every collector, curator, and breeder tracked the

388; FAO Commission on Plant Genetic Resources, "International Information System on Plant Genetic Resources," Provisional Agenda, December 1984, CPGR/85/6, www.fao.org/tempref/docrep/fao/meeting/015/aj375e.pdf.

[14] IBPGR, *Annual Report 1976* (Rome: IBPGR, 1977), p. 17.   [15] Ibid.

[16] IBPGR and IS/GR, *Descriptors for Wheat & Aegilops: A Minimum List* (Rome: IBPGR, March 1978), p. 1, https://hdl.handle.net/10568/73164.

[17] D. J. Rodgers, B. Snoad, and L. Seidewitz, "Documentation for Genetic Resources Centers," in Frankel and Hawkes, eds., *Crop Genetic Resources*, pp. 399–405.

[18] IBPGR, *Annual Report 1975* (Rome: IBPGR, 1976), p. 12.

[19] IBPGR, *Annual Report 1977* (Rome: IBPGR, 1978), p. 33.

same thirty-three essential items of information ("descriptors"), using the same scales and standardized responses ("descriptor states"), then it would be irrelevant where and by whom evaluations were done, at least with respect to interpreting the data. Plant height would always appear in centimeters and be calculated without including the awns. Kernel plumpness would be rated from 1 to 9, with 1 indicated "shrivelled" and 9 "plump." The number of spikelets per spike would be decided by averaging five spikes. And so on.[20]

Expert-developed and community-agreed descriptors like those created for wheat were meant to make collections "useful to workers other than those who have assembled them," overcoming institutional divisions of labor, as well as cultural divides.[21] Heretofore uncoordinated approaches to assessing phenotypic traits, which made it difficult to share and compare plant materials, would be aligned to a single standard – or, rather, a list of them. The first attempt to implement the new wheat descriptors in an international evaluation program revealed just how ambitious this goal was. Sets of 400 wheat samples from collections were sent to several sites, with instructions to grow and characterize them according to the agreed descriptors. Of the four institutions that returned results by 1980, none returned data for every descriptor. Very few descriptors were recorded across all institutions and one, drought resistance, was not recorded by any.[22] Researchers on the ground evidently lacked the time to assemble complete datasets. They may also have disagreed with top-level coordinators about which data were useful and which were not.

The aims of IBPGR's nascent descriptor program, and the obstacles to its realization, reflected the ambitions of the still-young CGIAR and the realities of international agricultural research in the 1970s. These were the heady years when a rapid extension of the Green Revolution through institution-building and technological innovation – especially innovation in crop varieties promising higher yields – seemed possible to funders like the Rockefeller Foundation and the World Bank. A system for shuttling the genetic resources considered essential to crop development from one site to another and one researcher to another would be a critical component of CGIAR's growing institutional network. If heterogeneous descriptions were an obstacle to the efficient transfer of material and information, then a group of experts could convene to decide standard ones, and all other researchers, whether at CGIAR centers or in national

[20] IBPGR and IS/GR, *Descriptors for Wheat and Aegilops*.
[21] IBPGR, *Annual Report 1976*, p. 17.
[22] IBPGR, *Annual Report 1980* (Rome: IBPGR, 1981), p. 67.

research institutions, would conform to this universal norm. The problem was that local circumstances – institutional, environmental, cultural, political – often resisted this centralizing, universalizing ambition to identify high-yield crops.

Descriptors' Ascendancy

CGIAR, through the new IBPGR, could and did build on the knowledge and expert communities that FAO had fostered since the 1940s as it began to coordinate the transit of breeding materials and information in the 1970s. Moreover, because it had no research capacity of its own, IBPGR depended on other institutions to achieve its objectives. These included CGIAR centers, national research institutes, and universities. This institutional positioning created circumstances in which the development of descriptor lists became the defining work of IBPGR through the 1980s with respect to its mission of facilitating cross-institutional exchange of genetic resources.

The intensified focus on descriptor lists as one of the defining contributions, if not *the* defining contribution, of IBPGR to international agricultural research in the 1980s followed a major disruption to the organization's communication and information program. The research group in Boulder that IBPGR had funded to develop its information management system, TAXIR/EXIR, was recruited in 1978 to lead the development of an information system for the US National Plant Germplasm System. On review, it looked as though IBPGR's significant financial investment – about $1.5 million between 1975 and 1978 – had mainly gone towards developing technical systems that did not serve most CGIAR centers' needs and expert knowledge that was now contracted to a different institution. Indeed, a panel assembled to evaluate these efforts deemed EXIR to be cost-ineffective, and IBPGR quickly abandoned its development of a centralized, universal computer system for genetic resources management across CGIAR. "Practical ad hoc adaptation" of any computerized data management system to existing local equipment was thought to promise faster, more sustainable results.[23]

Although IBPGR abandoned its aspirations for a unified approach to hardware and software, it intensified its goal of developing a universal language for recording and communicating information about breeding materials and crop varieties. The 1982 IBPGR *Annual Report* reiterated

---

[23] TAC Secretariat, "Report of the TAC Quinquennial Review of IBPGR," 22. See also TAC Secretariat, "Report of the TAC Mission to the IBPGR Programme at Boulder, Colorado"; TAC Secretariat, "Comments made by IBPGR on the Quinquennial Review Report," May 1980, https://hdl.handle.net/10568/118516.

that "the biggest and most difficult problem to solve" with respect to genetic resources remained accurate documentation. This was essential to nearly every task, from planning collecting expeditions to curating to sharing materials.[24] This in turn justified the accelerated production of lists of standardized descriptors and descriptor states for crops.

Over the next five years, IBPGR published dozens of descriptor lists in rapid succession. Its schedule ostensibly prioritized economically and socially important crops. In practice, priorities depended on the availability of existing information and relevant expertise, which in turn derived from previous research investments. Crops such as wheat or rice that had long histories in CGIAR centers, linked to their crucial role in industrial economies, provided obvious starting points for systematic information-gathering and discussion. The development of each descriptor list was supported by an advisory body that included biologists and agronomists with expertise in the crop at hand.[25] The twenty-one lists newly published or revised in 1985 alone included several staple grains (e.g., wheat, rye, oats, millets), fiber crops (cotton), oil plants (sunflower), pulses (lentil, chickpea, mung bean), beans (faba, tepary), fruits (apricot, cherry, peach, plum), multiple forages, and still others.[26] By 1991, IBPGR had published descriptors for seventy crops.[27]

The development of standard descriptor lists was accompanied by efforts to standardize across crops as well as within them. IBPGR introduced a new list format in 1982, identifying minimum information to be gathered by collectors and to be kept by curators on the status of samples maintained in a gene bank, as well as "standard numbering" and "standard descriptor states." This revised format also aimed to guide the production of more and better information at various points in the trajectory of a seed from farmer's field to gene bank to experimental site and back to the bank, not least by clearly demarcating collectors' responsibilities from those of curators. Collectors were further aided by the creation of standard collectors' forms, an intervention that was seen as resolving concerns about missing data, as well as inconsistencies in language and content.[28]

In attending to the publication of descriptors as its key contribution to research, IBPGR strove to streamline and standardize the characterization of seeds and other materials in the interest of efficient exchange and

---

[24] IBPGR, *Annual Report 1982* (Rome: IBPGR, 1983), p. ix.
[25] E. Gotor, A. Alercia, V. Ramanatha Rao et al., "The Scientific Information Activity of Bioversity International: The Descriptor Lists," *Genetic Resources and Crop Evolution* 55 (2008): 757–772.
[26] IBPGR, *Annual Report 1985* (Rome: IBPGR, 1986).
[27] Gotor et al., "Scientific Information Activity," 760.
[28] IBPGR, *Annual Report 1982*, pp. 75–76.

use. Its "minimum" lists sought to achieve maximum compliance by limiting the quantity of information required of hurried collectors, harried curators, and financially stressed research institutions. Yet these materials were scattered across institutions that deployed esoteric cataloguing systems and different computer software and hardware, and where researchers and curators spoke different languages. Institutions bore responsibilities for diverse crop species and responded to divergent cultural expectations for those crops, not all of which could be adequately captured in the standard descriptor list. The diversity of research made adherence to the minimalist ideal difficult.

This tension was apparent even in 1980, when external reviewers first formally advised IBPGR to drop the development of hardware and software and focus instead on descriptors themselves. The advisory panel urged against "over-elaborate descriptor lists," calling these "self-defeating" and recommending instead that lists be kept "as short as possible" by focusing on the institutional identifiers and "basic botanical characters."[29] IBPGR's ostensible emphasis on minimal descriptor lists would suggest that it acceded to the panel's admonitions as it reoriented its activities in the early 1980s – except that this emphasis was short-lived. By 1992, IBPGR descriptor lists were viewed not as minimal but as maximal and celebrated as providing "the widest number of descriptors that will assist with the characterization of the crop."[30] Consider the 1991 descriptor list for sweet potato, an early product of the new comprehensive approach. It included four categories of descriptors: passport (collectors' data), characterization (highly heritable, highly visible traits), preliminary evaluation (a limited number of traits "thought desirable" by many users consulted during list development), and further evaluation (basically, anything else considered useful in breeding). Users could record, in standardized form, collection data (e.g., site, collector, institution, environmental qualities), basic characteristics of the plant (vine color, leaf shape), more fine-grained details (root surface defects, flesh flavor), and a breathtaking array of evaluation data (data and location, soil taxonomy, root cracking, crude fiber content, keeping quality, drought tolerance, pest resistance).[31]

---

[29] TAC Secretariat, "Report of the TAC Quinquennial Review of IBPGR," 23.

[30] IBPGR, *Annual Report 1992* (Rome: IBPGR, 1993), p. 37.

[31] Zosimo Huamán, "Descriptors for the Characterization and Evaluation of Sweet Potato Genetic Resources," in *Exploration, Maintenance and Utilization of Sweet Potato Genetic Resources*, Report of the First Sweet Potato Planning Conference, February 1987 (Lima, Peru: International Potato Center, 1988), pp. 331–355. See also Helen Anne Curry, "Diversifying Description: Sweet Potato Science and International Agricultural Research after the Green Revolution," *Agricultural History* 97, no. 3 (August 2023): 414–447.

The curator at the CGIAR's International Potato Center (CIP) in Peru who oversaw this publication insisted that the newly expanded list was essential for improving management and use of sweet potato collections. The minimal list of sweet potato descriptors that had been agreed in 1981 by a small group of experts convened in South Carolina, USA and published by IBPGR had been revised and expanded almost immediately, after researchers attempted to apply it to collections in Fiji and Papua New Guinea. In 1986, when CIP launched an assessment of its 1,500 sweet potato accessions, the curator had expanded this already expanded minimal list still further. Yet, as he later reported, "even this expanded list was not adequate enough to describe all the morphologic variation shown in CIP's collection."[32]

The "single-language" vision of descriptors was abandoned, much as the single computer system had been. What had happened? It is tempting to suggest that the diversity of crops and crop researchers was just too great to be accommodated in universal standardized minimal lists. This is what the example of the sweet potato seems to indicate. However, looking outwards to the political debates and institutional wrangles in which IBPGR was involved in the 1980s suggests that these tussles were at least as important as the technical, biological, and cultural constraints encountered at the coalface of descriptor production. Throughout the decade, IBPGR was embroiled in a fight over the ownership of plant genetic resources that played out with particular fury within FAO (see Marianna Fenzi, Chapter 11, this volume). In response to accusations of its pirating seeds from farmers of the Global South to gene banks of the Global North and grossly mismanaging collections, IBPGR scrambled to show its commitment to maintaining open access to seeds and to rectifying perceived management issues.[33] Efforts at centralization and control were pushed aside in favor of inclusivity and inviting broader expertise, and IBPGR renewed its emphasis on data production and circulation.

The scrutiny of IBPGR in FAO forums and beyond prompted significant institutional change. The existing international system for collecting and conserving crop genetic materials, ostensibly overseen by IBPGR and therefore reporting to CGIAR, was heavily reliant on CGIAR centers and well-funded agricultural research institutions in a handful of industrialized

---

[32] Huamán, "Descriptors," p. 331.

[33] E.g., J. T. Williams, "A Decade of Crop Genetic Resources Research," in J. H. W. Holden and J. T. Williams, eds., *Crop Genetic Resources: Conservation and Evaluation* (London: Allen & Unwin, 1984), pp. 1–17; J. H. W. Holden, "The Second Ten Years," in Williams, ed., *Crop Genetic Resources*, pp. 277–285.

countries.[34] Critics wanted to see FAO placed in charge of such a system. FAO offered equal representation and voice to all member nations, whereas CGIAR in the 1980s was still chiefly an organization of donor countries and their scientist advisors. As part of their bid to undermine IBPGR, advocates of change pointed out that it had no clear legal standing: unlike other CGIAR centers, it had not been founded as an independent international institution via an agreement with a host country.[35] An initial attempt to resolve these concerns ultimately resulted in an institutional break between IBPGR and FAO and the establishment of IBPGR as an independent entity.

IBPGR's emphasis on decentralization and inclusivity in the creation of crop descriptor lists came during this period of institutional crisis. It took shape as part of a response to complaints about CGIAR's largely self-assumed – and to some critics unauthorized – management of global crop genetic resources. This suggests that maximal description was a political solution as much as a technical one. It attempted to improve the quality and usability of descriptors while also shoring up the perceived legitimacy of IBPGR.

Going Global

As the form of crop descriptors expanded, so too did their functions. The elaboration of new international frameworks for managing crop genetic resources, beginning with the International Undertaking on Plant Genetic Resources for Food and Agriculture agreed at FAO in 1983, made data generation and data norms and standards more important than ever before.[36] First conceived as a tool for the exchange of information about accessions to collections, and therefore the exchange of accessioned materials, descriptors were integrated into new international regimes for tracking and governing plant genetic resources. In the run-up to the 1992 Convention on Biological Diversity (CBD), for example, descriptor lists produced by IBPGR were portrayed as a tool for

---

[34] J. Hanson, J. T. Williams, and R. Freund, *Institutes Conserving Crop Germplasm: The IBPGR Global Network of Genebanks* (Rome: IBPGR, 1984). See also Peres, "Seed Banking as Cryopower"; Imke Thormann, Johannes M. M. Engels, and Michael Halewood, "Are the Old International Board for Plant Genetic Resources (IBPGR) Base Collections Available through the Plant Treaty's Multilateral System of Access and Benefit Sharing? A Review," *Genetic Resources and Crop Evolution* 66 (2019): 291–310.

[35] E.g., Pat R. Mooney, "The Law of the Seed: Another Development and Plant Genetic Resources," *Development Dialogue*, 1–2 (1983): 65–68.

[36] See, e.g., FAO Commission on Plant Genetic Resources, "International Information System on Plant Genetic Resources."

promoting information exchange as part of the technical and scientific cooperation mandated by the convention.[37]

Positioning descriptor lists as key tools to support international cooperation, thereby highlighting the technical contributions of CGIAR to global agricultural development, was of special strategic significance at the start of the 1990s. CGIAR had grown to encompass eighteen centers and was taxed by the complexity of managing this institutionally diverse and geographically dispersed network while also negotiating an expanded research remit.[38] CGIAR administrators grappled with pressing financial concerns, including both the extent of resources required to orchestrate work across various locations and the need to comply with the demands of funders while respecting the autonomy of each center. In addition, the United States Agency for International Development (USAID), which had provided most of the financial support for CGIAR since 1971, was increasingly reluctant to do so.[39] Retrospective accounts have characterized the 1990s as a period of "crisis" for CGIAR, during which it faced criticism for its inconsistent and uncoordinated portfolio, its inability to address emerging challenges as a result of cumbersome managerial and financial structures, and its exclusion of representatives from the Global South.[40]

During this period of institutional crisis, CGIAR took steps to shore up its central role in the international flow and management of genetic resources. IBPGR transitioned into a new, independent, and legally authorized CGIAR center, the International Plant Genetic Resources Institute (IPGRI) in 1994. IPGRI was tasked with serving the genetic resources needs of the other CGIAR centers, making its operation the first cross-institute initiative specifically focused on standards for general use. Among the functions that IPGRI assumed – in this case from both IBPGR and FAO – was that of maintaining an authoritative, comprehensive list of internationally accessible gene banks.[41]

The standards developed by IBPGR/IPGRI were seen as means to connect and coordinate the sprawling network of CGIAR centers, and to clarify their relations to other international initiatives, as well as to address concerns about a lack of inclusivity within CGIAR. At the technical level, one way to show support for a more diverse user base was to

[37] Secretariat of the Convention on Biological Diversity, *Convention on Biological Diversity: Text and Annexes* (Montreal, Canada: UNEP, 2011), Article 18.3, www.cbd.int/doc/leg al/cbd-en.pdf; see also Gotor et al., "Scientific Information Activity," 769.

[38] Selçuk Özgediz, *The CGIAR at 40: Institutional Evolution of the World's Premier Agricultural Research Network* (Washington, DC: CGIAR Fund, 2012), pp. 32–34.

[39] Ibid., p. 13.   [40] Ibid., pp. 31–54.

[41] IBPGR, *Annual Report 1991* (Rome: IBPGR, 1992), pp. 10–11.

emphasize the broad relevance of the standards produced and their inclusivity compared with other systems. In 1992, IBPGR had confirmed its crop descriptor lists as allowing the "widest number of descriptors." However, when it became apparent that comprehensive descriptors were cumbersome for breeders with fewer resources to deploy – implemented primarily by those at well-resourced institutions with the effect of excluding others, especially those working in the Global South – the pendulum swung back.[42] Around 1993, IBPGR/IPGRI began to resimplify descriptors. Comprehensive descriptors were not abandoned, but instead accompanied by a reduced, general list of "minimal," "highly discriminating" descriptors that could be applied across species and locations. This new format, first trialed with barley in 1994, was thought to "reduce redundancy" and again make descriptors more user-friendly.[43] IPGRI acknowledged that some descriptor lists were long but encouraged researchers "to utilize those that are important in their own situations."[44]

This solution was envisaged as cost saving, in that it would reduce the resources dedicated to implementing crop descriptors within each center. It also chimed with, and was subsumed into, a larger quest to develop common computational tools and infrastructure to support system-wide coordination within CGIAR. The early 1990s saw efforts to "solidify a network of computer systems" across the centers, under the guidance of the data communications firm CGNET International, as well as the installation of equipment and software for managing large databases at IBPGR/IPGRI.[45] In 1994, CGIAR launched the System-Wide Information Network for Genetic Resources, which aimed to facilitate data sharing by linking the independent genetic resources databases of twelve CGIAR centers. The quest for internal, system-wide compatibility of the data used to document and manage crop genetic resources sought to make these available both within and – crucially, given the controversies about accessibility of breeding materials to all users – outside the CGIAR system.[46]

The changing circumstances in funding to and governance of CGIAR in the 1990s included other efforts to redress the perceived imbalance of power in determining the direction of international agricultural research and development. Responding to concerns that national research institutions, though crucial to the success of most agricultural development

---

[42] Gotor et al., "Scientific Information Activity," 759.
[43] International Plant Genetic Resources Institute (IPGRI), *Annual Report 1993* (Rome: IPGRI, 1994).
[44] IPGRI, *Annual Report 1994* (Rome: IPGRI, 1995), p. 66.
[45] IBPGR, *Annual Report 1990* (Rome: IPGRI, 1991).
[46] IPGRI, *Annual Report 1995* (Rome: IPGRI, 1996), pp. 69–70.

objectives, had little voice in setting priorities, CGIAR and other international institutions such as the UN International Fund for Agricultural Development tried to create mechanisms that would amplify the voice and role of national agricultural research systems.[47] Consultative processes that engaged state-level organizations bore witness to their demands for better venues for transnational dialogue and cooperation. These processes led to the convening in 1996 of a Global Forum for Agricultural Research (GFAR) that was to encompass all stakeholders, from farmer organizations to national research systems to the World Bank, FAO, and other international actors. GFAR was charged with, among many things, reassessing the mandate of CGIAR.[48] By dint of the breadth of institutions included, GFAR convenings highlighted the disparity between CGIAR's central political and strategic influence on global agriculture and its relatively minor economic role. Aggregating across the many and varied institutions engaged in agricultural development, CGIAR represented "only 3% of the annual investment in research geared to agriculture in developing countries," and yet it played a crucial role in providing the means and standards for effective cooperation among agricultural organizations.[49]

This role included coordinating information about genetic resources, an area that GFAR had not actively targeted but was nonetheless of pressing concern for many participants. The 1992 CBD had made obvious the need for a binding international agreement on plant genetic resources, which eventually emerged as the 2001 International Treaty on Plant Genetic Resources for Food and Agriculture, or Seed Treaty. The Seed Treaty's power to shape global seed exchange depended on international strategy and consensus, but also on local organizations' willingness to adopt standards and monitor the movement of plant materials. In consultations over the Seed Treaty, which included the formulation of a Global Plan of Action for the Conservation and Sustainable Utilization of Plant Genetic Resources, CGIAR confirmed its position as a key provider of scientific and technical solutions for genetic resources management. It achieved this, in part, through its promotion of descriptor lists. The lists were already tethered to CGIAR centers' crop germplasm collections, held "in trust for humanity" by CGIAR on behalf of the FAO

[47] H. Gregersen, "The CGIAR and National Agricultural Research Systems (NARS): Concepts Note for TAC Deliberations on Collaborative Relationships and Comments," February 1999, https://hdl.handle.net/10568/118931.
[48] Global Forum on Agricultural Research (GFAR), "Terms of Reference for the Establishment of the Global Forum Steering Committee Secretariat," Discussion Paper 29, 1997; GFAR, "Establishment of a Donor Support Group to the Global Forum for Agricultural Research," Discussion Paper, October 1997.
[49] Özgediz, *CGIAR at 40*, p. 43; Gregersen, "CGIAR and National Agricultural Research Systems."

Plant Genetic Resources Commission. In 1996, IPGRI extended its crop-specific descriptor work to "multi-crop passport descriptors."[50] Working in collaboration with FAO, IPGRI sought to produce "consistent coding schemes for a number of key passport descriptors that can be used for all crops," which it imagined – rightly, as it turned out – would facilitate data exchange across national borders.[51]

These continued efforts to make descriptors as useful as possible, and as widely used as possible, paid off. A 1997 CGIAR survey of seed and gene bank curators revealed that 80 percent relied on standardized descriptors – and more than two-thirds used IPGRI-produced descriptors.[52] The survey underscored the usefulness of such work on the ground, across dispersed sites and diverse crops.[53] It also illustrated the willingness of national bodies and regional breeder organizations to adopt IPGRI-produced descriptors as guidelines for crop research management and related trade. IPGRI trumpeted this contribution to international collaboration on plant genetic resources, describing in 1999 that "IPGRI is making it easier for genetic resources workers to document and explore collections as well as to identify promising accessions, through development of crop descriptors." Descriptors also helped CGIAR carry out its mandated responsibilities for stewarding genetic resources, as they were used in the System-Wide Information Network for Genetic Resources to standardize databases across eleven CGIAR gene banks.[54]

The prominence that descriptors acquired during the 1990s resulted from internal and external policies developed at CGIAR as it sought to maintain relevance in a changing institutional and political landscape, as well as the novel technical demands that emerged from new cross-institution programs and international agreements. Initially set up as tools to enable the circulation of crop materials, descriptor lists became a concrete mechanism through which to foster cooperation and exchange among locations and an instrument for the international governance of plant genetic resources. As a largely autonomous entity, IPGRI could act

---

[50] Th. Hazekamp, J. Serwinski, and A. Alercia, "Multi-crop Passport Descriptors," in *Central Crop Databases: Tools for Plant Genetic Resources Management*, compiled by E. Lipman, M. W. M. Jongen, Th. J. L. van Hintum, T. Gass, and L. Maggioni (Rome: IPGRI/CGN, 1997), pp. 35–39.

[51] IPGRI, *Annual Report 1996* (Rome: IPGRI, 1997), p. 67.

[52] B. Laliberté, L. Withers, A. Alercia, and T. Hazekamp, "Adoption of IPGRI Crop Descriptors – IPGRI," in Lee Sechrest, Michelle Stewart, and Timothy Sickle, eds., *A Synthesis of Findings Concerning CGIAR Case Studies on Adoption of Technological Innovation* (Rome: IAEG Secretariat, 1999), pp. 80–87.

[53] It was also undoubtedly useful to those who commissioned it in apparently demonstrating the value of investments in CGIAR and IPGRI programs. See discussion of the survey in Gotor et al., "Scientific Information Activity."

[54] IPGRI, *Annual Report 1999* (Rome: IPGRI, 2000), p. 29.

as a reference point for institutions seeking technical standards to ground new forms of cooperation, regulation, and monitoring. By its nature, the scope of descriptor development extended well beyond the CGIAR network, connecting users such as breeders and crop scientists worldwide. It therefore enhanced the visibility and impact of CGIAR, and to some extent made outside researchers dependent on its continued activities. At the turn of the twenty-first century, descriptor lists were central to the global system of germplasm exchange, and CGIAR accrued prominence and legitimacy as their principal creator.

Expanding Scope

In the years leading up to the 2001 Seed Treaty, descriptor lists were established as key tools for the legal and institutional governance of plant genetic resources. At FAO and IPGRI, staff focused on ensuring that descriptor lists would retain their international credibility and be ready to facilitate compliance with the Seed Treaty. Among other things, this meant expanding existing descriptor lists and prioritizing crops included in Annex 1 – that is, the sixty-four crops for which genetic resources would be made available through the less restricted multilateral system.[55] The five years leading up to the Seed Treaty saw the second-highest number of lists ever produced, with a marked drop after 2001 (Table 10.1).

This work intersected with preparations for the formal launch of the multicrop passport descriptors list, also in 2001, which was coordinated

Table 10.1 *The annual production of descriptor lists between 1977 and 2006, including multiple publications for the same crop when published in different languages. Adapted from Gotor et al., "Scientific Information Activity."*

Year Interval	Number of Descriptor Lists Published	Percentage of Total
1977–81	12	8
1982–86	38	25
1987–91	20	13
1992–96	31	21
1997–2001	36	24
2002–06	15	9
Total (1977–2006)	150	100

[55] Gotor et al., "Scientific Information Activity," 761.

by FAO with CGIAR.[56] The technical labor of developing these standards involved their alignment not only with existing and forthcoming descriptor lists but also, in some cases, with regional data management systems. For example, the European Plant Genetic Resources Search Catalogue (EURISCO), which stored passport information on *ex situ* collections maintained in forty European countries, was developed on the basis of the multicrop passport descriptor standard.

The passport standards came to play a central role in agricultural research, in part thanks to the prominence acquired by genetic technologies and genome sequencing by the turn of the millennium. The promise of precision agriculture, which included a focus on innovation driven by genomic manipulation, directed new attention to transnational information systems. Genomic information could be readily digitized and shared, especially in comparison with the highly diverse and often intractable data linked to plant morphology. Meanwhile, efforts to expedite computerized data exchange were frustrated by the limits of information and communication technologies. The technological focus was therefore less on the general opportunities offered by comprehensive data collection and more on how to exploit new genetic technologies. This arguably led to a shift in the very concept of what constituted a descriptor, with novel descriptor types accepted as significant and complementary to the global circulation of crop germplasm – and, increasingly, the availability of genomic information about such germplasm. In 2004 the Genetic Marker Technologies list was launched, establishing genetic descriptors as important tools alongside those focused on plant morphology.[57]

At the same time, the entrenchment of descriptor lists, marker technologies, and related passport standards into global agricultural research and international trade made it ever more evident that decisions about whether and how to include crops in such systems would shape the recognition (or not) of those plants as socially, scientifically, or economically significant. This facet of international standard-setting was heightened by continued lack of agreement over intellectual property rights in plants. Many questions centered on so-called traditional or Indigenous knowledge about plants: whether such knowledge should be captured in databases, and to what extent this was possible given a system centered on

---

[56] A. Alercia and M. MacKay, "Contribution of Standards for Developing Networks, Crop Ontologies and a Global Portal to Provide Access to Plant Genetic Resources," IAALD 13th World Congress, Montpelier, 2010, http://iaald2010.agropolis.fr/final-paper/ALE RCIA-2010-Contribution_of_standards_to_networks,_ontology_and_portals_to_provi de_access_to_plant_genetic_resources_b.pdf.

[57] C. de Vicente, T. Metz, and A. Alercia, *Descriptors for Genetic Markers Technologies* (Rome: Bioversity, 2004), https://hdl.handle.net/10568/74490.

traits of relevance to "modern" agriculture and reliant on English as a *lingua franca*. The 1990s saw ethnobotany rise to new prominence, and ethnobotanical knowledge increasingly featured among potential sources of data for crop scientists.[58] IPGRI in turn developed standards to facilitate communication of contextual information about plants' lifecycles and uses.[59]

This aligned with a larger CGIAR agenda. In 1996, the CGIAR Chairman Ismail Serageldin's vision for future research emphasized local knowledge: "the CGIAR's research programs need to be guided . . . by the need for greater stakeholder participation in the research process. . . . Indigenous knowledge must be integrated with new science."[60] Within the realm of descriptor development, this meant new recognition for previously overlooked information. It ultimately led to the 2009 Descriptors for Farmers' Knowledge of Plants list, which set standards for integrating traditional knowledge into descriptor lists. Here characteristics such as "seed supply system," "plant uses," and "market traits" appeared alongside morphological, functional, and environmental ones.[61]

Meanwhile, IPGRI devoted increased attention to developing descriptor lists in languages other than English. Scarce funding, and the resulting need to focus on the widest possible audiences, meant that additional languages were nonetheless limited to Spanish, French, and Portuguese, thus producing descriptor lists that mapped onto each crop's colonial heritage (Table 10.2).

Environmental concerns provided an additional impetus to expand the remit of descriptors. The potential impact of climate change on agriculture fostered interest in environmental information, such as data on soil and climate. In addition, a major review of CGIAR in 1998 had recommended refocusing on the environmentally sustainable management of natural resources.[62] This led to a restructuring of CGIAR operations around heritage crops and the role of biodiversity in developing resilient

---

[58] Richard E. Schultes and Siri von Reis, *Ethnobotany: Evolution of a Discipline* (Portland, Oregon: Dioscorides Press, 1995).

[59] E.g., IPGRI, *Descriptors for Taro (Colocasia esculenta)* (Rome: IPGRI, 1999), https://hdl .handle.net/10568/73039.

[60] CGIAR, *CGIAR Annual Report 1996, Part One: The Year in Review*, https://hdl.handle .net/10947/5690.

[61] Bioversity International and The Christensen Fund, *Descriptors for Farmers' Knowledge of Plants* (Rome: Bioversity International; Palo Alto, CA: The Christensen Fund, 2009), https://hdl.handle.net/10568/74492.

[62] CGIAR System Review Secretariat, "The International Research Partnership for Food Security and Sustainable Agriculture," Third System Review of the CGIAR, October 8, 1998, https://library.cgiar.org/bitstream/handle/10947/1586/3SysRev.pdf.

Table 10.2 *The languages of the official descriptor lists, 1977 to 2006. Adapted from Gotor et al., "Scientific Information Activity."*

	1977–81	1982–86	1987–91	1992–96	1997–2001	2002–06	Total	Percentage of Total
English	11	36	14	19	14	9	101	67
Spanish	1	1	4	6	10	2	24	16
French	0	0	2	6	9	1	18	12
Portuguese	0	0	0	0	3	0	3	2
Arabic	0	0	0	0	0	1	1	>1
Chinese	0	1	0	0	0	0	1	>1
Russian	0	0	0	0	0	1	1	>1
Italian	0	0	0	0	0	1	1	>1
Total	**12**	**38**	**20**	**31**	**36**	**15**	**150**	

sources of food, and created space for interest in medicinal plants.[63] A drive to include the health of forests and wildlife within CGIAR's remit further expanded the focus beyond the usual staple crops.[64] The growing focus on measuring and fostering biodiversity within CGIAR included the rebranding of IPGRI as Bioversity International in 2006 and culminated in the launch of the Biodiversity for Food and Nutrition Project at the Convention on Biological Diversity in 2012. The project, which aimed to identify and promote biodiverse, nutrient-rich plant species, was coordinated by Bioversity and funded by the Global Environment Facility, a trust fund administered by the World Bank and financed by forty donor countries.[65]

The expertise and resources devoted by CGIAR to developing descriptors and other data standards sat at the technical epicenter of a global shift towards precision agriculture and environmental stewardship driven by diverse but standardized data about crops, cultures, and climates. At the same time, what should count as a descriptor, and how descriptor lists

[63] Özgediz, *CGIAR at 40*, pp. 48–52; F. Pank, "Experiences with Descriptors for Characterization of Medicinal and Aromatic Plants," *Plant Genetic Resources* 3, no. 2 (2005): 190–198; P. Quek, G-T. Cho, S-Y. Lee et al., "Introduction to Development of Electronic Descriptors of Medicinal Plants to Promote Information Exchange and Sustainable Uses of Plant Genetic Resources," in *International Conference of Medicinal Plants*, Conference Proceedings, KL, Malaysia, December 5–7, 2005.
[64] Centre for International Forestry Research (CIFOR), *A Year for Forests: Annual Report 2011* (Bogor Barat: CIFOR, 2012), www.cifor.org/knowledge/publication/3798.
[65] United Nations Environment Programme, "Mainstreaming Biodiversity Conservation and Sustainable Use for Improved Human Nutrition and Well-Being," Project Document (2011–16), www.b4fn.org/fileadmin/templates/b4fn.org/upload/documents/Project_TRs/BFN_Project_document.pdf; Özgediz, *CGIAR at 40*, p. 15.

could and should complement genetic data collection, became more contested as technological opportunities grew. The very expertise employed to provide feedback and input into descriptor lists shifted from the 1990s to early 2000s, with the gradual disappearance of the Crop Advisory Groups once selected by IPGRI to develop the lists, and the emergence of ad hoc, crop-specific collectives whose composition shifted depending on the type of crop and related funders and stakeholders.[66]

Bioversity signaled its continuing attention to descriptor lists as a core mechanism for facilitating transnational collaboration on plant genetic resources, including via the Seed Treaty, by launching a survey of the lists' users in 2006. A part of the "External Review" of Bioversity's Understanding and Managing Biodiversity program, the survey measured the usefulness of descriptors "in facilitating the establishment and development of databases; improving collaboration and information exchange among organizations; and finalizing the ambitious objective of building a Clearing-House Mechanism to assure a full implementation of the Convention on Biological Diversity."[67] The results of the survey supported a view of Bioversity descriptor lists as the best-known standard for descriptors in the world, relied on well beyond CGIAR and acclaimed by users as an effective tool for crop data collection and sharing. This spurred further work on multiple descriptor lists, which became the backbone of influential regional and global crop databases, including the Global Information System backed by the International Treaty on Plant Genetic Resources and the FAO/Bioversity List of Multicrop Passport Descriptors.[68]

Conclusion

From the founding of CGIAR until the early 2000s, descriptor lists occupied a central place within the network of institutions connected via CGIAR and beyond. Descriptors were a technical solution to facilitate the international exchange of breeding materials and information about them. Over time, descriptor lists became standards essential to the implementation of increasingly stringent mechanisms for the international governance of plant genetic resources. As global agriculture extended its focus from the appropriation of seeds and other plant germplasm materials towards the capture of molecular, environmental, and traditional knowledge about germplasm,

---

[66] Gotor et al., "Scientific Information Activity."   [67] Ibid., 769.
[68] A. Alercia, S. Duilgheroff, and M. MacKay, "FAO/Bioversity Multicrop Passport Descriptors V.2," 2012, https://hdl.handle.net/10568/91224.

descriptors proved essential to aggregating and linking disparate sources of data and relevant biological materials. Descriptor lists were therefore a key means for CGIAR, working especially through IBPGR and its successor institutions, IPGRI and Bioversity, to position itself as a central repository of scientific and technical know-how to sustain both agricultural development and global policy. Even as other closely related elements of CGIAR activities came under political fire, such as its management of seed banks and its environmental and social sustainability, descriptors served as a tool for demonstrating responsiveness to those critiques and willingness to reform.

Early ambitions for universalizing the standards and protocols for describing crops, and recording these descriptions so that all researchers could use and benefit from them, were repeatedly derailed. Although the gap between ambition and achievement could sometimes be traced to the limitations of technology or financial resources, the implementation of universal descriptors was more often stymied by the diversity – of crops, humans, institutions, and goals – encompassed in the international agricultural research community that descriptors sought to discipline.

Over the last decade, developments in digital "Big Data" technologies and curatorial standards have promised to finally encompass such diversity and therefore enable the implementation of descriptors in their original, idealized form without incurring losses, discrimination, or exclusions. One of the most significant recent expressions of this expectation is the GARDIAN database, set up in 2017 to power the CGIAR Big Data Platform that would facilitate – and monitor – the sharing of data across CGIAR centers.[69] In 2021, the Big Data Platform became a key element of CGIAR's restructuring as "One CGIAR," further highlighting the scale and ambition of the data integration effort envisaged and its perceived role in coordinating across CGIAR institutes. The digital platform of One CGIAR is meant to include all data produced by CGIAR centers and their collaborators, encompassing crops, pathogens, soil composition, climate, socioeconomic information about farming communities, and more.[70] Crop descriptors are essential to this

---

[69] CIAT and IFPRI (Centro Internacional de Agricultura Tropical and International Food Policy Research Institute), *Big Data Coordination Platform: Full Proposal 2017–2022*, Proposal to the CGIAR Fund Council (Cali, Colombia: CIAT and IFPRI, 2016), https://hdl.handle.net/10947/4450; T. Abell, M. Ambrosius, J. van den Berg et al., *Accelerating CGIAR's Digital Transformation: A High-Level Assessment of Digital Strategy across CGIAR* (CGIAR, 2019), https://hdl.handle.net/10568/101268.

[70] B. King, M. Devare, M. Overduin, et al., *Toward a Digital One CGIAR: Strategic Research on Digital Transformation in Food, Land, and Water Systems in a Climate Crisis* (Cali, Colombia: CIAT, 2021), https://hdl.handle.net/10568/113555.

data linkage system.[71] Their continued use defies concerns about the potential implications of such an extensive standardization and testifies to the power of naming standards – and by extension the institutions that control these – within an ever more digitalized system of global agricultural governance.[72]

---

[71] On the recent evolution of descriptors into bio-ontologies, see S. Leonelli, "Process-Sensitive Naming: Trait Descriptors and the Shifting Semantics of Plant (Data) Science," *Philosophy, Theory and Practice in Biology* 14 (2002): article 16.

[72] Leonelli and Williamson, "Towards Responsible Plant Data Linkage."

11   Crop Genetic Diversity under the CGIAR Lens

Marianna Fenzi

In 1967, at the Technical Conference on the Exploration, Utilization, and Conservation of Plant Genetic Resources organized at the headquarters of the United Nations Food and Agriculture Organization (FAO) in Rome, the term "genetic erosion" was used for the first time to raise the alarm about an urgent problem: the loss of genetic diversity in agricultural crop plants. As the record of that meeting declared:

> The genetic resources of the plants by which we live are dwindling rapidly and disastrously ... the reserves of genetic variation, stored in the primitive crop varieties which had been cultivated over hundreds or thousands of years ... have been or are being displaced by high-producing and uniform cultivars, and by forest plantations ... This "erosion" of our biological resources may gravely affect future generations which will, rightly, blame ours for lack of responsibility and foresight.[1]

This chapter is devoted to the genesis of plant genetic resources conservation as a scientific object and agricultural concern and its institutionalization inside FAO and the Consultative Group on International Agricultural Research (CGIAR). I present the efforts to conserve crop plant genetic resources prior to the establishment of a network of international agricultural centers, as well as the forces shaping the management of plant genetic resources inside CGIAR. I am especially interested in the imaginaries – the worldviews and expectations that produced and shaped conservation efforts – and epistemologies – the modes of knowledge creation – involved in this process.

Many people, both within and beyond CGIAR, have described the creation and operation of its centers' gene banks. These institutions collect, store, and distribute seeds or other plant genetic materials, often described today as "plant genetic resources." In the case of the largest CGIAR gene

---

[1] Erna Bennett, ed., Record of the 1967 FAO/IBP Technical Conference on the Exploration, Utilization, and Conservation of Plant Genetic Resources, PL/FO: 1967/M/12, David Lubin Memorial Library (hereafter DLML), FAO, Rome.

banks, curators aim to represent most, if not all, of the extant diversity in a crop species and its wild relatives and to make this available to breeders and other researchers on request. In 2022, there were eleven CGIAR gene banks, which together held more than 730,000 samples and had a "legal obligation to conserve and make available accessions of crops and trees on behalf of the global community."[2] Institutional histories illustrate the activities that precipitated the creation of these gene banks and the function of their collections within CGIAR.[3] Other accounts have discussed the scientific and political tensions that shaped plant genetic resources management both in CGIAR institutions and elsewhere.[4] For example, multiple studies highlight the geopolitics of distribution and access to plant genetic resources arising from their use in agro-industrial and biotechnological development.[5]

Another way to study the history of the CGIAR gene banks is to explore the ideas about genes, crop varieties, and agricultural change that underpin a common understanding of gene banks as possessors of valuable plant genetic resources. In the first half of the twentieth century, state-led agricultural modernization projects, tasked with developing more productive crop varieties, paved the way for the concept of genetic resources as "building blocks" for breeders.[6] Historians have shown how agricultural institutions in industrialized countries competed and collaborated in conducting systematic collections of these "raw materials" containing

---

[2] CGIAR Genebank Platform, www.cgiar.org/research/program-platform/genebank-plat form; CGIAR Genebank Platform, "Genebanks and Germplasm Health Units," www .genebanks.org/genebanks.

[3] See, e.g., Otto Herzberg Frankel and John Gregory Hawkes, eds., *Crop Genetic Resources for Today and Tomorrow* (Cambridge: Cambridge University Press, 1975); Donald L. Plucknett, Nigel J. H. Smith, J. Trevor Williams, and N. Murthi Anishetty, *Gene Banks and the World's Food* (Princeton, NJ: Princeton University Press, 1987); Johannes M. M. Engels and Andreas W. Ebert, "A Critical Review of the Current Global Ex Situ Conservation System for Plant Agrobiodiversity: I. History of the Development of the Global System in the Context of the Political/Legal Framework and Its Major Conservation Components," *Plants* 10, no. 8 (2021): 1557.

[4] Robin Pistorius, *Scientists, Plants and Politics: A History of the Plant Genetic Resources Movement* (Rome: IPGRI, 1997); Johanna Sutherland, "Power and the Global Governance of Plant Genetic Resources," Ph.D. dissertation, Australian National University (2000).

[5] See, e.g., Lawrence Busch, William B. Lacy, Jeffrey Burkhardt, Douglas Hemken, Jubel Moraga-Rojel, Timothy Koponen, and José de Souza Silva, *Making Nature, Shaping Culture: Plant Biodiversity in Global Context* (Lincoln, NE: University of Nebraska Press, 1995); Robin Pistorius and Jeroen van Wijk, *The Exploitation of Plant Genetic Information: Political Strategies in Crop Development* (Wallingford, UK: CABI Publishing, 1999); Jack R. Kloppenburg, *First the Seed: The Political Economy of Plant Biotechnology, 1492–2000*, 2nd edn. (Madison: University of Wisconsin Press, 2004).

[6] Christophe Bonneuil, "Seeing Nature as a 'Universal Store of Genes': How Biological Diversity Became 'Genetic Resources,' 1890–1940," *Studies in History and Philosophy of Science Part C: Studies in History and Philosophy of Biological and Biomedical Sciences* 75 (2019): 1–14.

useful traits for breeding.[7] Conservation practices were therefore entangled with national programs of crop development and seed production, which typically followed a logic of "purity" and sought the standardization of varieties.[8] In short, the ever-increasing value accorded to diverse plant genetic resources was tied up with agricultural research and production systems that sought, ever more successfully, to impose uniformity across crops and farms.

Grounded in a similar approach, this chapter looks at the factors that influenced how crop diversity conservation was and is conceived and managed, especially within CGIAR, and at the "epistemic cultures" mobilized in the process. Following Karin Knorr Cetina, I understand epistemic cultures as the historically specific arrangements of individuals, institutions, and ideas that form "cultures that create and warrant knowledge."[9] In this chapter, I ask: How was the conservation of crop diversity in CGIAR shaped by the epistemic culture of plant breeders, especially those from the Global North who dominated the early development of conservation strategies? How did their representation of crop diversity as a stock of raw material awaiting discovery in the Global South lead to the concept of genetic erosion and to the prioritization of conservation in gene banks? Rather than interpret the Green Revolution as a homogenizing force wiping out crop diversity, I embrace the need to "provincialize" or decenter the categories taken to define the Green Revolution and its impacts.[10] I explore how the concept of genetic erosion, far from being just a description of how agricultural transformations would affect local diversity, was shaped by the perspective of scientists involved in the Green Revolution programs who defined the problem and framed its operational aspects. I analyze the subsequent trajectory of plant genetic resources conservation to show how approaches to conservation

[7] See, e.g., Michael Flitner, "Genetic Geographies: A Historical Comparison of Agrarian Modernization and Eugenic Thought in Germany, the Soviet Union, and the United States," *Geoforum* 34, no. 2 (2003): 175–185; Tiago Saraiva, "Breeding Europe: Crop Diversity, Gene Banks, and Commoners," in Nil Disco and Edna Kranakis, eds., *Cosmopolitan Commons: Sharing Resources and Risks across Borders* (Cambridge, MA: MIT Press, 2013), pp. 185–212; Helen Anne Curry, "From Working Collections to the World Germplasm Project: Agricultural Modernization and Genetic Conservation at the Rockefeller Foundation," *History and Philosophy of the Life Sciences* 39, no. 2 (2017): 5.
[8] Christophe Bonneuil, "Producing Identity, Industrializing Purity: Elements for a Cultural History of Genetics," in *A Cultural History of Heredity IV: Heredity in the Century of the Gene*, Preprint 343, Max-Planck-Institut für Wissenschaftsgeschichte (2008), pp. 81–110.
[9] Karin Knorr Cetina, *Epistemic Cultures: How the Sciences Make Knowledge* (Cambridge, MA: Harvard University Press, 1999), p. 1.
[10] Marianna Fenzi, "'Provincialiser' la Révolution Verte: Savoirs, politiques et pratiques de la conservation de la biodiversité cultivée (1943–2015)," Ph.D. dissertation, L'Ecole des Hautes Etudes en Sciences Sociales (2017).

were modified as a result of changes in scientific, institutional, and political contexts, including the entry of new epistemic cultures whose tools and assumptions differed from those of an earlier period. Examining these elements of crop diversity conservation as it developed within CGIAR is essential to understanding today's debates on the management and preservation of crop diversity.

FAO's Global Seed Coordination Campaigns

During the 1930s and 1940s, various countries established collections and catalogs of diverse cultivated varieties of rice, wheat, maize, forages, and other crops. Leading agricultural research institutes led "imperial" plant-hunting expeditions around the world to stock these national collections with seeds or other genetic materials (Figure 11.1).[11] However, with the exception of the Institute of Plant Industry in Leningrad, established by the Russian geneticist Nikolai Vavilov, there were few general collections that ranged widely across crop species.[12] Instead, collections generally targeted specific national agricultural ambitions or breeding programs.[13]

After World War II, the FAO Plant Production and Protection Division began to play an important role in plant genetic resources management in collaboration with the Commonwealth Scientific and Industrial Research Organization (CSIRO) of Australia.[14] Together they established an international network for the exchange of breeding materials, especially through the *FAO Plant Introduction Newsletter* launched in 1957.[15] Yet FAO's efforts to make information about collections available took shape in a context where breeders were not very receptive to the idea of sharing materials and coordinating collection missions across borders.[16] In the late 1950s, most breeders in industrialized countries continued to rely on national collections; the FAO catalogs

[11] Bonneuil, "Seeing Nature as a 'Universal Store of Genes.'"
[12] Flitner, "Genetic Geographies."
[13] See, e.g., Calestous Juma, *The Gene Hunters: Biotechnology and the Scramble for Seeds* (Princeton, NJ: Princeton University Press, 1989); Plucknett et al., *Gene Banks and the World's Food*; Garrison Wilkes and J. T. Williams, "Current Status of Crop Plant Germplasm," *Critical Reviews in Plant Sciences* 1, no. 2 (1983): 133–181.
[14] R. O. Whyte, *Plant Exploration, Collection and Introduction*, FAO Agricultural Studies No. 41 (Rome: FAO, 1958).
[15] In 1971, the name was changed to *Plant Genetic Resources Newsletter*, reflecting intensifying efforts to conserve plant genetic resources.
[16] This was despite scientists' advocacy from at least the 1920s about the importance of an international collaboration to collect genetic resources; see Flitner, "Genetic Geographies" and Bonneuil, "Seeing Nature as a 'Universal Store of Genes.'"

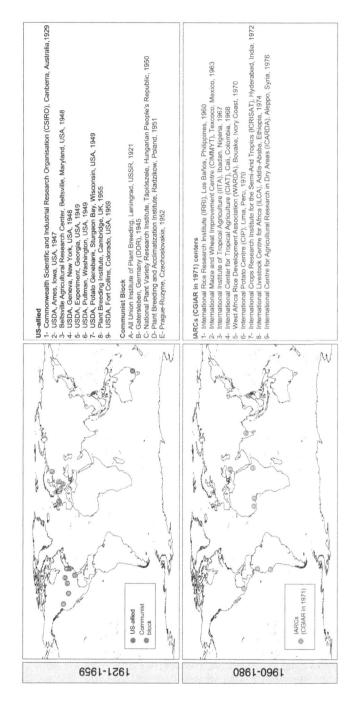

Figure 11.1 Key gene banks established between 1920 and 1980. The upper panel represents the main gene banks in US-allied countries and in the communist bloc established between 1921 and 1959. The lower panel represents the gene banks of the international agricultural research centers (associated with CGIAR from 1971) founded between 1960 and 1980.

were either not well known or not considered a useful tool by many breeders.

Towards the end of the 1950s and into the 1960s, the establishment of new international agricultural research centers such as the International Rice Research Institute (IRRI) and the International Maize and Wheat Improvement Center (CIMMYT) was associated with what the historian Jonathan Harwood characterizes as a transition from a "local strategy" to a "cosmopolitan strategy" in plant breeding – a quest for varieties that would perform well across many locations.[17] Other contributors to this volume characterize this strategy as the search for "widely adapted varieties" (see, e.g., Derek Byerlee and Greg Edmeades, Chapter 9, and Harro Maat, Chapter 6, this volume). This cosmopolitan approach was seen both to depend on and threaten the existence of farmers' varieties of rice, maize, wheat, and other crops that centers would seek to improve. The Rockefeller Foundation's collection and conservation of maize varieties, initially fostered in the 1940s through its agricultural program in Mexico and associated with efforts to extend "improved" maize varieties across Latin America, provided a model for how the international agricultural research centers could manage this dilemma – namely, by building up their own collections of farmers' varieties to ensure their future availability for breeding.[18]

Experts mobilized by FAO played an important role in parallel to that of the international agricultural research centers, constituting a specific framework for crop diversity management. For example, FAO oversaw actions concerning the exchange of breeding materials, working with breeding associations such as the European Association for Research on Plant Breeding (EUCARPIA) and the International Association of Plant Breeders for the Protection of Plant Varieties (ASSINSEL), and organizing initiatives like the World Seed Campaign in 1957.[19] As part of this initiative to encourage and coordinate exchange, the first FAO Technical Meeting on Plant Exploration and Introduction was held in July 1961. In 1967 a second conference was organized jointly by the FAO Plant Production and Protection Division and the International Biological Program; this was the Technical Conference on the Exploration,

---

[17] Jonathan Harwood, *Europe's Green Revolution and Others Since: The Rise and Fall of Peasant-Friendly Plant Breeding* (London: Routledge, 2011).

[18] Curry, "From Working Collections to the World Germplasm Project"; Fenzi, "'Provincialiser' la Révolution Verte."

[19] The FAO World Seed Campaign represented a turning point in the transfer of samples, experimentation, and dissemination of improved seeds; see FAO, *Nouvelles CMS*, no. 15 (April 1962): 13, copy available at DLML.

Utilization, and Conservation of Plant Genetic Resources.[20] In the aftermath of the 1967 meeting a new FAO team on genetic resources was created, the FAO Crop Ecology Unit. Together with representatives of the International Biological Program, this team constituted a heterogeneous expert group that started to frame "genetic erosion" as an international concern.[21]

Rising Awareness about Genetic Erosion

It was during the first Technical Meeting on Plant Exploration and Introduction in 1961 that FAO initially recorded concerns among breeders about the replacement of landraces – that is, farmers' locally adapted varieties – through the widespread adoption of "modern" varieties generated by professional breeders.[22] Even though the collection of genetic materials from regions considered remote and "less civilized" was already underway in many countries, the sense of urgency created by the assumption that local varieties were bound to vanish globally was not yet fully established. By the 1960s, this concern was increasingly felt. At FAO, experts deployed on its programs of seed dissemination and plant exploration began reporting changes in the distribution of landraces versus "modern" varieties. The 1961 and 1967 technical conferences were the first international events to specifically address declining genetic diversity as a "side effect" of efforts to deliver "improved" or "modern" varieties to farmers. Breeders gathered at FAO expressed concern about the consequences of expected success in disseminating these varieties, especially in regions that were hotspots of diversity. "Without the primitive crop races which are the raw materials of plant breeding, the continued production of high-yielding varieties is not possible," explained a 1969 editorial penned by staff of the FAO Crop Ecology unit.[23]

The alarm sounded at FAO on this issue presupposed a particular vision of crop diversity and agricultural change wherein "primitive" varieties – that is, landraces or farmers' varieties – would inevitably be replaced by "modern" ones. As the Australian wheat breeder and FAO consultant Otto Frankel summarized, "The crops of modern agriculture

---

[20] FAO, Record of FAO Technical Meeting on Plant Exploration and Introduction, Rome, Italy, July 10–20, 1961, PL 1961/8, DLML; Bennett, ed., "Record of the 1967 FAO/IBP Technical Conference on the Exploration, Utilization, and Conservation of Plant Genetic Resources."
[21] Marianna Fenzi and Christophe Bonneuil, "From 'Genetic Resources' to 'Ecosystems Services': A Century of Science and Global Policies for Crop Diversity Conservation," *Culture, Agriculture, Food and Environment* 38, no. 2 (2016): 72–83.
[22] FAO, Report of the Technical Meeting on Plant Exploration and Introduction.
[23] *Plant Introduction Newsletter*, no. 22 (July 1969), 2.

consist of varieties bred for high production and for uniformity. Over large areas of the world the same, or closely related, varieties are grown, uniformity is displacing the enormous variety of types."[24] The reports of the first FAO conferences on this subject show that participants took it as a fact that farmers would abandon local varieties once they had noticed the superiority of "modern" varieties, and that they would benefit from this transition. From this perspective, the first victims of varietal homogenization would be breeders, who would lose access to breeding materials, not farmers, who were imagined to be fully satisfied with the new varieties.

Under this dichotomy of "primitive" versus "modern," the complex and dynamic interplay between the circulation of new varieties and the disappearance of landraces was reduced to a clear-cut phenomenon of new replacing old. Just a decade earlier, by contrast, most breeders had been convinced that the poor economies and "backward" agricultural systems of countries that were the centers of origin of crops, and therefore hotspots of crop genetic diversity, would ensure the continuation of farmers' varieties. Frankel later suggested that no one could have imagined that local varieties in these places would be at risk of erosion.[25] He and many of his contemporaries saw cultivated biodiversity as a "primitive" product preserved in a natural state in the cradles of agriculture. For example, in neither the 1961 nor 1967 FAO technical conferences did participants explicitly observe that farmers' practices contribute to the conservation and evolution of crop genetic diversity. According to Frankel, "Plant breeders, searching the world for even more productive strains, must have genetic pools to provide 'building stones.' The plants of primitive agriculture and related wild plants are this treasury, now depleted by development."[26] For those who shared this view, crop genetic diversity was a "raw material" or "building stone," and not a product of human labor with nature. This perspective further suggested that it was up to professional plant scientists alone, and not farmers, to resolve the problem of conservation, given that farmers did not feature within this view as producers or managers of genetic diversity.

However, this was not the only perspective available. Within FAO, the issue of genetic erosion and proposals to cope with it sparked debates and divisions between two different epistemic cultures. Two figures can be taken as representative of these epistemic cultures: the breeder Otto

[24] Otto Frankel, "Survey of Crop Genetic Resources in Their Centres of Diversity," First Report, February 1973, FAO, DLML.

[25] Frankel and Hawkes, eds., *Crop Genetic Resources*, p. 106.

[26] Otto Frankel, "Guarding the Plant Breeder's Treasury," *New Scientist* 35 (1967): 538–540, at 538.

Figure 11.2 The plant geneticist Erna Bennett of the UN FAO Crop Ecology Unit in Greece, undated. Photographer unknown, republished from author's personal collection.

Frankel, and the population geneticist Erna Bennett (Figure 11.2). Frankel and Bennett played fundamental but antagonistic roles in getting the conservation of crop genetic resources onto the international agenda.[27] Frankel, along with others who shared his epistemic culture, conceived conservation as the sheltering of entities – in this case, genes. He focused on the managerial and technical aspects of conservation. He was interested in the development of *ex situ* or off-site conservation approaches, chiefly through storage in gene banks, and the standardized exchange of breeding materials (often referred to as germplasm) to enable breeders' activities. By contrast, within the second epistemic culture Erna Bennett and others aimed to maintain plants' interactions with their environment and all of the processes that generate diversity *in situ* – that is, in agro-ecosystems – "to preserve the evolutionary potential of local population-environment complexes."[28] Her vision was supported by scientists who were part of an evolutionary epistemic culture, including

[27] Fenzi, "'Provincialiser' la Révolution Verte."
[28] Erna Bennett, "Plant Introduction and Genetic Conservation: Genecological Aspects of an Urgent World Problem," *Scottish Plant Breeding Station Record* (1965): 27–113, at 91.

population geneticists, ecologists, and botanists.[29] Despite the support of some geneticists and ecologists for the latter approach, and thus the lack of a broad consensus across all actors interested in the conservation of crop diversity at FAO, the epistemic culture of the breeders won out. Only a system of *ex situ* conservation outside the plants' environment of origin was pursued and, as I discuss below, implemented. The gene bank approach, focused on providing breeders with the materials they needed, was considered tried and tested, and seen as easier to set up than *in situ* conservation programs, which lacked an operational plan.

The Constitution of a Global Conservation Network

The erosion of genetic resources was included on the global environmental agenda at the United Nations Conference on the Human Environment held in Stockholm in June 1972 – a crucial turning point in advocacy on this issue. The Stockholm conference represented the high point of a period of vigorous action on environmental protection and conservation, including the invention and definition of the "global environment."[30] The endangered future of agricultural development was illustrated at the conference by two problems: the first was genetic erosion, with purported evidence taken from FAO reports and conferences.[31] The second problem was the *Helminthosporium maydis* or southern corn leaf blight epidemic that caused serious losses to US hybrid maize between 1970 and 1971. An expert group assembled to investigate the disease outbreak stated that "[t]he key lesson of 1970 is that genetic uniformity is the basis of vulnerability to epidemics," and warned that American crop varieties were "impressively uniform genetically and impressively vulnerable."[32] The stark illustration of the dependence of industrialized agriculture of the Global North on "exotic" or foreign germplasm to shore up vulnerable crops fueled the growing sense of urgency about global coordination on genetic resources. Although strong

[29] The leading advocate of the evolutionary perspective in FAO, Bennett was perceived as an anomaly by many colleagues for personal reasons as much as scientific ones. Working in a predominantly male environment, she was communist, unmarried, and living with another woman. She was also a poet, journalist, and pacifist.

[30] Yannick Mahrane, Marianna Fenzi, Céline Pessis, and Christophe Bonneuil, "From Nature to Biosphere: The Political Invention of the Global Environment, 1945–1972," *Vingtième Siècle: Revue d'Histoire* 1, no. 113 (2012): 127–141.

[31] These included the conference proceedings published as Otto Frankel and Erna Bennett, eds., *Genetic Resources in Plants: Their Exploration and Conservation*, IBP Handbook No. 11 (Oxford: Blackwell, 1970) and the manuscript of a survey conducted for FAO by Frankel, *Survey of Crop Genetic Resources in Their Centres of Diversity*.

[32] National Academy of Science, *Genetic Vulnerability of Major Crops* (Washington, DC: NAS, 1972), p. 1.

evidence of genetic erosion was still lacking in 1972, surveys conducted by FAO and the *Helminthosporium maydis* epidemic's impact helped place genetic erosion among the global environmental problems of greatest concern recognized by the United Nations. The conservation of genetic resources was the subject of 7 out of 109 recommendations established in Stockholm.[33]

In the 1960s, FAO had tried to organize international management of crop genetic resources but failed, owing to a lack of interest and, crucially, resources. The Stockholm Conference created new possibilities for establishing a network of regional centers for collecting and conserving landraces and other crop varieties considered endangered, along with infrastructure "to grant all countries access to basic breeding materials."[34] As I describe below, in the 1970s, governments were invited to participate in collection campaigns and, in cooperation with FAO, to ensure the conservation of plant genetic resources in a global network of gene banks. Inventories of threatened genetic resources were compiled and registers of existing collections updated in an effort to monitor the progress of conservation on a global scale. A new phase in the conservation of crop genetic resources was thus inaugurated. However, despite the centrality of FAO expertise in preceding decades, its role in the conservation of plant genetic resources after 1972 gradually diminished. CGIAR, which was created in 1971 to extend the Green Revolution by perpetuating scientific research for agricultural development, was instead the institution in charge of the new network. CGIAR centers became the operational hubs for conservation activities like collection, evaluation, and storage in gene banks, and CGIAR, through its Technical Advisory Committee (TAC), took charge of political, managerial, and economic matters.

In 1973, FAO hosted another technical conference on genetic resources. This conference saw FAO involved for the last time as the legitimate leading institution on international genetic resources conservation. At the end of the conference, the task of establishing an international network of genetic resources centers was assumed by CGIAR.[35] FAO staff and consulting experts had developed an action plan for this network and published two manuals on technical aspects of conserving

---

[33] United Nations, "Report of the United Nations Conference on the Human Environment," Stockholm, June 5–16, 1972, UN Doc. A/CONF 48 General Assembly, 1972.

[34] Ibid., A/CONF 48/7, 48.

[35] Pistorius and van Wijk, *The Exploitation of Plant Genetic Information*, pp. 96–100; Frankel and Hawkes, eds., *Crop Genetic Resources*.

crop diversity since 1966.[36] However, in the midst of international attention to the Green Revolution and anticipation of further agricultural transformation, CGIAR and its international research centers were able to present themselves as the institutions best positioned to guide the conservation and use of genetic resources.[37] CGIAR promoted the commitment to plant genetic resources of such emblematic figures of the Green Revolution as Norman Borlaug and Monkombu Swaminathan. Borlaug famously received the Nobel Peace Prize in 1970 for his work on wheat at CIMMYT. Swaminathan, meanwhile, was celebrated as the master of making India self-sufficient in grain and became the director of IRRI in 1982.

In 1974, CGIAR created the International Board for Plant Genetic Resources (IBPGR) as an institution independent of the United Nations but headquartered at FAO in order to benefit from FAO's diplomatic role in the Global South. Although located within FAO, at the policy and operational levels IBPGR was centered more within the CGIAR network and operated as a CGIAR institution alongside the other international agricultural research centers. Ultimately, the conservation of genetic resources in gene banks became a branch of the centers' research on the improvement of wheat, rice, maize, and other crops, and was guided by the imperatives of crop productivity and agricultural "modernization" associated with the Green Revolution.[38]

FAO scientists and their collaborators thus succeeded in raising the issue of genetic erosion to the level of an international problem – and in generating action – within a decade. However, with the entry of CGIAR, their scientific and decision-making power disappeared. Some considered this situation a defeat, including Bennett, who resigned in 1982. Others kept a certain influence, including Frankel, who maintained a consultative role, and the young botanist Trevor Williams, who had participated in FAO collecting missions in the 1970s and became IBPGR's executive secretary in 1978.

Just as the institutions of the Green Revolution won out over FAO, so too did the epistemic culture of the Green Revolution – namely, that of plant breeders – win out over approaches from disciplines such as population genetics and ecology. As a result, the complex dynamic between

---

[36] Frankel and Hawkes, eds., *Crop Genetic Resources*. Together with Frankel and Bennett, eds., *Genetic Resources in Plants*, this work formalized the theoretical basis for *ex situ* conservation.

[37] Accounts that describe the transition of coordinating responsibility from FAO to CGIAR include Pistorius, *Scientists, Plants and Politics*; Curry, "From Working Collections to the World Germplasm Project."

[38] See discussion in D. L. Plucknett and N. J. Smith, "Agricultural Research and Third World Food Production," *Science* 217, no. 4556 (1982): 215–220.

"primitive" varieties and "modern" varieties, or farmers' varieties and professional breeders' varieties, was reduced to the problem of genetic erosion, disregarding evolutionary processes unfolding through farmers' practices in local environments. The mission of conserving genetic materials for breeding came to be both the dominant approach to the study of crop diversity and the organizing principle of actions to conserve it. Among other outcomes, a geographical distribution of conservation activities came to be formalized in which the "poorly equipped" South supplied crop diversity, and the North, with its techno-scientific power, managed it.[39] The scientific debate in FAO, which included the viewpoints of botanists and population geneticists, was displaced by a massive technical routine consisting of lists of collecting priorities, databases of plant materials, and jars of seeds in gene banks (Figure 11.3). In short, under IBPGR, conservation entered a "chronic alert" phase, where the problem of erosion was managed by creating new collections.

Ex situ Conservation between Routine and Crisis

Unlike the other CGIAR centers, IBPGR was not a research institute. As a 1979 policy document described, "it is a service organization, whose primary purpose is to assist plant breeders."[40] It began its activity in 1974 with a modest budget of about $250,000, an amount that gradually increased over the following years, reaching nearly $3.8 million in 1982. Funding remained at around this level over the following two decades.[41] IBPGR continued FAO's work of collecting and making materials available for plant improvement programs. However, it abandoned the scientific discussion about how to conserve – specifically whether *in situ* or *ex situ* approaches were more appropriate – and for what purpose, topics on which FAO experts had led. As set out by its technical mission, IBPGR developed new lists of accessions, an international system of descriptors for genetic resources (as described by Helen Anne Curry and Sabina Leonelli,

---

[39] Kloppenburg, *First the Seed.*
[40] "Policies of the Board 1974–1978," in IBPGR, *A Review of Policies and Activities 1974–1978 and of the Prospects for the Future* (Rome: IBPGR, 1979).
[41] The overall sum spent on genetic resources research within CGIAR reached $55 million in 1982 and remained close to this figure throughout the 1980s. More than half of this sum supported gene banks located in industrialized countries, particularly the United States. Around 14 percent was distributed among genebanks in the Global South, 17 percent among CGIAR international agricultural research centers, and the rest to various bilateral aid initiatives and UN agency projects. See "Budgets and Expenditures of IBPGR since 1974," in IBPGR, *A Review of Policies and Activities.* For approximations of spending in the 1980s and 1990s, see C. P. Fowler and P. R. Mooney, *Shattering: Food, Politics and the Loss of Genetic Diversity* (Tucson: University of Arizona Press, 1990) and Plucknett et al., *Gene Banks and the World's Food.*

Figure 11.3 Accessions stored in the gene bank of the International Maize and Wheat Improvement Center (CIMMYT), Mexico, 2018. Photo by Luis Salazar/Crop Trust. By permission of Global Crop Diversity Trust.

Chapter 10, this volume), "minimum standard" protocols for plant exploration, and information-sharing activities.[42] These actions were directed by five ad hoc committees for the crops with the greatest economic importance: wheat, maize, rice, sorghum, and a single combined committee for millets and beans.[43] As Erna Bennett later reflected, these patterns confirmed that IBPGR followed the orientation dictated by the plant breeders' epistemic culture:

Landraces with no commercial value but that are important in local diets would not be present in these collections. Many centers, despite their strategic position in regions of great genetic diversity, deal exclusively with collecting material from a single species. Moreover, these single-species collections are also not representative of genetic variability [within each species], and there were serious lacunae in the collections. This system reflected the CGIAR's favoured approach based on major crops.[44]

In 1976, IBPGR established a first action plan to bring in species deemed insufficiently represented in its activities. This program involved various

---

[42] John Gregory Hawkes, "Plant Genetic Resources: The Impact of the International Agricultural Research Centers," CGIAR Research Study Paper, no. CGR3, CGIAR and World Bank, 1985.

[43] The activities of these committees were reported in *Plant Genetic Resources Newsletter*. See issues published in 1976–80.

[44] Erna Bennett, personal communication with author, 2011.

regions (the Mediterranean, southern Asia, West Africa, Ethiopia, Central America, and Brazil) and identified fifty-eight crop species assigned to three different priority levels.[45] Despite this attempt to prioritize other species, conservation policies remained linked to the agendas of the international agricultural research center system. The primary focus therefore continued to be on maize, wheat, and rice, and on the productivity goals firmly anchored in the traditional pathway to improvement, which Bennett described as "the search for major genes and homogeneity."[46] From the mid 1980s, failures of the system of *ex situ* gene banks (which I describe below) reduced the money allocated for plant exploration and collection activities, especially those targeting minor crop species.

In the late 1970s, the "global network" of genetic resources conservation included eight international centers and fifty-four regional centers, of which twenty-four had long-term storage systems that conformed to IBPGR's technical standards.[47] With the opening of more regional gene banks, the global network continued to grow. By 1983, there were forty-eight gene banks compliant with international standards for long-term storage at sites around the world. Thirty of these were officially enrolled in the IBPGR network, operating in twenty-four countries and covering the main crop species.[48] IBPGR also had regional offices, for example in Aleppo, Bangkok, Nairobi, and Cali.[49] In the 1980s, IBPGR had as many as 600 scientists, working in 100 countries. They were supplied by hundreds of collection missions in which a veritable army of collectors sought new samples of landraces and other plant genetic materials across sixty-two countries.[50] IBPGR was also able to "capture" collections produced by outside organizations and governments, which were invited to join its global conservation effort.

[45] Priorities for action on crops (Annex III) in IBPGR, *A Review of Policies and Activities 1974–1978*.

[46] Erna Bennett, personal communication with author, 2011.

[47] J. T. Williams et al., *Seed Stores for Crop Genetic Conservation* (Rome: IBPGR, 1979).

[48] J. T. Williams, "A Decade of Crop Genetic Resources Research," in J. H. W. Holden and J. T. Williams, eds., *Crop Genetic Resources: Conservation and Evaluation* (London: Allen & Unwin, 1984), pp. 1–17; Jean Hanson, J. T. Williams, and R. Freund, *Institutes Conserving Crop Germplasm: The IBPGR Global Network of Genebanks* (Rome: IBPGR, 1984), p. 2; Plucknett et al., *Gene Banks and the World's Food*, p. 203.

[49] Donald L. Plucknett, Nigel. J. H. Smith, J. Trevor Williams, and N. Murthi Anishetty, "Crop Germplasm Conservation and Developing Countries," *Science* 220, no. 4593 (1983): 163–169.

[50] IBPGR, *Annual Report 1979* (Rome: IBPGR, 1980); *Annual Report 1981* (Rome: IBPGR, 1982); *Annual Report 1984* (Rome: IBPGR, 1985); these and other reports are available at https://cgspace.cgiar.org/collections/44e7ddf6-b69d-4075-8c80-a7aab65495af. See also Williams, "A Decade of Crop Genetic Resources Research."

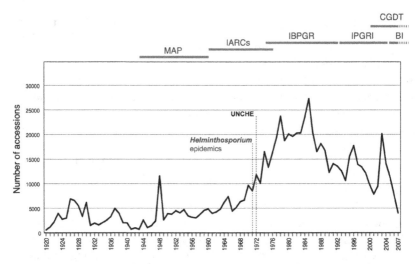

Figure 11.4 Annual number of accessions to selected gene banks, 1920–2007, including those of CGIAR centers. Adapted from United Nations Food and Agriculture Organization (FAO), *Second Report on the World's Plant Genetic Resources for Food and Agriculture* (Rome: FAO, 2010), 57. Reproduced with permission of FAO.

Thanks to these collection campaigns, the number of samples held in CGIAR gene banks and other key international institutions rapidly grew.[51] The annual number of samples accessioned into collections remained roughly stable from the 1920s through the 1960s (Figure 11.4), with a peak in 1948 linked to collections created through the Rockefeller Foundation's agricultural program in Mexico (labeled MAP in the figure). It then started to rise in parallel with the creation of the first international agricultural research centers (IARCs) in the late 1960s. After the *Helminthosporium maydis* epidemic and the Stockholm conference (UNCHE) in the early 1970s, and the formation of IBPGR in 1974, the annual number of accessions leaped higher. Then, starting in the mid 1980s, new accessions progressively decreased, but with two exceptions: the late 1990s, in correspondence with the activities of the International Plant Genetic Resources Institute (IPGRI), and in 2004, with the establishment of the Global Crop Diversity Trust (GCDT).[52]

---

[51] William L. Brown, "Genetic Diversity and Genetic Vulnerability: An Appraisal," *Economic Botany* 37, no. 1 (1983): 4–12.

[52] The latter became famous for conserving crop diversity inside the emblematic Svalbard Global Seed Vault in Norway.

As CGIAR's accessions rose for more than ten years during the 1970s, new concerns emerged regarding the efficiency of *ex situ* conservation. In 1978 the US National Academy of Sciences published *Conservation of Germplasm Resources: An Imperative*, which reported the challenges faced by *ex situ* conservation of microorganismal, marine, plant, and animal germplasm. These challenges ranged from technical issues, such as the maintenance of the original genetic base, to organizational ones including inventories, evaluation, and quality control.[53] This report spurred CGIAR to commission the agronomist Donald Plucknett to conduct a detailed survey of the state of its collections.[54] However, it was only the beginning of a period of questioning and concern. In September 1981, an article in the *New York Times* described the precarious situation of the main long-term gene bank in the United States, located in Fort Collins, Colorado: "In the chilly seed storage rooms here, sacks of seeds are piled on the floors, overflowing the laboratory's facilities."[55] Conservation failures in one of the most important gene banks in the world prompted the IBPGR Secretariat to take a greater interest in problems affecting *ex situ* conservation – and therefore its entire network of gene banks.[56] A new report published in 1983 highlighted losses in gene banks resulting from lack of personnel, negligence, malfunctioning equipment, fires, and still other factors. IBPGR attributed most of the responsibility for these problems to the curators: "[gene bank] curators may well contribute more to loss of valuable material than might have occurred in the field."[57] Despite recognition of the failure of many gene banks with respect to their core mission, no general reevaluation of *ex situ* approaches came to pass. IBPGR instead sought to manage the crisis by imposing more rigorous technical and procedural approaches in seed storage centers.

At the same time, the orientation of breeding programs towards the rapid release of commercial products discouraged breeders from using local varieties or their wild relatives. In the 1980s, assessments internal and external to CGIAR showed that most of the diversity in gene banks

---

[53] National Research Council, *Conservation of Germplasm Resources: An Imperative* (Washington, DC: National Academy of Sciences, 1978).

[54] Plucknett et al., *Gene Banks and the World's Food*; Plucknett et al., "Crop Germplasm Conservation and Developing Countries."

[55] A. Crittenden, "US Seeks Seed Diversity as Crop Assurance: A World to Feed US," *New York Times*, September 21, 1981, A1.

[56] IBPGR, *Practical Constraints Limiting the Full and Free Availability of Genetic Resources* (Rome: IBPGR, 1983). The recommendations are found in a complementary report: D. R. Marshall, "Practical Constraints Limiting the Full and Free Availability of Genetic Resources," Consultant report AGPG: IBPGR/84/20, Rome, 1983.

[57] This conclusion of the report is cited in Hanson, Williams, and Freund, *Institutes Conserving Crop Germplasm*, p. 1.

remained unused, if not totally abandoned.[58] The measures taken in response to this perceived concern indirectly affected field research and collection activities, in particular limiting the development of the collection of so-called "orphan" crops and crop wild relatives. It would be another ten years before the problems of increasing wild species and "neglected and underutilized species" accessions in gene banks would be treated as an important subject in the major conservation institutions.[59] CGIAR took the position that what was needed, more than new accessions, was to maintain what it had in good condition – and what it had in good condition were collections of the world's key agricultural commodity crops. As a 1984 assessment confirmed, the collections that were most representative of extant diversity and kept in the best conditions were those of the most economically important species: potato at the International Potato Center (CIP); wheat in European banks, the Vavilov Institute, CIMMYT, and the International Center for Agricultural Research in the Dry Areas (ICARDA); maize at CIMMYT; rice at IRRI; barley at ICARDA; and sorghum at the International Crops Research Institute for the Semi-Arid Tropics (ICRISAT).[60]

In sum, once an international system for crop genetic conservation had been established in the early 1970s, associated with the epistemic culture of plant breeders and premised on the idea of collecting endangered breeding materials before their inevitable replacement, the focus of conservation moved to technical improvements in storage. The dominant conceptual and scientific approach simplified the problem of conservation in order to make it immediately operational, and it persisted even as evidence of its shortcomings accumulated.

Genetic Resources through the Prism of Geopolitical Tensions

The geneticists who organized the system of exchanges and collections within FAO in the 1960s and 1970s contributed to the emergence of the idea that plant genetic resources are a common heritage of humanity. For example, Erna Bennett used the phrase "human heritage" and Otto

---

[58] RAFI, "A Report on the Security of the World's Major Gene Banks," *RAFI Communiqué*, July 1987; IBPGR, *Progress on the Development of the Register of Genebanks* (Rome: IBPGR, 1987).
[59] Reem Hajjar and Toby Hodgkin, "Using Crop Wild Relatives for Crop Improvement: Trends and Perspectives," in N. Maxted et al., eds., *Crop Wild Relative Conservation and Use* (Wallingford, UK: CABI, 2008), pp. 535–548.
[60] Judith Lyman, "Progress and Planning for Germplasm Conservation of Major Food Crops," *Plant Genetic Resources Newsletter* 60 (1984): 3–19.

Frankel spoke of a "genetic estate" comprising the "biological heritage, the genetic endowment of organisms now living."[61] IBPGR, appropriating the notion of common heritage, strove to construct its own image as a "catalyst of genetic resources flows" between countries, applying principles of free exchange and fair distribution.[62] However, observers from the 1970s onward increasingly condemned this activity as a raiding of the genetic resources of the Global South by greedy Northern interests. Pat Mooney, a Canadian activist and author of the influential 1979 book *Seeds of the Earth*, observed of the global network of gene banks, "the Third World is being invited to put all its eggs in someone else's basket."[63] Scientists' work on genetic resources, which was dominated by CGIAR through IBPGR, was gradually confronted with a critical discourse emerging at the international level, articulated by nongovernmental organizations (NGOs) such as the Rural Advancement Foundation International (RAFI) and the International Coalition for Development Action.[64]

In the same years that FAO and CGIAR worked to establish international genebanks with a long-term conservation mission, problems had emerged that could not be contained by technical solutions. The blossoming of Third World alliances in the 1970s transformed genetic resources into a new field of tensions between Global North and South. Some tensions arose from the expansion of intellectual property rights in plant varieties, which implied limitations on the free circulation of varieties among breeders and institutions and new restrictions on farmer seed-saving. (For a detailed discussion of intellectual property concerns see David J. Jefferson, Chapter 12, this volume.) In the 1970s and 1980s, the International Union for the Protection of New Varieties of Plants (UPOV), established in 1961 to enable breeders' intellectual property claims in plant varieties, was updated to adapt it to the patent system for commercial innovations in genetic engineering and biotechnology.[65]

---

[61] Bennett, "Plant Introduction and Genetic Conservation," 93; Otto H. Frankel, "Genetic Conservation: Our Evolutionary Responsibility," *Genetics* 78, no. 1 (1974): 53–65, at 53.

[62] S. Jana, "Some Recent Issues on the Conservation of Crop Genetic Resources in Developing Countries," *Genome* 42, no. 2 (1999): 562–569.

[63] P. R. Mooney, *Seeds of the Earth: A Private or Public Resource?* (Ottawa: Inter Pares, 1979).

[64] Plucknett et al., *Gene Banks and the World's Food*, p. 143; Pat Roy Mooney, "The Law of the Seed: Another Development and Plant Genetic Resources," *Development Dialogue* 1–2 (1983): 1–173, at 79; Kloppenburg, *First the Seed*, p. 165; José Esquinas-Alcázar, Angela Hilmi, and Isabel López Noriega, "A Brief History of the Negotiations on the International Treaty on Plant Genetic Resources for Food and Agriculture," in M. Halewood, I. L. Noriega, and S. Louafi, eds., *Crop Genetic Resources as a Global Commons* (London: Routledge, 2013), pp. 135–149.

[65] UPOV was established by a convention in 1961 and revised in 1972, 1978, and 1991. UPOV 1991 grants breeders at least twenty years of rights over novel, distinct, uniform,

Facing restrictive new seed regulations, attitudes towards the sharing of genetic resources shifted in many countries in the Global South.[66] Other tensions emerged from political restrictions on access to supposedly global collections held in trust in national gene banks. For example, embargoes prevented researchers in Afghanistan, Albania, Cuba, Iran, Libya, the Soviet Union, and Nicaragua from accessing materials held in US collections.[67] Some countries of the Global South began in turn to impose restrictions on trade in species with a strategic national economic role: Ethiopia over coffee, Ecuador over cocoa, and so on.[68]

The "seed wars," in which countries struggled to assert control over plant genetic materials, reached a peak at the Twenty-First FAO Conference in November 1981. Backed by the G77, the developing-country coalition within the United Nations, the Mexican delegation proposed a "new international genetic order," independent of CGIAR, in what was later designated Resolution 6/81.[69] The aim of Mexico's proposed resolution was to bring global collections of crop genetic resources back under the aegis of FAO, granting it full control over a new international gene bank. Under the proposal, FAO was to ensure the conservation and circulation not only of landraces and crop wild relatives, but also the breeders' lines produced in public and private research centers "without restrictive practices that limit their availability" to countries in the Global South.[70] This resolution struck FAO – whose staff were not prepared, and probably did not want, to take on this responsibility – like a meteorite. Resolution 6/81 was one of the most highly debated in FAO history.[71]

Several industrialized countries, particularly the United States, Australia, and the United Kingdom, opposed the proposal, initially arguing that building a new gene bank would be too expensive. Other concerns proved more potent. Following the conference, ASSINSEL alerted

and stable varieties. Under UPOV regulation, protected seeds cannot be sold or exchanged, eventually only saved, and reused only under specific national agreements.

[66] Henk Hobbelink, *New Hope or False Promise? Biotechnology and Third World Agriculture* (Brussels: International Coalition for Development Action, 1987).

[67] William B. Lacy, "The Global Plant Genetic Resources System: A Competition-Cooperation Paradox," *Crop Science* 35, no. 2 (1995): 335–345, at 338.

[68] C. Fowler, *Unnatural Selection: Technology, Politics and Plant Evolution* (Yverdon, Switzerland: Gordon and Breach, 1994), p. 181.

[69] Resolution 6/81 of the Twenty-First Session of the FAO Conference, November 1981. The description "new international genetic order" is from Kloppenburg, *First the Seed*.

[70] Resolution 6/81, point 1.

[71] Giacomo T. Scarascia-Mugnozza and Pietro Perrino, "The History of Ex Situ Conservation and Use of Plant Genetic Resources," in J. Engels et al., eds., *Managing Plant Genetic Diversity* (Rome: IPGRI-CABI, 2002), pp. 1–22; Robin Pistorius, *The Environmentalization of the Genetic Resources Issue: Consequences of Changing Conservation Strategies for Agricultural Research in Developing Countries* (Copenhagen: Centre for Development Research, 1993), p. 80.

the UPOV Council about the risks that the proposal's provisions on the circulation of breeders' lines posed to their activities. Maintaining established intellectual property protections became the primary focus of industrialized countries' objections. In the wake of Resolution 6/81, IBPGR continued to defend its image as a good manager and "catalyst" of initiatives promoting the conservation of genetic resources. However, the main donors to IBPGR were countries and institutions in the Global North that strongly opposed the resolution.[72] Under pressure from IBPGR, FAO succeeded in orienting the supporters of Resolution 6/81 towards the establishment of a network of collections instead of a single gene bank under FAO management.[73] The resolution was transformed into a proposed International Undertaking on Plant Genetic Resources, which stipulated that member countries must make their genetic resources available without restriction, including lines developed by breeders, as part of a "common heritage."

The eleven-article undertaking mandated that samples of plant genetic material "be made available free of charge, on the basis of mutual exchange or on mutually agreed terms."[74] All existing conservation institutions were asked to adhere to new standards, implemented and overseen by FAO, as part of this global agreement.[75] Among other provisions, the undertaking called for "the equitable and unrestricted distribution of the benefits of plant breeding." including the circulation of "special genetic stocks (including elite and current breeders' lines and mutants)."[76] In other words, the "common heritage" framework was intended to allow the Global South to obtain access to protected lines. The undertaking thus potentially implied a substantial revision of plant breeders' rights. Although it did not gain the support of key industrialized nations or international agricultural institutions, 103 countries signed a revised version of the agreement in November 1983.[77] The victory was more symbolic than material. In 1989, after long negotiations, FAO

---

[72] IBPGR, *Annual Report 1983* (Rome: IBPGR, 1984).
[73] Mooney, "The Law of the Seed," 33–34; Pistorius, *The Environmentalization of the Genetic Resources Issue*, p. 80.
[74] FAO Resolution 8/83, International Undertaking on Plant Genetic Resources, Article 5.1, Twenty-Second Session of the FAO Conference, 1983.
[75] The agreement involved norms for the management of collections and the transfer of germplasm, a code for biotechnology, and a global plan of action for conservation. Its management architecture included three networks: 1) the World Information and Early Warning System on Plant Genetic Resources for Food and Agriculture, which would identify risks to collections and enable immediate international action; 2) a network of gene banks; 3) a network of areas for *in situ* and on-farm conservation.
[76] FAO Resolution 8/83, Annex to Resolution 8/83, Article 2.1.a.v.
[77] Membership was not unconditional; some participating Northern countries declared that they would apply restrictions.

and IBPGR signed a "Memorandum of Understanding" that formalized a new relationship set in motion by the undertaking but also abandoned the original political project.[78] As part of these negotiations, FAO stipulated that "'plant breeders' rights as provided for under UPOV are not incompatible with the International Undertaking."[79]

Towards New Approaches to the Conservation of Genetic Resources

With the rising demands of Indigenous communities and peasant associations, the opening of new spaces of socioenvironmental struggle, and growing criticism of globalization, the landscape of biodiversity conservation grew more complex in the 1980s and 1990s.[80] Within the arena of crop conservation, broader confrontation with nongovernmental actors pushed established institutions towards new approaches and ultimately the incorporation of new epistemic cultures. To use a formulation from the sociology of social problems, breeders were no longer the sole "owners" of the problem of agricultural biodiversity.[81]

The most important result of the negotiations first set in motion by FAO Resolution 6/81 was the 1983 creation of the Commission on Plant Genetic Resources within FAO.[82] This commission aimed, among other things, to better represent the countries of the Global South in agreements and to represent "farmers' rights" to use and share seeds for the first time. In 1991, 127 countries participated in the Fourth Conference of the Commission on Plant Genetic Resources, which defined the distribution of responsibilities between FAO and IBPGR. FAO would focus on *in situ* conservation, favoring an ecological approach for species outside the sphere of CGIAR. Meanwhile IBPGR would be the main institution for *ex situ* conservation. However, in the 1990s both institutions increasingly confronted the emergence of alternative and more ecological approaches to conservation. A new global governance for biodiversity, inaugurated by the 1992 Convention on Biological Diversity (CBD),

---

[78] FAO Resolution 4/89, Agreed Interpretation of the International Undertaking, Twenty-Fifth Session of the FAO Conference, 1989.

[79] FAO Resolution 5/89, Twenty-Fifth Session of the FAO Conference, 1989.

[80] Arturo Escobar, "Whose Knowledge, Whose Nature? Biodiversity, Conservation, and the Political Ecology of Social Movements," *Journal of Political Ecology* 5, no. 1 (1998): 53–82; Jean Foyer, *Il était une fois la bio-révolution: Nature et savoirs dans la modernité globale* (Paris: Presses Universitaires de France, 2010).

[81] Joseph R. Gusfield, "Constructing the Ownership of Social Problems: Fun and Profit in the Welfare State," *Social Problems* 36, no. 5 (1989): 431–441.

[82] Resolution 9/83, Establishment of a Commission on Plant Genetic Resources, Twenty-Second Session of the FAO Conference, 1983.

directly affected crop conservation.[83] Recommendations of Agenda 21, the nonbinding UN action plan on sustainable development, on the CBD implied the creation of a World Information and Early Warning System on Plant Genetic Resources for Food and Agriculture, the implementation of a Global Plan of Action for *ex situ* and *in situ* conservation, and the recognition of the farmers' rights agenda.

Throughout the 1990s, the Commission on Plant Genetic Resources and FAO tried to implement CBD recommendations. The Commission's fourth technical conference, held in Leipzig in 1996, represented an important moment in the advancement of those proposals. At the conference, a Global System for the Conservation and Utilization of Plant Genetic Resources was approved to combine *ex situ* and *in situ* conservation strategies for the sustainable use of plant genetic resources.[84] The conference also encouraged the production of 155 national reports, which formed the basis for the first FAO report on genetic resources published in 1998. These activities culminated in November 2001 with the International Treaty on Plant Genetic Resources for Food and Agriculture, also known as the Seed Treaty. Under the Seed Treaty, which entered into force on June 29, 2004, *ex situ* collections, including CGIAR gene banks, were made available through a multilateral system, and benefits generated from using genetic resources (e.g., in commercial crop varieties) were supposed to be shared through a collective funding system.[85] Based on voluntary decisions, the collective funding system struggled to materialize. However, responding to a more mutualist logic, where genetic resources were considered public goods, the Seed Treaty became the privileged space for discussing farmers' rights, farmers' seed systems, and alternative approaches to conservation.[86]

This opening of new institutional spaces and dialogues in the 1990s allowed for the assertion of new epistemologies and practices within the plant genetic conservation regimes of CGIAR. Thanks to the institutional critiques and upheavals of the 1980s, IBPGR transitioned into a new arrangement as IPGRI from 1993 to 2006, an operation later renamed Bioversity International. The changes of the 1980s and 1990s also prompted an expanded network of experts, range of knowledge, and list of collaborating agencies at IPGRI. The challenge increasingly faced by IPGRI and its successor organizations was no longer just to conserve

---

[83] Secretariat of the Convention on Biological Diversity, *Convention on Biological Diversity: Text and Annexes* (Montreal, Canada: UNEP, 2011), www.cbd.int/doc/legal/cbd-en.pdf.

[84] FAO, Global Plan of Action for the Conservation and Sustainable Utilization of Plant Genetic Resources for Food and Agriculture and the Leipzig Declaration, 1996, Rome.

[85] See FAO, "The Multilateral System," www.fao.org/plant-treaty/areas-of-work/the-multi lateral-system/landingmls/en.

[86] Esquinas-Alcázar et al., "A Brief History of the Negotiations on the International Treaty."

genetic resources for breeders, but to include other actors, such as farmer organizations and NGOs, and to involve local communities through participatory approaches like participatory plant breeding and community seed banks.[87] Inside these institutions as well as in national research programs, crop diversity was increasingly reconceived in the context of agro-ecosystems and the cultures from which it had originated (Figures 11.5 and 11.6). The dynamics of crop diversity now had to be explored from multiple angles: social, ecological, and agronomic.

In the 1980s, certain branches of biology and botany, coupled with studies in anthropology, had already developed analytical tools that could be practically applied to *in situ* genetic resources conservation. The work of ethnobotanists, ethnobiologists, and anthropologists studying agricultural biodiversity in its centers of origin played a determining role in reframing the concept of genetic resources to integrate evolutionary processes and farmers' practices.[88] These disciplines, together with participatory plant breeding, contributed to the development, within IPGRI/ Bioversity and elsewhere, of *in situ*/on-farm conservation approaches that in the 1990s sought to sustain the farming practices and social contexts that create and maintain agricultural diversity.[89]

Once integrated into the mainstream conservation landscape, *in situ* conservation was reconceived as providing services that would enable the adaptation of agriculture amid global change. It newly positioned farmers as "guardians" of globally important diversity. At the same time, *in situ* approaches contributed to the development of new practices and values in which crop diversity and farmers were more than service providers for industrial agriculture. In this understanding, crop diversity was also seen as fundamental for the flourishing of farmers in the Global South. This in turn suggested that the major challenge for conservation was not storing

---

[87] O. T. Westengen, K. Skarbø, T. H. Mulesa, and T. Berg, "Access to Genes: Linkages between Genebanks and Farmers' Seed Systems," *Food Security* 10 (2018): 9–25; O. Westengen, T. Hunduma, and K. Skarbø, "From Genebanks to Farmers: A Study of Approaches to Introduce Genebank Material to Farmers' Seed Systems," *Noragric Report* 80 (2017).

[88] Devra Jarvis, Christine Padoch, and H. David Cooper, eds., *Managing Biodiversity in Agricultural Ecosystems* (New York: Columbia University Press, 2007); Hugo R. Perales, "Landrace Conservation of Maize in Mexico: An Evolutionary Breeding Interpretation," in N. Maxted, M. E. Dulloo, and B. V. Ford-Lloyd, eds., *Enhancing Crop Genepool Use: Capturing Wild Relative and Landrace Diversity for Crop Improvement* (Wallingford, UK: CABI, 2016), pp. 271–281.

[89] Stephen B. Brush, *Genes in the Field: On-Farm Conservation of Crop Diversity* (Rome: IPGRI, 2000); Margery L. Oldfield and Janis B. Alcorn, "Conservation of Traditional Agroecosystems," *BioScience* 37, no. 3 (1987): 199–208; Miguel A. Altieri and Laura Merrick, "In Situ Conservation of Crop Genetic Resources through Maintenance of Traditional Farming Systems," *Economic Botany* 41, no. 1 (1987): 86–96.

Figure 11.5 A maize granary in Yucatan, Yaxcaba, Mexico represents on-farm (or *in situ*) conservation of crop diversity, 2013. Photo by Marianna Fenzi.

and providing genetic materials to breeding companies but providing opportunities and solutions to farmers and their farming systems. In this scenario, gene banks and breeding programs finally had a role in directly supporting farmers.[90]

Conclusion

In 2018, the United Nations Declaration on the Rights of Peasants and Other People Working in Rural Areas declared that these individuals "have the right to seeds, including ... the right to save, use, exchange and sell their farm-saved seed or propagating material." It called on states to "recognize the rights of peasants to rely either on their own seeds or on other locally available seeds of their choice" and to "take appropriate measures to support peasant seed systems and promote the use of peasant

---

[90] Louafi Sélim, Mathieu Thomas, Elsa T. Berthet et al., "Crop Diversity Management System Commons: Revisiting the Role of Genebanks in the Network of Crop Diversity Actors," *Agronomy* 11, no. 9 (2021): 1893.

Figure 11.6 Maize seeds from a farmers' seed fair in Mérida, Mexico, 2014. Photo by Marianna Fenzi.

seeds and agrobiodiversity."[91] Even though these principles are hardly fully translated into national agricultural policies, their declaration signals the important institutional changes driven in part by a shifting balance of power between different epistemic cultures in crop conservation and use alongside hard-fought political struggles. Many contemporary under-standings of crop diversity contradict the long-dominant view of land-races as a stock of raw materials stored in "wild" landscapes. Because scholars have recognized farmers' practices as important to the evolution, improvement, and conservation of cultivated biodiversity, institutions now confront the need to change seed regulations, as well as conservation methods.

Despite important shifts, the concept of genetic erosion and the accom-panying notion of crop diversity as a resource to be mined by breeders for value still shape scientific practices, conservation actions, and policies. As my history of the conservation of plant genetic materials in and beyond CGIAR makes clear, this early and influential perspective largely ignored the role of farmers and took for granted the spread of breeders' innov-ations. In 2022 the adoption of "improved" varieties is considered as

[91] United Nations Declaration on the Rights of Peasants and Other People Working in Rural Areas, September 28, 2018, A/HRC/RES/39/12, Article 19.

inexorable as ever, and the reversion to *ex situ* conservation nearly always prevails over options that integrate farmers' practices and knowledge into plans for enhancing and improving cultivated biodiversity. Within the CGIAR system, and despite its encompassing very different scientific souls, the epistemic culture that sustained the Green Revolution's approaches and vision is as vigorous as it ever was.[92]

Meanwhile, other ways of knowing and understanding crop diversity continue to issue challenges to this predominating culture. Without denying an overall pattern in which crop diversity is diminishing over time, it is nonetheless also possible to observe that commercial varieties are often unable to fulfill the needs of heterogeneous smallholder agriculture. For this reason, many of the world's farmers cannot completely rely on commercial varieties and still sow seeds that they produce themselves.[93] They actively work on crop diversity and, especially in the Global South, they still grow landraces, introduce "modern" varieties, make crosses, select for valued traits, and exchange seeds and knowledge. The classic framework of genetic erosion did not take into account farmers as a powerful evolutionary force that is still active and capable of participating in the search for solutions to ever-changing agricultural needs.[94] The work of many scholars over many decades has made clear that genetic resources can no longer be considered a "raw" product constantly under threat. The goal should not be to permanently conserve the same genetic configuration, as CGIAR gene banks sought to do for much of their existence, but to reconnect diversity, conservation practices, and farming systems.

---

[92] See for example CGIAR's recent partnership with the Alliance for a Green Revolution in Africa (AGRA) funded by the Bill & Melinda Gates Foundation and the Rockefeller Foundation.

[93] C. J. Almekinders and N. P. Louwaars, "The Importance of the Farmers' Seed Systems in a Functional National Seed Sector," *Journal of New Seeds* 4, nos. 1–2 (2002): 15–33.

[94] Mauricio R. Bellon, Alicia Mastretta-Yanes, Alejandro Ponce-Mendoza et al., "Evolutionary and Food Supply Implications of Ongoing Maize Domestication by Mexican Campesinos," *Proceedings of the Royal Society B: Biological Sciences* 285, no. 1885 (2018): 20181049; Perales, "Landrace Conservation of Maize in Mexico," pp. 271–281.

12 When Public Goods Go Private
The CGIAR Approach to Intellectual Property, 1990–2020

David J. Jefferson

For most of the twentieth century, intellectual property was of little relevance for public agricultural research. When the Consultative Group on International Agricultural Research (CGIAR) was established in 1971, its centers considered the privatization of research products to be antithetical to the network's mission, which endeavored to promote food security in developing countries through sustainable agriculture. To realize this mission, CGIAR scientists distributed the products of their research, such as new crop varieties, directly to farmers, free of charge, through extension services provided in collaboration with public national agricultural research systems. In contrast, private agricultural firms generally focused on commercializing products in high-income countries where industrialized agriculture was common and intellectual property operated to secure market exclusivity for new products.

Beginning in the 1980s, several changes unsettled the public–private balance in agricultural science and provoked a reimagination of the role of intellectual property in the research and development process. Various factors help to explain these shifts, including developments in science (e.g., advent of new genetic transformation techniques), the law (e.g., expansion of intellectual property systems), and politics (e.g., decrease in governmental support for research). The ability to claim a broader range of agrarian inventions as property, coupled with the rethinking of how public institutions should leverage exclusive rights, have raised the stakes of agricultural science and ignited tensions that affect the work of many institutions worldwide, including CGIAR.

In the 1990s, agricultural experts – including agronomists, plant scientists, economists, lawyers, and development policy specialists,

Acknowledgments: I am grateful for feedback that members of the Harnessing Intellectual Property to Build Food Security research group at the University of Queensland provided on early drafts of this chapter, as well as for the support of Helen Anne Curry and Timothy W. Lorek as editors of this volume.

but notably not farmers – working within or as consultants for CGIAR developed at least three approaches to how the network and its centers should respond to the global expansion of intellectual property in agriculture. I describe the first approach as maximalist, based on an understanding that formalized intellectual property ownership could provide an important means for centers to augment the impact of their technologies for target beneficiaries. In contrast, I characterize the second approach as adaptationist. Adherents expressed skepticism about the appropriateness of claiming intellectual property rights, but they also recognized that sooner or later CGIAR would need to modify its existing practices to accommodate a reality in which many products of science were regarded as proprietary objects. Finally, I portray the third approach as rejectionist. Proponents claimed that intellectual property was antithetical to CGIAR's mission and its historical focus on small-scale, sustainable agriculture. In this chapter, I argue that over the thirty years from 1990 to 2020, the adaptationist approach crystallized as the overarching approach to intellectual property within CGIAR, as internal debates stabilized and internal governance structures developed and matured.

When intellectual property first emerged as a matter of concern for CGIAR, activists and researchers aligned with organizations that rejected privatization under any circumstance – including Via Campesina, Third World Network, and Genetic Resources Action International (GRAIN) – clashed with industry representatives who thought the centers should maximize the benefits of a capitalist approach to technology dissemination – such as those from the International Association of Plant Breeders for the Protection of Plant Varieties (ASSINSEL) and the International Seed Trade Federation. Over time, CGIAR found ways to accommodate both perspectives to some extent, with each center still able to exercise autonomy over technology management and private-sector partnerships. As of 2020, the centers operated along a continuum, such that some regularly engaged with intellectual property systems while others rarely sought patents, plant variety protection, or other forms of exclusive ownership for their inventions. However, and although CGIAR formally retained its focus on the production of "global public goods,"[1] by the end of the second decade of the new millennium it was clear that across the

---

[1] In the CGIAR context, global public goods (now officially termed "international public goods") are products of scientific research whose social returns on investment exceed any potential private returns. In theory, global public goods are freely available to all (nonexcludable) and not diminished by use (nonrivalrous). However, according to the current CGIAR conceptualization, intellectual property may be justified to render certain technologies not freely available to all (excludable), where doing so increases value for society

global research partnership that CGIAR represents[2] ignoring the influence of intellectual property was no longer tenable. The ascendancy of the adaptationist approach was evident in the fact that responses to the growth of proprietary science had been thoroughly woven into the research and technology development practices of all the centers and CGIAR itself.

Although the need to respond to the expansion of intellectual property led to the alteration of certain CGIAR activities, doing so did not produce the effects that many experts initially expected. Throughout the 1990s and early 2000s, while opponents of privatization feared that the pursuit of intellectual property rights in the form of patents and plant variety protection would undermine CGIAR's mission, proponents foresaw the potential to incentivize partnerships with commercial entities and to provide an alternative source of revenue in an era of diminished public funding. By 2020, neither of these visions had been actualized. The possibility that centers might obtain intellectual property for their creations did not substantially alter their research agendas, lead to a dramatic increase in proprietary claims for CGIAR technologies, or directly generate significant revenue through the commercialization of protected technologies.

Instead, intellectual property had subtle and diffuse effects on the activities of individual centers, and on how they relate to one another as members of the CGIAR global partnership. The expansion of proprietary science also transformed how some centers interact with private-sector partners, especially agribusiness firms. During the early 2000s, all centers adopted institutional policies to deal with the potential effects of intellectual property, and all hired personnel to resolve questions related to the ownership of research results and the commercialization of CGIAR technologies. Furthermore, intellectual property played a role in the structure and internal governance standards of the CGIAR network as a whole, providing both a justification for centralization (e.g., through juridical harmonization and the consolidation of legal services across the network)

as a whole. See D. G. Dalrymple, "International Agricultural Research as a Global Public Good: Concepts, the Global Experience, and Policy Issues," *Journal of International Development* 20 (2008): 347–379, at 350–351.

[2] In 2019, CGIAR announced a major reform known as "One CGIAR," which was driven by a "need for collaboration to become more systemic to better capture strategic opportunities and synergies across the organization." The aim is to create better integration among CGIAR partners and enhance the impacts of CGIAR research. While this transformation will no doubt result in significant effects, as of the time of writing in 2022, it has not resulted in a dramatic shift in CGIAR's intellectual property policies or practices. "Toward Greater Impact: A CGIAR Engagement Framework for Partnerships & Advocacy," Global Director, Partnerships and Advocacy, 4, March 29, 2022, https://storage.googleapis.com/cgiarorg/2022/03/CGIAR-Engagement-Framework-29-March-2022.pdf.

and a platform for individuation (e.g., by allowing each center to define its own operational approach to intellectual property). This chapter focuses on the period of 1990 to 2020, when numerous discussions and concrete changes occurred in reaction to the increasing influence of intellectual property on agricultural research worldwide. Over the course of these three decades, CGIAR leaders and consultants engaged in debates, produced reports, and drafted, adopted, and harmonized policies, leading to a systematized approach to intellectual property governance that is now shared across the global partnership. The chapter draws on internal documents and consultants' reports to recount the history of the consolidation of a coordinated CGIAR approach to intellectual property. It shows that the debates sustained between different experts mirrored discussions about agricultural science and the commercialization of research products that were ongoing in other institutions, including universities and national government agencies, during the same period. Notwithstanding the ambitions and concerns of proponents and opponents of privatization and commercialization, a radical shift away from the global public goods model did not occur. Instead, the formal endorsement of the adaptationist approach to intellectual property precipitated subtler transformations to CGIAR research administration.

Historical and Institutional Context

A series of scientific, economic, and legal developments that occurred in the latter part of the twentieth century led to the expansion of formal intellectual property norms into many domains of agricultural research and plant breeding. As national and international laws were created or expanded, researchers in fields such as molecular biology, genetics, and plant sciences could more easily claim proprietary rights in their creations. In parallel, the locus of plant varietal improvement shifted from the public to the private sector in many countries, while firms trading in seeds, fertilizers, and other farming inputs consolidated through a multitude of mergers and acquisitions.[3] As agricultural science and technology development became increasingly intertwined with intellectual property laws and with globalized capitalism, debates surged about the privatization of seeds and other plant materials, which international legal systems historically had treated as the common heritage of humankind.[4]

[3] S. C. Price, "Public and Private Plant Breeding," *Nature Biotechnology* 17, no. 10 (1999): 938; R. Tripp and D. Byerlee, "Plant Breeding in an Era of Privatisation," *Natural Resource Perspectives* 57 (2000): 1–4; P. H. Howard, "Visualizing Consolidation in the Global Seed Industry: 1996–2008," *Sustainability* 1, no. 4 (2009): 1266–1287.

[4] J. R. Kloppenburg, Jr. and D. L. Kleinman, "Property versus Common Heritage," in J. R. Kloppenburg, Jr., ed., *Seeds and Sovereignty: Debate over the Use and Control of Plant*

Many of these discussions were characterized by certain assumptions. These included the idea that the availability of the exclusive rights provided by intellectual property regimes should incentivize innovation in agricultural science and plant breeding, which in turn was expected to benefit farmers, for example by making the seeds of improved crop varieties more broadly available.[5] However, a competing assumption held that some farmers – including smallholders and Indigenous cultivators, especially in the Global South – would be harmed by the expansion of intellectual property in agriculture. The assumption was that the increased privatization of public research products and the corresponding prioritization of maximizing economic returns would lead to a neglect of crop species and varieties for which large markets do not exist, while proscribing customary cultivation practices such as the saving, exchange, and local sale of farm-saved seeds.[6]

It was inevitable that as the largest public agricultural research system in the world, CGIAR would need to contend with intellectual property issues. While debates over the use of proprietary legal vehicles to claim agricultural technologies became common in research institutions worldwide in the 1980s and 1990s, such discussions had unique features within CGIAR. This may be partially explained by the complex character of the network. At the time when intellectual property became a matter of concern for agricultural science, CGIAR operated simultaneously as a loose affiliation of individual research centers – each with their own missions, governance models, and scientific orientations – and as a centralized institution in its own right. The variegated nature of CGIAR meant that it had to both accommodate centers' diverse responses to intellectual property, and harmonize local approaches to create a coherent, system-wide strategy. In this way, CGIAR needed to transcend the dichotomous thinking that characterized many late twentieth-century debates about the global expansion of intellectual property in agriculture.

The formation of CGIAR in 1971 forged a formal link between institutions that had emerged independently from post–World War II, country-specific agricultural programs. In part because certain centers predated CGIAR, tension between centralization and autonomy imbued the network from the time of its establishment. Competing

---

Genetic Resources (Durham, NC: Duke University Press, 1998), pp. 173–203. Here "common heritage" is defined as when plants and seeds are viewed as a common good for which no payment is necessary or appropriate.

[5] L. R. Helfer, *Intellectual Property Rights in Plant Varieties: International Legal Regimes and Policy Options for National Governments*, FAO Legislative Study No. 85 (Rome: FAO, 2004).

[6] N. P. Louwaars, R. Tripp, D. Eaton et al., *Impacts of Strengthened Intellectual Property Rights Regimes on the Plant Breeding Industry in Developing Countries* (Wageningen, Netherlands: World Bank, 2005).

interests that alternately advocated for unification or independence contributed to divergent views about the appropriate role of intellectual property throughout the 1990s and early 2000s. For example, there was tension between efforts to establish universal policies, performance standards, and decision-making protocols for resource allocation, and the need to safeguard individual centers' capacities to innovate and set appropriate internal governance standards.[7]

Economic factors also underpinned the intellectual property debates that emerged in the 1990s. In CGIAR's early years, the centers were mainly funded through donations from national and international governmental agencies. More recently, however, financial support from governments became increasingly scarce. While private philanthropy stepped in to fill some gaps, the number of "public–private partnerships" with for-profit firms also grew.[8] Reliance on associations with profit-driven entities required that CGIAR reconcile its nonproprietary global public goods model with the commercialization strategies of multinational agribusinesses, which typically were grounded in the protection of research products as intellectual property for the purpose of securing market exclusivity. This dynamic was further compounded by scientific developments, such as the emergence of new agricultural biotechnologies (e.g., transgenic plants), and the global expansion of patent and plant variety protection laws. Thus, at the dawn of the 1990s, a series of international scientific, economic, and legal developments brought intellectual property to the fore within CGIAR.

Intellectual Property Becomes a Matter of Concern, 1990 to 1996

The first formal review of the implications that intellectual property could have for CGIAR was initiated in 1982, but by then certain centers, most notably the International Rice Research Institute (IRRI), had already obtained patents for their inventions.[9] As the 1990s commenced, all

---

[7] D. Byerlee and J. K. Lynam, "The Development of the International Center Model for Agricultural Research: A Prehistory of the CGIAR," *World Development* 135 (2020): 105080.

[8] From 2011 to 2022, the Bill & Melinda Gates Foundation contributed the second-highest amount to the CGIAR Trust Fund ($990.6 million), behind only the United States Agency for International Development (USAID) (US$1,474.1 million); see CGIAR, "CGIAR Trust Fund Contributions," www.cgiar.org/funders/trust-fund/trust-fund-cont ributions-dashboard. On public–private partnerships, see D. J. Spielman, F. Hartwich, and K. von Grebmer, *Sharing Science, Building Bridges, and Enhancing Impact: Public–Private Partnerships in the CGIAR*, IFPRI Discussion Paper 00708 (Washington, DC: IFPRI, 2007).

[9] W. E. Siebeck, D. L. Plucknett, and K. Wright-Platais, "Privatization of Research through Intellectual Property Protection and Its Potential Effects on Research at the International

CGIAR center directors "accepted that the legal protection of inventions and intellectual property" had become standard practice in modern agricultural science, particularly for research involving the use of novel biotechnologies.[10] Although the directors expressed confidence that the growth of intellectual property could be accommodated without abandoning the global public goods model, they also acknowledged the "clear need" for expert guidance on patent and plant variety protection issues. They argued that CGIAR centers should be shielded from any detrimental effects associated with the increased utilization of intellectual property in agricultural research but should also be able to "take advantage of potential benefits," including "the promotion of collaborative arrangements" and "the facilitation of access to technologies."[11]

The directors presented a draft paper on intellectual property at a 1990 meeting of the CGIAR leadership, where their proposals generated "considerable discussion."[12] Shortly afterward, the CGIAR chairman convened a consultation that brought together twenty-eight experts from national governmental agencies, universities, and nongovernmental development organizations (NGOs) to "think creatively about a CGIAR strategy for the 1990s."[13] Consultation participants represented the United Nations Food and Agriculture Organization (FAO), United Nations Development Programme (UNDP), World Bank, Rockefeller Foundation, and several European and North American government agencies and universities. Industry representatives were not invited. Nevertheless, and notwithstanding their public-sector affiliations, some consultants favored greater engagement with businesses, highlighting that the private sector encompassed "a wider universe ... than just the multinational companies" and that the centers could play an important role in supporting small industries in rural areas in the countries where they were located.[14] However, others were skeptical of partnering with industry, querying, "Could the CGIAR hurt

---

Centers," in D. R. Buxton et al., eds., *International Crop Science I* (Madison: Crop Science Society of America), pp. 499–504. Early IRRI patents covered inventions including extracts from rice plants used as insecticides (PH 12554) and herbicides (PH 13021), a seed plate planter (PH 13473), a process of rice seedling production (PH 13550), a reaper (PH 14108), and a chemical compound used for flavoring foods (US 4522838).

[10] CGIAR Center Directors Committee, "Biotechnology in the International Agricultural Research Centers of the Consultative Group on International Agricultural Research: A Statement by Center Directors," CGIAR Mid-Term Meeting, the Hague, the Netherlands, May 21–25, 1990, 5, https://hdl.handle.net/10947/201.

[11] Ibid.   [12] Ibid.

[13] CGIAR Ad Hoc Strategy Consultation, Synthesis Report, February 1992, encl. in Letter from CGIAR Chairman V. Rajagopalan, letter to Heads of CGIAR Delegations, February 24, 1992, 1, https://hdl.handle.net/10947/718.

[14] Ibid., 30.

itself in some ways in some countries if its relationship with private companies is too close?"[15]

In summarizing the discussion, economist and Stanford University professor Walter Falcon, who served as moderator, noted that "[s]trong anti-private sector sentiments exist in several circles related to CGIAR."[16] Correspondingly, many stakeholders would likely oppose the future utilization of intellectual property laws to protect CGIAR technologies, because "for some persons and donors, intellectual property rights are a political issue, at least in part, while they are moral or ethical issues for others."[17] Despite this, Falcon concluded that intellectual property issues, particularly in relation to patents, plant variety protection, and material transfer agreements would almost certainly figure more prominently in the centers' work in the future. CGIAR "must learn how to handle these questions effectively."[18]

The conversation continued to gather momentum at a 1992 meeting, where CGIAR leadership debated the recently released "Suggested Principles for a Future CGIAR Policy on Intellectual Property Rights" and a discussion document on "Intellectual Property, Biosafety and Plant Genetic Resources." The latter identified several situations that might justify centers' use of intellectual property, including "to prevent preemptive protection by others, which might restrict the availability of those inventions, especially to ... developing countries."[19] Intellectual property ownership could also give centers leverage in negotiations for the use of third parties' technologies, where a cross-licensing or similar arrangement could be brokered. However, the leadership concluded that centers should not pursue intellectual property for economic reasons, and any financial returns generated from licensing or commercializing technologies that centers owned would need to be used for the direct benefit of developing countries.[20]

Although the discussion document was unanimously adopted at the 1992 meeting, divergent views on intellectual property persisted. One year later, during another leadership conference, some experts rejected the idea that CGIAR should adopt a formal intellectual property policy, while others wanted to unambiguously encourage collaboration with private-sector partners.[21] Further complicating matters, two major shifts in the international legal landscape occurred in the early 1990s that

---

[15] Ibid.   [16] Ibid.   [17] Ibid., 31.   [18] Ibid., 38.
[19] CGIAR Discussion Document on Intellectual Property, Biosafety, and Plant Genetic Resources, Mid-Term Meeting, May 18–22, 1992, 2, https://hdl.handle.net/10947/648.
[20] Ibid.
[21] W. E. Siebeck, "Intellectual Property Rights and CGIAR Research – Predicament or Challenge?" in *CGIAR Annual Report 1993–1994* (Washington, DC: CGIAR Secretariat, 1994), pp. 17–20.

created uncertainty about intellectual property governance within CGIAR. The changes were the entry into force of the Convention on Biological Diversity (CBD) in December 1993, and the signing of the Agreement on Trade-Related Aspects of Intellectual Property Rights (TRIPS) of the World Trade Organization (WTO) in April 1994. These international agreements, which in some ways were in tension with one another, led the new CGIAR chairman, the Egyptian scientist and economist Ismail Serageldin, to convene a panel on intellectual property rights in 1994.[22] Given the reforms anticipated in the wake of the CBD and the TRIPS Agreement, the panel urged CGIAR to analyze how changes to national intellectual property laws might affect the dissemination of agricultural technologies in the Global South.[23]

The Indian geneticist and Green Revolution plant breeder M. S. Swaminathan chaired the panel, which comprised center directors and experts from academic, governmental, and philanthropic institutions. For the first time an industry representative joined the conversation: the CEO of the pharmaceutical and agrochemicals company Zeneca (now AstraZeneca). Panel experts agreed on several points, including the circumstances that would justify the use of patents to protect CGIAR inventions and the principle that if a center obtained a patent for one of its inventions, it should provide royalty-free licenses to developing countries.[24] Panelists also agreed that questions such as where to apply for patent protection and how to share intellectual property ownership rights under collaborative research agreements should be determined case by case.[25]

The panel further recommended the establishment of pooled technical and legal services to enable centers to understand intellectual property issues and develop "common operational approaches" to technology management.[26] Finally, and revealing of the longstanding tension between centralization and autonomy, some panelists endorsed the idea that CGIAR should have independent legal personality. This would formalize the ad hoc funder–center partnership structure and enable

[22] C. Lawson and J. Sanderson, "The Evolution of the CBD's Development Agenda That May Influence the Interpretation and Development of TRIPS," in J. Malbon and C. Lawson, eds., *Interpreting and Implementing the TRIPS Agreement: Is It Fair?* (Cheltenham, UK: Edward Elgar, 2008), pp. 131–158.

[23] CGIAR Intellectual Property Rights Panel and M. S. Swaminathan, "Report of the Intellectual Property Rights Panel," September 30, 1994, i, https://hdl.handle.net/1094 7/1094.

[24] Ibid., ii. Justifiable circumstances included preventing appropriation by third parties, ensuring further product development and delivery to farmers, and negotiating access to other proprietary technologies.

[25] Ibid., ii–iii.   [26] Ibid., iii.

CGIAR to act on behalf of individual centers, for instance when filing for patent protection.[27]

The panel also considered how CGIAR should approach plant variety protection as a form of intellectual property alternative to patents. This was especially relevant considering the 1991 reform of the Convention of the Union for the Protection of New Varieties of Plants (UPOV Convention) and the TRIPS Agreement.[28] The latter treaty required all members of WTO, including developing countries, to enact plant variety protection laws.[29] At the time, activists critical of the TRIPS Agreement interpreted this requirement as an implicit endorsement of the UPOV Convention.[30] In this context, panelists "strongly support-[ed]" the recognition of exceptions to plant variety protection, which would allow farmers to save and exchange seeds, and permit protected varieties to be used for research.[31] Panel experts additionally suggested that CGIAR should co-sponsor the formation of a standardized approach to plant variety protection, in collaboration with the governments of developing countries, which could operate as an alternative to the UPOV Convention.[32]

The CGIAR leadership reviewed the panel's report during a meeting in December 1994 and "broadly accepted" its recommendations, endorsing another round of consultation that aimed to develop a system-wide intellectual property policy.[33] After two years of research and discussions, the "Guiding Principles for the CGIAR Centers on Intellectual Property and Genetic Resources" were released at a 1996 leadership meeting. Like earlier policy statements, these principles emphasized that centers should continue to prioritize the full disclosure of research results and release products into the public domain, except where seeking intellectual property protection "is needed to facilitate technology transfer or otherwise

---

[27] Ibid., 4.
[28] The 1991 Act of UPOV substantially expanded the scope of intellectual property available to plant breeders. For example, it enabled a broader set of plant materials to be claimed and lengthened the periods of exclusivity, while also limiting certain exemptions that had been previously recognized.
[29] Notably, the TRIPS Agreement exempted "least developed countries" that are WTO members from implementing the agreement until 2006, which was later extended until July 2034 at the earliest. See WTO, "WTO Members Agree to Extend TRIPS Transition Period for LDCs until 1 July 2034," June 29, 2021, www.wto.org/english/news_e/news21_e/trip_30jun21_e.htm.
[30] V. Shiva, "Agricultural Biodiversity, Intellectual Property Rights and Farmers' Rights," *Economic and Political Weekly* 31, no. 25 (1996): 1621–1631, at 1628.
[31] "Report of the Intellectual Property Rights Panel," iii. [32] Ibid.
[33] CGIAR Secretariat, *CGIAR International Centers Week, Washington, DC, October 24–28, 1994: Summary of Proceedings and Decisions* (Washington, DC: CGIAR, December 1994), p. 48, https://hdl.handle.net/10947/273.

protect the interests of developing nations."[34] Furthermore, CGIAR institutions should not view exclusive rights as a means to secure monetary returns. However, the principles also indicated that if a center did benefit financially from intellectual property commercialization, the center would need to ensure that the funds were used to further its public goods mandate and the overall objectives of CGIAR.[35] These examples demonstrate that by the time the guiding principles were released in 1996, CGIAR had largely consolidated a standardized approach to intellectual property.

A System-Wide Policy Is Consolidated, 1996 to 2012

Although by 1996 it appeared that CGIAR was ready to enact a system-wide intellectual property policy, its leadership decided that the guiding principles should continue to operate as nonbinding working guidelines until ongoing legal questions were resolved.[36] In the meantime, chairman Serageldin formed a panel on proprietary science and technology, which conducted interviews with administrators and scientists from seven centers, in addition to intellectual property managers at five major US land-grant universities and five multinational agricultural companies.[37] Timothy Roberts, a British chemist and former intellectual property manager of ICI Seeds (now AstraZeneca) chaired the panel, reflecting Serageldin's growing belief that the private sector would be an essential part of future CGIAR strategy.

The panel presented its final report at a leadership meeting in 1998. The document was notable for its consideration of issues that had received little attention in prior deliberations. For instance, the report identified risks that could arise if intellectual property were sought for CGIAR inventions, including the substantial expenditures associated with the preparation, filing,

[34] "Guiding Principles for the Consultative Group on International Agricultural Research Centers on Intellectual Property and Genetic Resources," principle 7, published in CGIAR Center Directors Committee and CGIAR Committee of Board Chairs, "CGIAR Center Statements on Genetic Resources, Intellectual Property Rights, and Biotechnology," May 1999, https://hdl.handle.net/10947/253.
[35] Ibid., principle 8.
[36] These questions included the potential impact of the International Treaty on Plant Genetic Resources for Food and Agriculture (then still in negotiation) and likely reforms to national intellectual property laws. See CGIAR Secretariat, *The CGIAR at 25: Into the Future: ICW96 Summary of Proceedings and Decisions*, CGIAR International Centers Week 1996: Summary of Proceedings and Decisions (Washington, DC: CGIAR, January 1997), p. 67, https://hdl.handle.net/10568/119103.
[37] "Mobilizing Science for Global Food Security," Report of the CGIAR Panel on Proprietary Science and Technology, SDR/TAC:IAR/98/7.1, April 20, 1998, www.fao.org/3/w8425e/w8425e00.htm.

and maintenance of patent and plant variety protection applications.[38]
Obtaining proprietary rights could also skew the centers' research agendas.
For instance, centers might begin to focus more attention on investigations
that could lead to the development of marketable products while neglecting
research on questions with limited commercial applications. On the other
hand, benefits that could result from intellectual property utilization
included the possibility of facilitating technology transfer to target beneficiaries, the ability of centers that partnered with external entities to reserve
rights to jointly owned intellectual property for humanitarian use, and the
potential to attract local investments and enable "capital formation" in
countries where centers operated.[39]

   Panel members acknowledged that any revenues derived from licensing
a protected technology would constitute a potential benefit for the center
that owned it, but they disagreed about the extent to which CGIAR
should engage in commercial activities in the first place. While the majority thought that generating income should never be the main reason to
seek intellectual property, the minority "strongly" believed that not protecting certain technologies would be tantamount to "wast[ing] useful
resources."[40] Although panel members generally concurred that CGIAR
should establish a set of mission-driven criteria to guide decision-making,
discord permeated the report because participants "disagree[d] markedly
as to what an ideal situation should be."[41] The panel was particularly
divided over the question of "whether CGIAR should campaign against
all intellectual property on life-forms, or whether it should promote
extension of [intellectual property] to promote innovation, transfer and
adoption of useful technologies."[42]

   Deliberations over how CGIAR should practice science in relation to
intellectual property were manifested in three approaches or viewpoints.
The first approach, which I describe as maximalist, was endorsed by some
panel members, who "believe[d] strongly that advanced biotechnology
and the development of transgenic crop varieties are central to the goal of
increasing food production in developed and developing countries, and
that only in the context of strengthened intellectual property regimes will
these proceed efficiently."[43] The panel's report overtly referenced
a "Statement on Biotechnology and the Agri-Food Industry" by the
International Agri-Food Network as representative of this approach.[44]
Although the report did not specifically mention which panelists
endorsed the viewpoint that I term maximalist, it is likely that they
included at least Robert Horsch of Monsanto and Bernard Le Buanec

---

[38] Ibid., section 3.2.   [39] Ibid.   [40] Ibid.   [41] Ibid., section 6.   [42] Ibid.
[43] Ibid., section 6.2.1.   [44] Ibid.

of the International Seed Trade Federation and ASSINSEL, whose institutions had also endorsed the International Agri-Food Network's Statement.[45]

Proponents of maximalism championed the widespread utilization of intellectual property by at least some centers. Maximalists also thought that CGIAR should endorse the ratcheting-up of international intellectual property regimes, including by broadening the scope and reach of the TRIPS Agreement. Adherents to maximalism were "gravely concerned" about the notion that CGIAR should act as a "voice for the poor," believing that enabling aid recipients to express their views would "inevitably polarize the CGIAR's supporters; put at risk its scientific credibility; and undermine its ability to continue its enormously valuable technical contribution to the welfare of the poor."[46] In other words, they maintained that centers should continue to deliver new technologies to poor farmers but that they should not empower farmers politically, because doing so might offend CGIAR donors and partners or make the centers appear unscientific.

Other panel members supported an approach that I characterize as adaptationist, according to which most of CGIAR's core work could proceed without major changes to centers' customary lack of engagement with intellectual property laws. The adaptationist viewpoint recognized that "the increasing use of proprietary property in agricultural research and development is a fact of life, whether regrettable or beneficial," so centers should acclimate while continuing to focus on furthering their missions. In adjusting to this new reality, "the very substantial costs of increasing CGIAR capacity to manage intellectual property must be weighed carefully against potentially competing needs of an arguably underfunded CGIAR system."[47] Likewise, centers should judiciously consider the opportunity costs of using limited resources to enforce their intellectual property rights in the event of infringement. Adaptationists also expressed concern that patents owned by third parties were unreasonably constraining research. Therefore, they recommended that CGIAR advocate for a clearer definition of the "research exemption" to patents, which effectively limits the scope of exclusive rights to commercial uses rather than investigative or experimental activities.[48]

Finally, some panel members embraced an approach that I term rejectionist, resisting the idea that the most "advanced" agricultural science was the kind associated with industrial biotechnology and the development of transgenic plant varieties. Instead, a truly advanced approach would pursue "the better understanding, improvement, and adaptation to various developing country conditions of sustainable, diversity-based

---

[45] Ibid., appendix B.   [46] Ibid., section 6.2.1.   [47] Ibid., section 6.2.2.   [48] Ibid.

Figure 12.1 Protesters in the Philippines, 2010s, take a stand against Golden Rice, genetically modified organisms (GMOs), transnational corporations (TNCS), and the International Rice Research Institute (IRRI). IRRI is represented by the bespectacled, white-coated puppet at back right. By permission of MASIPAG.

agricultural systems, and the related management of genetic, crop, soil, and other agricultural resources."[49] According to the report, examples of this viewpoint could be found in the Thammasat Resolution, a 1997 declaration by representatives of Indigenous, peasant, nongovernmental, academic, and governmental organizations including Via Campesina, the Third World Network, and GRAIN, as well as in a statement issued by the prominent agro-ecologist Miguel Altieri (Figure 12.1).[50]

Rejectionists believed that intellectual property should have little relevance for the centers' work. Instead of becoming involved with commercialization, CGIAR "should only make research investments in technologies that the private sector is not investing in, and for which the only 'market' is the poor."[51] The rejectionist viewpoint argued that CGIAR should actively oppose a proposed expansion of the TRIPS Agreement, which would have required all WTO member countries to recognize patents for inventions based on animals and plants.

---

[49] Ibid., section 6.2.3.   [50] Ibid., appendices D-5 and D-6.   [51] Ibid., section 6.2.3.

Simultaneously, rejectionists advocated for "alternative" intellectual property regimes that would support CGIAR's mission of making plant varieties freely available to poor farmers in developing countries.[52]

Unlike the maximalists and adaptationists who participated in the 1998 panel on proprietary science and technology, rejectionists believed that CGIAR was in a position to actively shape rather than merely passively respond to changes driven by techno-legal developments and the spread of global capitalism. As Altieri argued in an appendix to the panel's report, "[i]t is time for the CGIAR to play a more active role in defining the future [intellectual property rights] scenarios so as to prevent that the free exchange of knowledge and resources does not give way to a monopoly vested in those who control capital and hence the resources for research."[53]

Notwithstanding the discrepancies between the maximalist, adaptationist, and rejectionist approaches, few significant changes were made to the official CGIAR stance on intellectual property after the report was presented. The guiding principles that were first introduced in 1996 continued to provide a system-wide framework until an updated policy, "CGIAR Principles on the Management of Intellectual Assets" (hereafter Intellectual Assets Principles) was finally adopted in 2012. Over this fifteen-year period, the adaptationist approach to intellectual property came to dominate. Meanwhile, certain centers, most notably IRRI and the International Maize and Wheat Improvement Center (CIMMYT), became increasingly maximalist by deepening engagements with private-sector partners and seeking intellectual property for inventions that could prove commercially viable. Other centers continued to avoid making proprietary claims for their technologies, maintaining a rejectionist approach. However, by 2012 it was clear that at the system level, the rejectionist viewpoint, with its advocacy for strengthening local, customary farming systems as an alternative to advanced biotechnologies, had been formally marginalized.

The dismissal of the rejectionist approach and its adherents' advocacy for CGIAR to take a more active role in shaping global agricultural research practices might be partially explained by the fascination with the "gene revolution" that pervaded agriscience discourse in the 1990s. When the 1998 CGIAR system review report highlighted that genetic "breakthroughs" were typically only achieved by the private sector, it also indicated that "CGIAR's challenge is to create a new form of public–private partnership that will protect intellectual property while bringing the benefits of this research to the poorest nations."[54] In his opening

---

[52] Ibid., section 6.2.3.   [53] Ibid., appendix D-5.
[54] CGIAR System Review Secretariat, *The International Research Partnership for Food Security and Sustainable Agriculture*, Third System Review of the Consultative Group on

remarks at a 2000 leadership meeting, Chairman Serageldin used even starker language to describe the situation:

CGIAR now faces a future characterized by make-or-break challenges, and make-or-break opportunities . . . The implicit bargain among the developing countries – the possessors of germplasm – the advanced research organizations, the main producers of new science, and international institutions working with national agricultural research systems . . . is becoming more and more difficult to maintain, as scientific developments become increasingly subject to private control. The private sector is now at the head of most developments in the field of science and, to recoup the billions of dollars it invests on research, is expanding the application of patents and intellectual property rights. We cannot remain indifferent to what goes on beyond the parameters of that bargain.[55]

In 2000, a CGIAR working group on intellectual property rights and the private sector echoed Serageldin, noting that "CGIAR must negotiate from a position of strength. Its leverage is strengthened when its own [intellectual property] is of interest to partners. It must be a trusted and respected player."[56] In other words, the working group insisted that to remain both scientifically relevant and economically viable, at minimum centers would need to speak the language of agribusiness, conceptualizing their own technologies as CGIAR intellectual property.

As the first years of the new millennium unfolded, the CGIAR approach to intellectual property stabilized. Proprietary science issues were debated with far less frequency in internal documents published between 2000 and 2010 in comparison with the previous decade.[57] Simultaneously, the organization's leadership shifted its focus to bolstering CGIAR as a centralized entity, while harmonizing the various centers' internal policy frameworks. For instance, following a series of meetings in 2005, the CGIAR genetic resources policy committee generated a template intellectual property policy statement, which was intended to promote consistency in centers' practices.[58] Despite these efforts, a 2008 independent review found that although some centers had already adopted internal policies and hired professional

International Agricultural Research (Washington, DC: CGIAR), September 1998, p. viii, https://hdl.handle.net/10947/1586.

[55] CGIAR Secretariat, *Charting the CGIAR's Future: A New Vision for 2010*, Summary of Proceedings and Decisions, Mid-Term Meeting 2000, Dresden, Germany, May 21–26, 2000 (Washington, DC: CGIAR, July 2000), p. 16, https://hdl.handle.net/10947/300.

[56] Ibid., 28.

[57] This assessment is based on keyword searches of documents housed in the CGSpace repository in August 2021, https://cgspace.cgiar.org.

[58] CGIAR Genetic Resources Policy Committee, "Summary Report of the Genetic Resources Policy Committee (GRPC) Meetings Held in 2005," appendix 2, https://hdl.handle.net/10947/3935.

staff to resolve intellectual property questions, the majority had not, and they "tend[ed] to deal with these issues on an ad hoc basis, often reacting to crisis."[59] A frank warning accompanied this assessment: "CGIAR cannot ignore or causally handle issues of intellectual property protection."[60]

Although all fifteen CGIAR centers[61] had already adopted intellectual property policy statements at the time the independent review was conducted, only six had established in-house units or offices dedicated to intellectual property management. Furthermore, while the review found that scientists working at the centers were increasingly aware of the relevance of intellectual property to their research, they lacked an understanding of pertinent international and national legal regimes.[62] Another issue was the fact that most centers did not allocate resources to intellectual property management in their annual budgets.[63]

The recommendations issued by the independent review – and indeed, the initial rationale for its formation – reflected the consolidation of the adaptationist and maximalist approaches. The institutional response to the deficiencies that the review identified was the 2012 adoption of the Intellectual Assets Principles. These principles espoused a commitment to the "sound management" of intellectual property as a means to advance the "CGIAR Vision" of a "world free of poverty, hunger and environmental degradation."[64] The policy formally articulated CGIAR's conceptualization of research results as global public goods and embraced a commitment to the "widespread diffusion and use of these goods to achieve the maximum possible access, scale, scope of impact and sharing of benefits to advantage the poor, especially farmers in developing countries." Simultaneously, the principles outlined CGIAR's commitment to the "prudent and strategic

---

[59] Elizabeth McAllister, Keith Bezanson, G. K. Chadha et al., *Bringing Together the Best of Science and the Best of Development: Independent Review of the CGIAR System: Technical Report* (Washington, DC: CGIAR, November 2008), p. 5, https://hdl.handle.net/10947/4949.

[60] Ibid., p. 4.

[61] The number of CGIAR centers has fluctuated over time. While there were fifteen centers in 2012, since that time Bioversity International and the International Center for Tropical Agriculture have formed an alliance, reducing the total number of centers to fourteen in 2022.

[62] Ibid., 250. The six centers that as of 2008 had established in-house intellectual property units were the International Crops Research Institute for the Semi-Arid Tropics (ICRISAT), the International Livestock Research Institute (ILRI), IRRI, the International Center for Tropical Agriculture (CIAT), Bioversity International, and CIMMYT.

[63] Ibid., 251.

[64] CGIAR System Management Office, "CGIAR Principles on the Management of Intellectual Assets," March 7, 2012, principle 5.1, https://hdl.handle.net/10947/4486.

use" of intellectual property, including requirements that centers manage their technologies with "integrity, fairness, equity, responsibility, and accountability," and that they engage in due diligence to ensure that they do not infringe third-party proprietary rights.[65]

One year after the Intellectual Assets Principles were adopted, CGIAR issued a set of implementation guidelines that provided additional information and examples to facilitate understanding and ensure coherent intellectual property management across the centers.[66] The implementation guidelines clarified that when centers consider whether to seek formal intellectual property protection, they should follow an internal evaluation procedure to ensure that doing so is necessary. The culmination of this procedure should typically entail the preparation of a written report that describes the strategy for technology development, dissemination, and commercialization, the reasons for filing the application, the benefits expected to result from protection, and the risks that may result from declining to seek protection, among other issues.[67] The standardization of these internal evaluation procedures is but one example[68] of how a culture of intellectual property had permeated CGIAR's operations by the second decade of the 2000s, even as the actual number of applications for patents and plant variety protections that centers filed remained low.

Lessons from the Intellectual Assets Reports

Every year since the adoption of the Intellectual Assets Principles in 2012, CGIAR has published an "intellectual assets report" on centers' technology management activities, including claims made under intellectual property laws. The first report indicated that although CGIAR institutions did not file a single application for patents or plant variety protection in 2012, intellectual property was already shaping their cultures and practices. For instance, by the end of that year, all centers had developed legal and intellectual property expertise in the form of in-house or external personnel, in contrast to what the 2008 independent review had found. Centers had responded to CGIAR's prioritization of intellectual property capacity development by enrolling staff

---

[65] Ibid., principle 1, principle 6.4.1, principle 5.2, and principle 5.3.
[66] CGIAR System Management Office, "Implementation Guidelines for the CGIAR IA Principles on the Management of Intellectual Assets," June 14, 2013, background, https:// hdl.handle.net/10947/4487.
[67] Ibid., IP rights (article 6.4).
[68] Another example can be found in the CGIAR Intellectual Property Community of Practice, a system-wide forum launched in 2013 whose aim is to promote the effective management of intellectual property across all CGIAR institutions. S. Cummings et al., eds., "Open for Business: Pathways to Strengthen CGIAR's Responsible Engagement with the Private Sector," 2022, 32, https://hdl.handle.net/10568/119305.

in technical seminars, recruiting additional legal experts, organizing workshops and training activities for researchers and administrators, and mobilizing resources to support local intellectual property management units.[69] Furthermore, ten of the fifteen centers had already reviewed and modified their policies to ensure compliance with the Intellectual Assets Principles.[70]

Notwithstanding these activities, even today the privatization of CGIAR technologies remains rare. The intellectual assets reports from 2012 to 2021 indicated that in any given year, few centers sought formal intellectual property protection. Over this ten-year period, a total of fifty-two patent filings were made, while only seven applications for plant variety protection were submitted.[71] Furthermore, the actual number of technologies that these filings represented was lower than the figures suggest. Many of the patent applications were reported multiple times across different years, for example when an application claiming a particular invention converted from a provisional to an international filing made under the Patent Cooperation Treaty, and subsequently progressed to national phase applications in specific countries.[72]

Between 2012 and 2021, nine centers lodged at least one intellectual property application.[73] However, one center accounted for the majority of the filings: IRRI made thirty-eight of the fifty-nine applications (64 percent) lodged during this period. The center with the second-highest number was the International Crops Research Institute for the Semi-Arid Tropics (ICRISAT), with six submissions. All other centers submitted a smaller number of applications throughout these ten years, indicating the relative infrequency with which formal intellectual property protection was sought across the CGIAR network. It is also notable that of the fifty-two patent applications that CGIAR institutions submitted between

---

[69] CGIAR Consortium Legal Counsel, "CGIAR Intellectual Assets (IA) Report for 2012," August 2013, 3–4, https://hdl.handle.net/10947/2887.

[70] Ibid., 8–9.

[71] These data were compiled from the annual intellectual assets reports for 2012–19, available at CGIAR, "Intellectual Assets Reports," www.cgiar.org/food-security-impact/intellectual-assets-reports.

[72] For example, the 2013 intellectual assets report disclosed that IRRI had lodged six provisional patent applications in the United States. The following year, five of these provisional applications were converted into international filings made through the Patent Cooperation Treaty (PCT), while one became the subject of a US utility patent application. Subsequently, in 2015, one of these PCT filings advanced to national phase applications in seven countries (Brazil, China, India, the Philippines, Thailand, the United States, and Vietnam). Therefore, although cumulatively this activity appears as nineteen patent filings, a single invention accounted for nine of the applications lodged.

[73] The centers that made intellectual property filings during this period were IRRI, CIMMYT, the International Center for Agricultural Research in the Dry Areas (ICARDA), the International Institute of Tropical Agriculture (IITA), ICRISAT, ILRI, the International Potato Center (CIP), CIAT, and Bioversity International.

US010999986B2

(12) **United States Patent**
Jena et al.

(10) **Patent No.:** **US 10,999,986 B2**
(45) **Date of Patent:** **May 11, 2021**

(54) INCREASING HYBRID SEED PRODUCTION THROUGH HIGHER OUTCROSSING RATE IN CYTOPLASMIC MALE STERILE RICE AND RELATED MATERIALS AND METHODS

(71) Applicant: **International Rice Research Institute**, Los Baños (PH)

(72) Inventors: **Kshirod K. Jena**, Cuttack '(IN); **Balram Marathi**, Hyderabad (IN); **Joie Ramos**, Los Baños (PH); **Reynaldo Diocton, IV**, Samar (PH); **Ricky Vinarao**, Los Baños (PH); **G. D. Prahalada**, Sira (IN); **Sung-Ryul Kim**, Ulsan (KR)

(73) Assignee: **International Rice Research Institute**, Los Baños (PH)

(*) Notice: Subject to any disclaimer, the term of this patent is extended or adjusted under 35 U.S.C. 154(b) by 12 days.

(21) Appl. No.: **15/579,247**

(22) PCT Filed: **Jun. 5, 2016**

(86) PCT No.: **PCT/IB2016/053294**
§ 371 (c)(1),
(2) Date: **Dec. 4, 2017**

(87) PCT Pub. No.: **WO2016/193953**
PCT Pub. Date: **Dec. 8, 2016**

(65) **Prior Publication Data**
US 2018/0160638 A1   Jun. 14, 2018

Related U.S. Application Data

(60) Provisional application No. 62/171,524, filed on Jun. 5, 2015.

(51) **Int. Cl.**
A01H 1/04	(2006.01)
C12Q 1/6895	(2018.01)
A01H 5/10	(2018.01)
A01H 1/02	(2006.01)
A01H 4/00	(2006.01)
C12N 15/82	(2006.01)

(52) **U.S. Cl.**
CPC *A01H 1/04* (2013.01); *A01H 1/02* (2013.01); *A01H 4/008* (2013.01); *A01H 5/10* (2013.01); *C12Q 1/6895* (2013.01); *C12N 15/8289* (2013.01); *C12Q 2600/13* (2013.01); *C12Q 2600/156* (2013.01)

(58) **Field of Classification Search**
None
See application file for complete search history.

(56) **References Cited**

U.S. PATENT DOCUMENTS

4,764,643 A	8/1988	Calub
2012/0240285 A1	9/2012	Jinushi et al.

FOREIGN PATENT DOCUMENTS

CN	102333439	1/2012
WO	WO 2016/193953	12/2016
WO	WO 2018/224861	12/2018

OTHER PUBLICATIONS

Miles et al (2008, "Quantitative Trait Locus (QTL) Analysis" , Nature Education 1(1):208).*
Taillebois et al (1986. "Improving Outcrossing Rate in Rice (*Oryza sativa* L.)", Proceedings of the International Symposium on Hybrid Rice, pp. 175-180).*
Miles et al (2008, "Quantitative Trait Locus (QTL) Analysis" , Nature Education 1(1):208; p. 4, 2nd paragraph).*
(Taillebois et al 1986, Proceedings of the International Symposium on Hybrid Rice, pp. 175-180).*
Communication Pursuant to Article 94(3) EPC dated Jan. 18, 2019 From the European Patent Office Re. Application No. 16729639.1. (4 Pages).
Examination Report dated Feb. 5, 2018 From the Ministry of Science and Technology of the Socialist Republic of Vietnam Re. Application No. 1-2018-00038 and Its Summary in English. (2 pages).
International Preliminary Report on Patentability dated Dec. 14, 2017 From the International Bureau of WIPO Re. Application No. PCT/IB2016/053294. (7 Pages).
International Search Report and the Written Opinion dated Dec. 1, 2017 From the International Searching Authority Re. Application No. PCT/IB2017/053363. (16 Pages).
International Search Report and the Written Opinion dated Aug. 8, 2016 From the International Searching Authority Re. Application No. PCT/IB2016/053294.
Angeles-Shim et al. "Molecular Analysis of *Oryza latifolia* Desv. (CCDD Genome)-Derived Introgression Lines and Identification of Value-Added Traits for Rice (*O. sativa* L.) Improvement", The Journal of Heredity, 105(5): 676-689, Advance Access Published Jun. 17, 2014.
Causse et al. "Prospective Use of *Oryza longistaminata* for Rice Breeding", Rice Genetics II—Proceedings of the Second International Rice Genetics Symposium, IRRI, Manila, Philippines,May 14-18, 1990, XP002760153, p. 81-89, May 14, 1990. p. 87, Para.2.
Dayun et al. "Preliminary Report on Transfer Traits of Vegetative Propagation From Wild Rice species to *Oryza sativa* Via Distant Hybridization and Embryo Rescue", The Kasetsart Journal Natural Sciences, 34(1): 1-11, Jan.-Mar. 2000.
Endo et al. "Molecular Breeding of A Novel Herbicide-Tolerant Rice by Gene Targeting", The Plant Journal, 52(1): 157-166, Oct. 2007.

(Continued)

Primary Examiner — Stuart F Baum

(57) **ABSTRACT**

Methods for increasing hybrid seed production are provided. Increased hybrid seed production is achieved through higher outcrossing rates in cytoplasmic male sterile (CMS) lines of rice by introgressing the long stigma trait of *Oryza longistaminata*. CMS lines having higher outcrossing rates capable of high hybrid seed set are also provided.

8 Claims, 17 Drawing Sheets
(15 of 17 Drawing Sheet(s) Filed in Color)

Specification includes a Sequence Listing.

Figure 12.2   In 2021, the US government granted a patent to IRRI for a method of increasing the production of hybrid rice seed. US Patent no. 10,999,986 B2, granted May 11, 2021 to the International Rice Research Institute, Los Baños, Philippines.

Figure 12.3 A worker cares for a sample of *Oryza longistaminata* at IRRI in 2009. This type of rice was used in the hybrid seed production method outlined in US Patent no. 10,999,986 B2, granted to IRRI in 2021. Photo by Ariel Javellana/IRRI and reprinted by permission of IRRI.

2012 and 2021, by 2022 only three had been approved. One patent, granted in the United States to IRRI, covers methods for increasing seed production in hybrid rice lines, as well as rice plants obtained by employing the claimed methods (Figures 12.2 and 12.3).[74] The other two patents were granted in the United States and Europe for the same ICRISAT invention, a DNA construct comprising a pigeon-pea gene, as well as plants whose genome contains the claimed DNA construct.[75]

[74] IRRI, "Increasing hybrid seed production through higher outcrossing rate in cytoplasmic male sterile rice and related materials and methods," US patent 10,999,986, filed June 5, 2016 and issued May 11, 2021. Patents were also filed for this invention in Australia, China, Brazil, and Europe, but the applications have been discontinued, while an application filed in the Philippines was still pending at the time of writing.

[75] ICRISAT, "Cytoplasmic male sterility gene ORF147 of pigeon pea, and uses thereof," US patent 11,060,106, filed December 1, 2017 and issued July 13, 2021; ICRISAT, "Cytoplasmic male sterility gene ORF147 of pigeon pea, and uses thereof," European patent 3,548,505, filed December 1, 2017 and issued January 27, 2021. Patent applications were also lodged in Canada and Australia for this invention. At the time of writing, the Canadian application was still pending, while the Australian application had been discontinued.

Notwithstanding the relatively small number of formal intellectual property claims that centers have made in the past decade, in recent years CGIAR institutions have worked to deepen engagement with the commercial sector. One manifestation of this effort is the proliferation of limited exclusivity agreements,[76] the vast majority of which have been executed between centers and private firms.[77] Between 2012 and 2021, a total of 302 of these kinds of contract were signed, amounting to more than five times the number of intellectual property filings made by centers during the same period.[78]

When the Intellectual Assets Principles were enacted, agreements granting limited exclusivity in the use of CGIAR technologies were, like intellectual property claims, seldom pursued. This began to change in 2017. Between 2017 and 2021 alone, 273 limited exclusivity agreements were signed (90 percent of the total). Although most of these contracts were between CIMMYT and seed company partners, the rise of agreements allowing third parties to exclusively use CGIAR technologies indicates the extent to which certain centers have begun to collaborate with commercial entities. In this way, adaptationist policies have accommodated an approach that intellectual property maximalists advocated in the mid 1990s.

By the end of the second decade of the new millennium, all CGIAR centers had enacted their own institutional policies to implement the Intellectual Assets Principles, and all had allocated part of their budgets to salaries for in-house or external intellectual property personnel, while also regularly training staff in intellectual property management. Nevertheless, despite decades of efforts to centralize and harmonize, each center continued to operate with substantial independence. In the future, it is possible that center autonomy in intellectual property management will be curtailed under the One CGIAR strategy, which was

---

[76] Limited exclusivity agreements are contracts through which CGIAR or the centers grant third parties exclusive rights to commercialize CGIAR "intellectual assets." These exclusive rights must be necessary for the further improvement of the intellectual assets or to enhance the scale or scope of impact on target beneficiaries, and as limited as possible in duration, territory, and/or field of use. Limited exclusivity agreements provide that CGIAR intellectual assets must remain available for noncommercial research by public-sector organizations and in the event of food security emergencies. CGIAR System Management Office, "CGIAR Principles on the Management of Intellectual Assets," principle 6.2.

[77] For example, in 2017 CIMMYT granted twenty-three licenses through limited-exclusivity agreements to partner institutions, 17 percent of which were public-sector institutions and parastatals, and 73 percent of which were private seed companies. CGIAR System Organization, *CGIAR Intellectual Assets Management Report 2017* (Montpellier, France: CGIAR System Organization, 2018), https://hdl.handle.net/105 68/102281.

[78] These data were compiled from the annual intellectual assets reports for 2012–21.

launched in 2019 and aims to achieve unified governance and institutional integration across all centers.[79]

Consistent with the One CGIAR approach, a 2022 special report recommended centralization in intellectual property management, stating that
CGIAR should become a "one-stop-shop" for engagement with private-
sector enterprises.[80] Recognizing that the absence of transversal mechanisms
to deal with intellectual property rights has posed a barrier to engagement
with businesses, especially multinational firms, the report recommended that
CGIAR should develop system-wide approaches to intellectual property
ownership that would enhance partnership with the private sector.[81]

Notwithstanding the ongoing drive towards centralization, there are
good reasons for CGIAR centers to retain some flexibility in defining
their approaches to intellectual property management. The centers
vary significantly in size and scope, and in the extent to which their
work is compatible with technoscientific and capitalistic agricultural
practices. For instance, it is logical that centers such as IRRI and
CIMMYT would be the most prolific users of patents, plant variety
protection, and limited exclusivity agreements, given that their
research priorities focus on rice, and wheat and corn, respectively.
These crops are the three most widely grown in the world, and they
also form the core of intellectual property portfolios owned by the
largest multinational agricultural corporations. Conversely, pursuing
intellectual property claims may be less relevant for centers such as
World Agroforestry or WorldFish, given these institutions' emphases
on ecological approaches to agriculture and aquaculture. Such
methods may be less compatible with privatization and industrialization, making them unlikely targets for corporate investment.

Reflecting the impact of broader scientific, economic, and legal shifts
that have occurred over the past three decades, CGIAR policies and
practices now formally regard all inventions made by the centers as
potentially protectable "intellectual assets." While many CGIAR technologies may still be distributed directly to farmers in the Global South, it
is increasingly possible that at least some centers will seek to develop and
commercialize their inventions in partnership with agribusinesses.

Given the diversity of the centers' research agendas, geographical locations, budgetary circumstances, and local administrative cultures, the
adaptationist approach of the Intellectual Assets Principles appears to
operate as a fair compromise. The principles established that CGIAR
institutions should generally avoid intellectual property claims, allowing

---

[79] See discussion of One CGIAR in note 2 above.
[80] Cummings et al., "Open for Business," 10.   [81] Ibid., 36.

rejectionist centers to continue to focus on nonproprietary forms of technology dissemination. Simultaneously, the policy held that, where appropriate, intellectual property ownership may "lead to the broadest possible impact on target beneficiaries in furtherance of [the] CGIAR Vision,"[82] providing a justification for maximalist centers to embrace entrepreneurial practices. Nevertheless, examining the history of intellectual property debates within CGIAR from 1990 to 2020 reveals that the rejectionist viewpoint was formally marginalized during this period. While some CGIAR administrators and scientists may continue to eschew intellectual property, they must do so in the context of an institutional culture that since 1990 has increasingly internalized a global capitalist approach to agricultural science.

---

[82] CGIAR System Management Office, "CGIAR Principles on the Management of Intellectual Assets," principle 6.

Index

Acosta, Ricardo, 209
Afghanistan, 29, 30, 33, 278
Africa, 12, 27, 29, 48, 49, 125, 126, 129,
    131, 133, 135, 145, 157, 160, 165,
    169, 171, 181, 185, 186, 187, 189,
    190, 209, 210, 213, 217, 220, 221,
    222, 223, 225, 227, 228, 231, 232,
    233. *See also* Middle East and North
    Africa (MENA)
East Africa, 110, 126, 164, 175, 231
sub-Saharan Africa, 104, 158–180, 226,
    227, 231
West Africa, 109, 126, 130, 135–157,
    187, 212, 227, 273
Africa Rice Center, 137, 166
African Agricultural Technology
    Foundation, 231
African animal trypanosomiasis (AAT). *See*
    trypanosomiasis
agribusiness, 6, 14, 75, 76, 78, 85, 87, 125,
    199, 288, 291, 301, 308 *See also* private
    sector; seed industry
agricultural economics, 11, 39, 81, 82, 120,
    123, 167
agricultural research for development
    (AR4D), 159, 179
agro-ecology, 11, 19, 27, 28, 31, 33, 40, 41,
    42, 49, 152, 165, 186, 188, 221, 267,
    282, 299
al-Assad, Bashar, 17, 41
al-Assad, Hafez, 23, 35, 36, 37, 41
Albania, 278
Algeria, 20, 30, 31, 32, 33
Alliance for Progress, 75, 77, 183, 198
Altieri, Miguel, 299, 300
Andean region, 69
Arab-Israeli war, 23
Arafat, Yasser, 37
Argentina, 40, 189, 190, 210
Arid Land Agricultural Development pro-
    gram (ALAD), 25, 33
arid regions. *See* drylands

Asia, 6, 18, 46, 48, 49, 50, 72, 80, 104, 124,
    126, 131, 149, 181, 185, 190, 205,
    210, 213, 221, 224, 232
Central Asia, 24, 42
South Asia, 97, 273
Southeast Asia, 93, 186
Western Asia, 21, 41. *See also* Middle
    East and North Africa (MENA)
Asian Development Bank, 119, 129
Australia, 129, 186, 278

Ba'ath Party, 23, 35
bananas, 80, 145, 202
Bangladesh War, 52, 60
Bänziger, Marianne, 227
barley, 24, 26, 39, 40, 104, 249, 276
Beachell, Henry, 138
beans, 11, 65, 83, 127, 181–206, 244, 272
    bean breeding, 11, 181, 185, 186, 188,
    189, 190, 195, 197, 200, 201, 205
broad beans, 39
velvet beans, 108, 109, 112
Belgium, 129
Bell, David, 84, 85
Bellagio meetings, 115, 116, 118, 119, 120,
    121, 122, 123, 124, 126, 128, 162
Bennett, Erna, 267, 268, 270, 272, 276
Bigirwenkya, Z.H.K., 31
Bill & Melinda Gates Foundation, 227,
    230, 285
biotechnology, 173, 229, 230, 232, 253,
    260, 277, 292, 297, 298
    transgenic animals, 159, 170, 174
    transgenic plants, 231, 291, 297,
    298, 306
Bioversity International, 12, 13, 255, 256,
    257, 281, 282, 302, 304 *See also*
    International Board for Plant Genetic
    Resources (IBPGR); International
    Plant Genetic Resources Institute
    (IPGRI)
Boerma, Addeke Hendrik, 24, 122, 127

310

Printed in the United States
by Baker & Taylor Publisher Services